International Review of Science

Physical Chemistry
Series Two

Consultant Editor
A. D. Buckingham, F.R.S.

Publisher's Note

The International Review of Science is an important venture in scientific publishing presented by Butterworths. The basic concept of the Review is to provide regular authoritative reviews of entire disciplines. Chemistry was taken first as the problems of literature survey are probably more acute in this subject than in any other. Biochemistry and Physiology followed naturally. As a matter of policy, the authorship of the Review of Science is international and distinguished, the subject coverage is extensive, systematic and critical.

The Review has been conceived within a carefully organised editorial framework. The overall plan was drawn up and the volume editors appointed by seven consultant editors. In turn, each volume editor planned the coverage of his field and appointed authors to write on subjects which were within the area of their own research experience. No geographical restriction was imposed. Hence the 500 or so contributions to the Review of Science come from many countries of the world and provide an authoritative account of progress.

The publication of Physical Chemistry Series One was completed in 1973 with thirteen text volumes and one index volume; in accordance with the stated policy of issuing regular reviews to keep the series up to date, volumes of Series Two will be published between the middle of 1975 and early 1976; Series Two of Organic Chemistry will be published at the same time, while Inorganic Chemistry Series Two was published during the first half of 1975. Volume titles are the same as in Series One but the articles themselves either cover recent advances in the same subject or deal with a different aspect of the main theme of the volume. In Series Two an index is incorporated in each volume and there is no separate index volume.

Butterworth & Co. (Publishers) Ltd.

PHYSICAL CHEMISTRY SERIES TWO

Consultant Editor
A. D. Buckingham, F.R.S.,
Department of Chemistry University of Cambridge

Volume titles and Editors

1 **THEORETICAL CHEMISTRY**
Professor A. D. Buckingham, F.R.S., *University of Cambridge* and Professor C. A. Coulson, F.R.S., *University of Oxford*

2 **MOLECULAR STRUCTURE AND PROPERTIES**
Professor A. D. Buckingham, F.R.S., *University of Cambridge*

3 **SPECTROSCOPY**
Dr. D. A. Ramsay, F.R.S.C., *National Research Council of Canada*

4 **MAGNETIC RESONANCE**
Professor C. A. McDowell, F.R.S.C., *University of British Columbia*

5 **MASS SPECTROMETRY**
Professor A. Maccoll, *University College, University of London*

6 **ELECTROCHEMISTRY**
Professor J. O'M. Bockris, *The Flinders University of S. Australia*

7 **SURFACE CHEMISTRY AND COLLOIDS**
Professor M. Kerker, *Clarkson College of Technology, New York*

8 **MACROMOLECULAR SCIENCE**
Professor C. E. H. Bawn, C.B.E., F.R.S., *formerly of the University of Liverpool*

9 **CHEMICAL KINETICS**
Professor D. R. Herschbach *Harvard University*

10 **THERMOCHEMISTRY AND THERMO-DYNAMICS**
Dr. H. A. Skinner, *University of Manchester*

11 **CHEMICAL CRYSTALLOGRAPHY**
Professor J. M. Robertson, C.B.E., F.R.S., *formerly of the University of Glasgow*

12 **ANALYTICAL CHEMISTRY —PART 1**
Professor T. S. West, *Macaulay Institute for Soil Research, Aberdeen*

13 **ANALYTICAL CHEMISTRY —PART 2**
Professor T. S. West, *Macaulay Institute for Soil Research, Aberdeen*

INORGANIC CHEMISTRY SERIES TWO

Consultant Editor
H. J. Emeléus, C.B.E., F.R.S.
Department of Chemistry University of Cambridge

Volume titles and Editors

1 **MAIN GROUP ELEMENTS—HYDROGEN AND GROUPS I–III**
Professor M. F. Lappert, *University of Sussex*

2 **MAIN GROUP ELEMENTS—GROUPS IV AND V**
Dr. D. B. Sowerby, *University of Nottingham*

3 **MAIN GROUP ELEMENTS—GROUPS VI AND VII**
Professor V. Gutmann, *Technical University of Vienna*

4 **ORGANOMETALLIC DERIVATIVES OF THE MAIN GROUP ELEMENTS**
Professor B. J. Aylett, *Westfield College, University of London*

5 **TRANSITION METALS—PART 1**
Professor D. W. A. Sharp, *University of Glasgow*

6 **TRANSITION METALS—PART 2**
Dr. M. J. Mays, *University of Cambridge*

7 **LANTHANIDES AND ACTINIDES**
Professor K. W. Bagnall, *University of Manchester*

8 **RADIOCHEMISTRY**
Dr. A. G. Maddock, *University of Cambridge*

9 **REACTION MECHANISMS IN INORGANIC CHEMISTRY**
Professor M. L. Tobe, *University College, University of London*

10 **SOLID STATE CHEMISTRY**
Dr. L. E. J. Roberts, *Atomic Energy Research Establishment, Harwell*

ORGANIC CHEMISTRY SERIES TWO

Consultant Editor
D. H. Hey, F.R.S., *formerly of the Department of Chemistry, King's College, University of London*

Volume titles and Editors

1 **STRUCTURE DETERMINATION IN ORGANIC CHEMISTRY**
Professor L. M. Jackman, *Pennsylvania State University*

2 **ALIPHATIC COMPOUNDS**
Professor N. B. Chapman, *University of Hull*

3 **AROMATIC COMPOUNDS**
Professor H. Zollinger, *Eidgenössische Technische Hochschule, Zürich*

4 **HETEROCYCLIC COMPOUNDS**
Professor K. Schofield, *University of Exeter*

5 **ALICYCLIC COMPOUNDS**
Professor D. Ginsburg, *Technion-Israel Institute of Technology, Haifa*

6 **AMINO ACIDS, PEPTIDES AND RELATED COMPOUNDS**
Professor H. N. Rydon, *University of Exeter*

7 **CARBOHYDRATES**
Professor G. O. Aspinall, *York University, Ontario*

8 **STEROIDS**
Dr. W. F. Johns, *G. D. Searle & Co., Chicago*

9 **ALKALOIDS**
Professor K. Wiesner, F.R.S., *University of New Brunswick*

10 **FREE RADICAL REACTIONS**
Professor W. A. Waters, F.R.S., *University of Oxford*

BIOCHEMISTRY SERIES ONE

Consultant Editors
H. L. Kornberg, F.R.S.
Department of Biochemistry University of Leicester and
D. C. Phillips, F.R.S., *Department of Zoology, University of Oxford*

Volume titles and Editors

1 **CHEMISTRY OF MACRO-MOLECULES**
Professor H. Gutfreund, *University of Bristol*

2 **BIOCHEMISTRY OF CELL WALLS AND MEMBRANES**
Dr. C. F. Fox, *University of California, Los Angeles*

3 **ENERGY TRANSDUCING MECHANISMS**
Professor E. Racker, *Cornell University New York*

4 **BIOCHEMISTRY OF LIPIDS**
Professor T. W. Goodwin, F.R.S., *University of Liverpool*

5 **BIOCHEMISTRY OF CARBO-HYDRATES**
Professor W. J. Whelan, *University of Miami*

6 **BIOCHEMISTRY OF NUCLEIC ACIDS**
Professor K. Burton, F.R.S., *University of Newcastle upon Tyne*

7 **SYNTHESIS OF AMINO ACIDS AND PROTEINS**
Professor H. R. V. Arnstein, *King's College, University of London*

8 **BIOCHEMISTRY OF HORMONES**
Professor H. V. Rickenberg, *National Jewish Hospital & Research Center, Colorado*

9 **BIOCHEMISTRY OF CELL DIFFER-ENTIATION**
Professor J. Paul, *The Beatson Institute for Cancer Research, Glasgow*

10 **DEFENCE AND RECOGNITION**
Professor R. R. Porter, F.R.S., *University of Oxford*

11 **PLANT BIOCHEMISTRY**
Professor D. H. Northcote, F.R.S., *University of Cambridge*

12 **PHYSIOLOGICAL AND PHARMACO-LOGICAL BIOCHEMISTRY**
Dr. H. K. F. Blaschko, F.R.S., *University of Oxford*

PHYSIOLOGY SERIES ONE

Consultant Editors
A. C. Guyton,
Department of Physiology and Biophysics, University of Mississippi Medical Center and
D. F. Horrobin,
Department of Physiology, University of Newcastle upon Tyne

Volume titles and Editors

1 **CARDIOVASCULAR PHYSIOLOGY**
Professor A. C. Guyton and Dr. C. E. Jones, *University of Mississippi Medical Center*

2 **RESPIRATORY PHYSIOLOGY**
Professor J. G. Widdicombe, *St. George's Hospital, London*

3 **NEUROPHYSIOLOGY**
Professor C. C. Hunt, *Washington University School of Medicine, St. Louis*

4 **GASTROINTESTINAL PHYSIOLOGY**
Professor E. D. Jacobson and Dr. L. L. Shanbour, *University of Texas Medical School*

5 **ENDOCRINE PHYSIOLOGY**
Professor S. M. McCann, *University of Texas*

6 **KIDNEY AND URINARY TRACT PHYSIOLOGY**
Professor K. Thurau, *University of Munich*

7 **ENVIRONMENTAL PHYSIOLOGY**
Professor D. Robertshaw, *University of Nairobi*

8 **REPRODUCTIVE PHYSIOLOGY**
Professor R. O. Greep, *Harvard Medical School*

International Review of Science

Physical Chemistry
Series Two

Volume 12

Analytical Chemistry — Part 1

Edited by **T. S. West**
The Macaulay Institute for Soil Research, Aberdeen

BUTTERWORTHS
LONDON - BOSTON
Sydney - Wellington - Durban - Toronto

THE BUTTERWORTH GROUP

ENGLAND
Butterworth & Co (Publishers) Ltd
London: 88 Kingsway, WC2B 6AB

AUSTRALIA
Butterworths Pty Ltd
Sydney: 586 Pacific Highway, NSW 2067
Also at Melbourne, Brisbane, Adelaide and Perth

CANADA
Butterworth & Co (Canada) Ltd
Toronto: 2265 Midland Avenue, Scarborough, Ontario M1P 4S1

NEW ZEALAND
Butterworths of New Zealand Ltd
Wellington: 26–28 Waring Taylor Street, 1

SOUTH AFRICA
Butterworth & Co (South Africa) (Pty) Ltd
Durban: 152–154 Gale Street

USA
Butterworths (Publishers) Inc
Boston: 19 Cummings Park, Woburn, Mass. 01801

Library of Congress Cataloging in Publication Data

Main entry under title:
Analytical chemistry.

(Physical chemistry, series two; v. 12–13) (International review of science)
Includes index.
1. Chemistry, Analytic. I. West, Thomas Summers.
II. Series. III. Series: International review of science.
QD450.2.P59 vol. 12–13 [QD75.2] 541′.3′08s [543]
ISBN 0 408 70612 0 75–19422

Typeset, printed and bound in Great Britain by
REDWOOD BURN LIMITED
Trowbridge & Esher

Consultant Editor's Note

The International Review of Science was conceived as a comprehensive, critical and continuing survey of progress in research. The difficult problem of keeping up with advances on a reasonably broad front makes the idea of the Review especially appealing, and I was grateful to be given the opportunity of helping to plan it.

Physical Chemistry Series One was published in 1972/1973. Its success was assured by the very great distinction of its editors and authors. Our need for critical reviews at a high level has not diminished. In the rather difficult times being experienced in most parts of the world, research workers should seek to broaden the range of their expertise; it is hoped that this Series will be of use in this connection.

Like its forerunner, Series Two consists of thirteen volumes covering Physical and Theoretical Chemistry, Chemical Crystallography and Analytical Chemistry. Each volume has been edited by a distinguished chemist; in several cases the person responsible for Series One has acted again as editor. The editors have assembled a strong team of authors, each of whom has assessed and interpreted recent progress in a specialised field in terms of his own experience. I believe that their efforts have again produced useful and timely articles which will help us in our efforts to keep abreast of progress in research.

It is my pleasure to thank all those who have collaborated in this venture —the volume editors, the authors and the publishers.

Cambridge A. D. Buckingham

ERRATUM

Library of Congress Cataloging in Publication Data are incorrect and should be replaced by

Library of Congress Cataloging in Publication Data

Main entry under title:
Analytical chemistry.

(Physical chemistry, series two; v. 12) (International review of science)
Includes index.
1. Chemistry, Analytic. I. West, Thomas Summers. II. Series. III. Series: International review of science.
QD450.2.P59 vol. 12 [QD75.2] 541'.3'08s [543]
ISBN 0 408 70611 2 75–19422

Preface

In Series One of these volumes selected topics in analytical chemistry were received such as ESCA, X-ray fluorescence, atomic absorption and fluorescence spectrosopy, ion-selective electrode potentiometry, catalytic and kinetic methods, thermoanalytical techniques, etc.

In these two volumes of Series Two, others that necessarily could not be covered previously are dealt with in a similar manner. Radiochemical techniques including the application of computers are reviewed at some length; main-stream techniques such as infrared analysis and complexometry; separation techniques such as the electrophoretic ones and ion-exchange chromatography; surface examination techniques such as electron microprobe and particle size analysis; multi-disciplinary, mission-oriented areas such as forensic analysis, air-pollution analysis and ultra-microanalysis are also reviewed. In addition, ion-selective electrode potentiometry has again been reviewed because of the continued developmental activity in this area.

As before, the authors are clearly identified as researchers of international repute and again they have been chosen on a geographical basis, as a secondary consideration, to derive as international a spread of viewpoint as possible. Each author has been given full freedom to develop the subject matter within the agreed framework of the topic and the limitations of available space.

There are, of course, many techniques and topics not reviewed here or in Series One. Their omission is no reflection on their vitality or relative importance, but rather an indication of the vastness of modern analytical chemistry and, of course, the arbitrary selection of the editor.

Aberdeen T. S. West

Contents

1
Activation Analysis — Part 1

R. DAMS, F. DE CORTE, J. HERTOGEN, J. HOSTE and W. MAENHAUT
Rijksuniversiteit Gent

and

F. ADAMS
Universitaire Instelling Antwerpen

1.1 INSTRUMENTAL NEUTRON ACTIVATION ANALYSIS

1.1.1 Introduction

It is customary to make a clear distinction between neutron activation analysis based on chemical separations and instrumental neutron activation analysis (INAA) which usually applies high-resolution γ-ray spectrometry to measure complex radioactive samples directly. This distinction is valid because both methods have clearly different advantages and disadvantages. The instrumental technique only very rarely approaches the ultimate sensitivity that can be reached through the judicious use of chemistry. It requires complex and costly equipment which the chemical separation method does not necessarily require. The most important features of the instrumental technique are the multi-element character of the trace analysis, the relative ease with which the technique can be adapted to different matrices, the possibility of automating the measurements and the data reduction. Large numbers of samples of widely differing composition can be analysed in a routine way as long as they do not give rise to considerable matrix activities.

When the matrix activities interfere with sensitive multi-element determinations, a simple group separation technique can often be applied. One example is the multi-element analysis of biological materials which is hindered by the formation of an excessively high ^{24}Na activity. An easy separation of the latter activity is possible, e.g. through the use of HAP (hydrated antimony pentoxide). Strictly speaking, in such cases the analysis is no longer instrumental. Nevertheless, all the essential features of the INAA approach are followed, with one or more chemical group separations between the end of the irradiation and the measurement. To stress this, and the fact that the 'instrumental' technique depends practically exclusively on the use of high-resolution Ge(Li) γ-ray spectrometry, the name 'nuclear (or neutron) activation spectrometry' might be used[1].

High-resolution γ-ray spectrometers became available in the early 1960s with the development of the lithium-drifted germanium detector, but it is only recently that large detectors with stable and optimum characteristics have become available. The main properties are an energy resolution down to 1.8 keV (FWHM) for the ^{60}Co radiation at 1333 keV, with a detection efficiency 5–10 times smaller than that of a sodium iodide detector with dimensions of 7.5 × 7.5 cm. Together with the neutron flux of the nuclear reactor, these data will determine what is feasible with the technique. It is now very unlikely that these specifications will improve as drastically in the future as they consistently did during the first years of the development of Ge(Li) detectors: the resolution approaches a fundamental lower limit based on statistics and the construction of larger size detectors would rapidly become uneconomical. The development of newer, mostly hyperpure germanium and compound semiconductors which could serve as the starting material for constructing newer classes of detector with superior characteristics has made rather slow progress. The latest achievements in this field have been reviewed by Bertolini *et al.*[2].

1.1.2 The high-resolution γ-ray spectrometer for activation analysis

The quality of analytical results, especially their precision and specificity, depends to a considerable extent on the characteristics of the γ-ray spectrometer. It is thus important to review the essential specifications for use in instrumental neutron activation analysis. Each application of semiconductor devices emphasises a number of different parameters, such as energy resolution, energy range of interest, detection efficiency, detector configuration and, possibly, coincidence timing characteristics.

1.1.2.1 Energy resolution

The resolution and the peak symmetry depend on the inherent properties of the Ge(Li) detector on the one hand and on the performance of the amplification and pulse-shaping circuitry on the other. With the FET pre-amplifiers now generally used and with amplifiers having built-in pole-zero cancellation

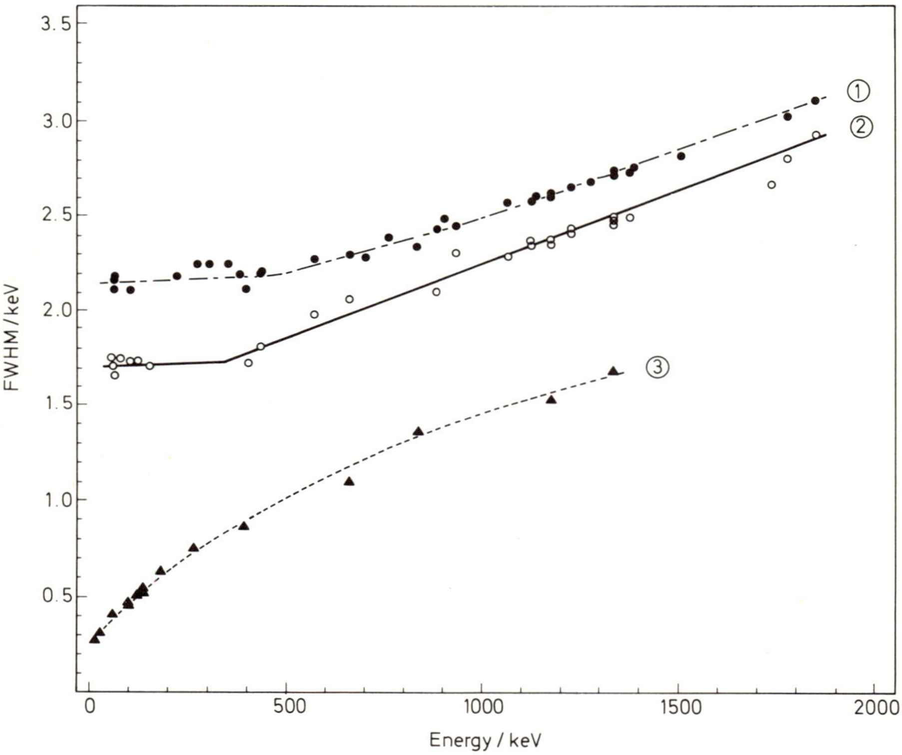

Figure 1.1 Resolution (FWHM) as a function of energy for typical spectrometers. The measurements were performed under common counting conditions, i.e. at moderate counting rates, without special precautions against pile-up effects: 1, coaxial Ge(Li), 53 cm^3 useful volume; 2, coaxial Ge(Li), 78 cm^3 useful volume; 3, planar Ge(Li) (LEPD), 78 mm^2 $\times$ 5 mm, 0.4 cm^3 useful volume

and base-line restoration, the resolution is determined mainly by the detector for energies above *ca.* 100 keV and counting rates up to 10 000 cps (counts per second). The resolution (FWHM) as a function of energy for typical spectrometers is displayed in Figure 1.1. Under optimum conditions, i.e. at amplifier shaping time constants of 2–10 μs, an energy resolution down to 1.8 keV (FWHM) at 1333 keV can be reached with large high-quality detectors. These 'optimum' conditions are, however, seldom compatible with a reasonable counting rate capability in instrumental analysis. A considerable gain in resolution for low-energy radiation below a few hundred keV is obtained with small planar 'low-energy photon Ge(Li) detectors' (LEPD) of the type described by Palms *et al.*[3]. In this case, one has to sacrifice some detection efficiency.

Most of the recent developments in pulse processing circuitry deal with the following two aspects of γ-ray spectrometry: (i) the preservation of good resolution characteristics at higher counting rates; and (ii) the improvement of the electronic resolution at low energies where it may become the dominant contribution to line-broadening. Interest in high-counting-rate Ge(Li) γ-ray spectrometry is apparent from recent publications on this subject. Johnson and Heath[4] have reported excellent results with a direct-coupled spectrometer and a small planar detector for input counting rates up to 100 000 cps. Significant results have also been obtained by Kandiah[5] with a high-rate system utilising a 40 cm^3 coaxial detector for input rates exceeding 160 000 cps. Dunn *et al.*[6] described the performance of a system consisting of a large detector, a direct-coupled recycling pre-amplifier and a unipolar Gaussian filter amplifier. The use of a pile-up rejector, inserted as an extra element in commonly used amplification chains, may also result in considerable improvements at high counting rates and this topic will be discussed more extensively in a later section. An important innovation in the field of low-energy spectrometry involves pre-amplifiers based on optical feed-back techniques. With d.c. optoelectronic feed-back[7,8], resolutions down to 145 eV are reported for a Si(Li) X-ray detector at 5.9 keV. Unfortunately, the main improvement is achieved at the cost of rather long amplifier time constants (8–16 μs). Better performances at higher counting rates have been obtained by means of the more recent pulsed-optical feed-back pre-amplifiers[9].

This technique makes use of pulsing a charge feed-back path to reset, at an appropriate point, the voltage ramp produced by the accumulation of radiation-induced charge pulses on the feed-back capacitor. These new circuits were originally proposed for energy dispersive X-ray fluorescence analysis, and it is still uncertain at the moment whether they will readily replace the conventional FET pre-amplifiers used nowadays in activation analysis.

1.1.2.2 Detection efficiency and peak-to-Compton ratio

The detector size and shape is important for two reasons. In the first place, the detection efficiency depends directly on these characteristics, although there is not necessarily a direct proportionality between intrinsic volume and efficiency. The detection efficiency is the product of the source–detector solid angle and the intrinsic detection efficiency. For detectors with a comparable sensitive volume, this illustrates the importance of detectors with large cross-sectional areas located as close to the outer cryostat as possible, thus maximising the solid angle. A few detection efficiency curves are showed in Figure 1.2.

The maximum efficiency for a given size of the detector is obtained with a well-type detector for which the solid angle is optimised by introducing the sample into the hollow core of the detector. This is not easily realised in the rather small size Ge(Li) detectors which are cooled to liquid-nitrogen temperatures and consequently need to be held in vacuum. A cross-section of a well-type Ge(Li) configuration is shown in Figure 1.3[10]. The best results

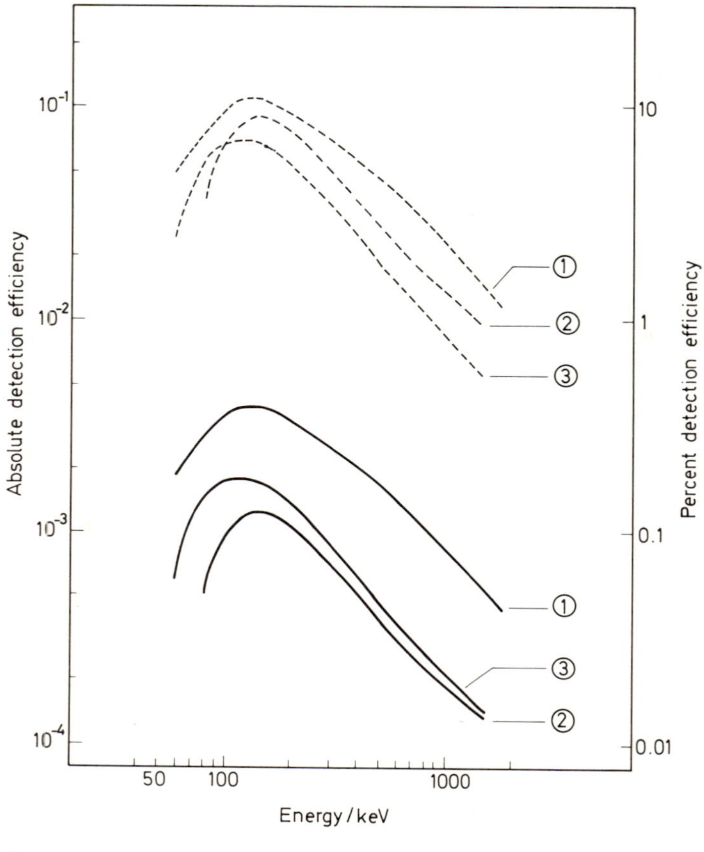

Figure 1.2 Full-energy peak efficiency curves for large coaxial Ge(Li) detectors ('point sources'). Full lines are experimentally determined curves: 1, 78 cm^3 useful volume, distance 12.3 cm; 2, 50 cm^3 useful volume, distance 18.7 cm; 3, 35 cm^3 useful volume, distance 16.2 cm. Dashed lines are *approximate* efficiency curves for closest geometry and are extrapolated from the foregoing (the ^{54}Mn 835 keV peak was used for the determination of the conversion factors) (Experimental data from Van der Linden[29], Bruggeman[372] and Maenhaut[30] (unpublished))

obtained so far were reported by Fowler and Toone[11] for a detector of 75 cm^3 useful volume with a 31 mm central hole.

It should be realised that although the detector efficiency is increased by a factor of three, at least in the well-type geometry, there is not necessarily a gain in sensitivity. This is due to the inevitable loss in resolving power caused indirectly by the increase of the capicitance of the device and in the second place by the decrease in peak-to-Compton ratio caused by the less than optimal compactness.

The peak-to-Compton ratio is, indeed, a very important characteristic for Ge(Li) detectors. It is defined as the ratio of the peak height to Compton-edge height for a given γ-ray, rather than as the counting rate of the full-energy peak normalised to the integral counting rate, because this ratio is so much easier to determine. The better the energy resolution for a given efficiency, the higher the peak height. The peak height and the peak shape are the two most essential parameters which have a bearing on the detectability of weak γ-rays in a spectrum. The ratio for the ^{60}Co 1333 keV radiation is commonly

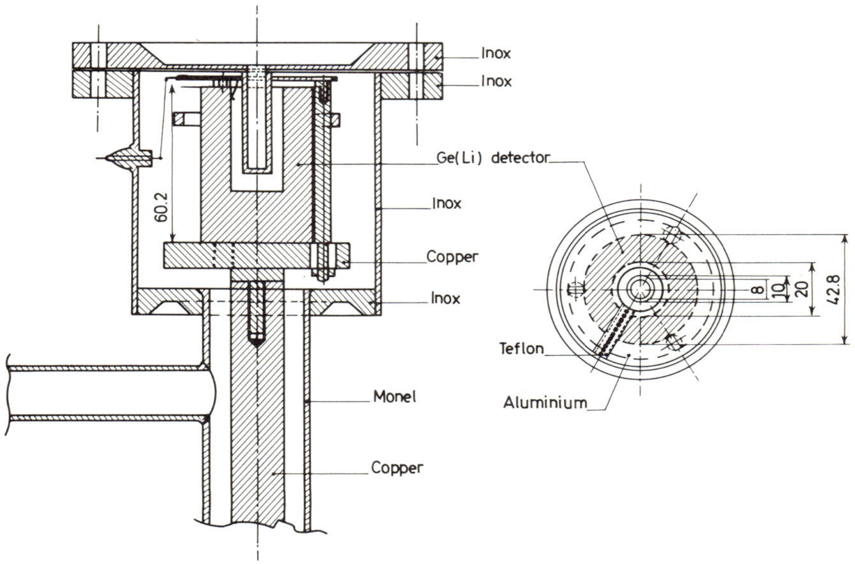

Figure 1.3 Cross-section of Ge(Li) well-type detector and upper part of cryostat; all dimensions in mm

given as a measure of performance. Typical values of 30:1 for a size of 50 cm³ are readily achieved. For smaller detectors this ratio will approach 10:1 only. Wagner *et al.*[12] have stated that there exists a general and widespread problem of declining efficiency as a function of time, owing to an unknown shortcoming of the raw germanium ingots. For a group of 50 germanium ingots, eight resulted in detectors whose efficiencies declined at rates greater than 8% per year. A substantial fraction of the cases studied displayed relative decreases of 10–20% over a period of six months. As a direct result of the latter paper, a number of spectrometers currently in use in the authors' laboratory were checked for their long-term stability. The results are presented in Figure 1.4. One of the detectors showed, indeed, a decrease of efficiency. It is still uncertain whether the phenomenon is as widespread as

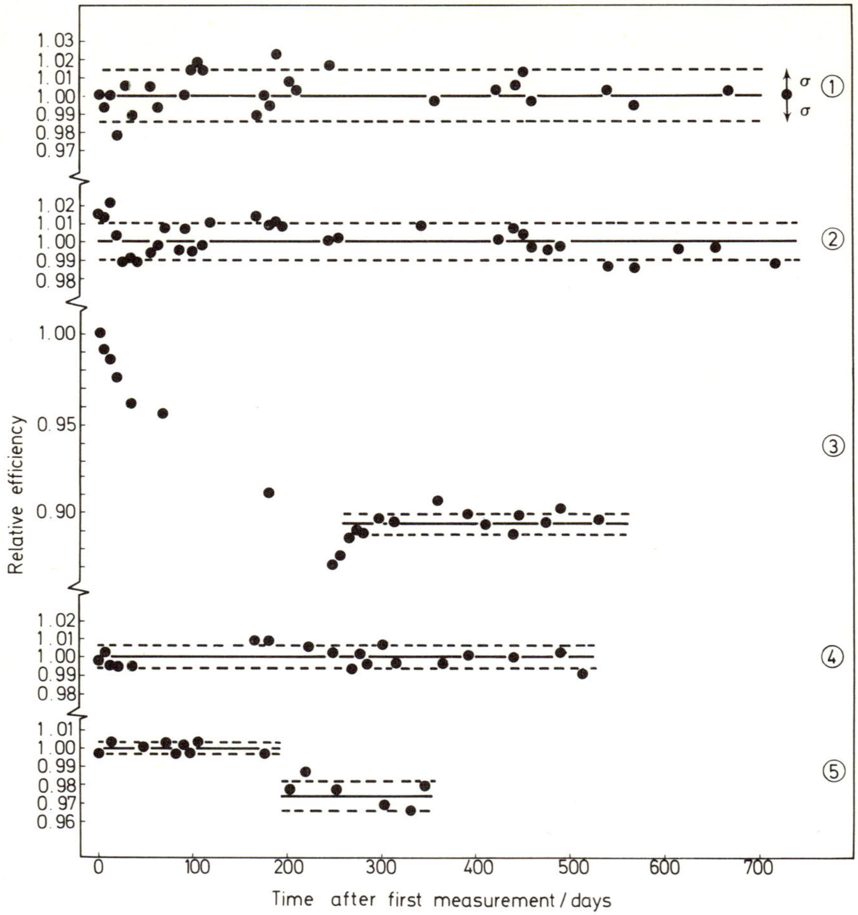

Figure 1.4 Relative variation of the detection efficiency of coaxial Ge(Li) detectors as a function of time, monitored with the ^{60}Co 1333 keV photopeak. Useful volumes of the detectors: 1, 35 cm^3; 2, 50 cm^3; 3, 53 cm^3; 4, 70 cm^3; 5, 78 cm^3

suggested by Wagner *et al.*[12]; however, great care should be exercised when using a calibrated detector in analysis and frequent recalibration of the equipment is recommended.

1.1.2.3 Figure-of-merit

Several authors have proposed a definition for a useful figure-of-merit by which the sensitivities of various detectors can be compared. Cline[13] suggests the use of the peak-to-total ratio divided by the resolution. The figure-of-merit proposed by McKenzie *et al.*[14] uses the ^{137}Cs peak height above a ^{60}Co

Compton continuum. An equation has been derived by Cooper[15] which should provide a quantitative means to evaluate the sensitivity of Ge(Li) γ-ray spectrometers. It is related to the five parameters that have a bearing on the sensitivity: resolution, peak efficiency, background interference, counting rate and counting time. This figure-of-merit is a measure of the detector's ability to detect a γ-ray of energy E_1 on the Compton continuum of a higher-energy γ-ray E_2, and is related to the peak efficiency of γ-ray (1), $\varepsilon_p(E_1)$, the resolution at energy E_1, $R(E_1)$, and the Compton efficiency at energy E_1 owing to γ-ray (2), $\varepsilon_{C2}(E_1)$, by the equation:

$$F_2(E_1) = \varepsilon_p(E_1)/[R(E_1)\cdot\varepsilon_{C2}(E_1)]^{\frac{1}{2}}$$

where $F_2(E_1)$ is the figure-of-merit for measuring γ-ray (1) in the presence of the higher-energy γ-ray (2). Equations similar to the above can be useful in the evaluation of various detectors because they provide a means of expressing quantitative improvements in various spectrometer parameters. For example, an improvement in resolution from 2.4 keV to 2.0 keV as might be expected by cooling the FET pre-amplifier would improve the sensitivity of well-resolved peaks on a smooth background by only a very small amount. This situation is, of course, entirely different when peaks are unresolved.

Figures-of-merit such as these do not, however, consider a number of important characteristics of the detector such as full-energy peak symmetry, dead or imperfectly compensated regions which limit or make impossible measurements at low energies, timing characteristics of the detector, etc.

1.1.2.4 Compton suppression spectrometers

Despite the excellent selectivity and sensitivity for the measurement of complex radioisotope mixtures, numerous problems still remain when the full-energy peaks of the less abundant radionuclides are obscured by the Compton distribution associated with the γ-radiation of the major radionuclides. In the past few years, a large number of sophisticated instruments have been developed to reduce this interference. They can be divided in two categories:

(i) Compton scattering spectrometers which use two Ge(Li) detectors, or a Ge(Li) di:ode[16, 17] operated in the sum-coincidence mode. Although these spectrometers greatly reduce the Compton continuum, they are seldom used in instrumental neutron activation analysis because they give rise to a considerable reduction in detection efficiency.

(ii) Compton suppression spectrometers based on anticoincidence shielding with a scintillator annulus around the Ge(Li) detector. They do not give rise to a reduction in detection efficiency if the source can be placed inside the shield, as long as the γ-ray of interest is not part of a cascade decay. In the latter case, the efficiency may drop by a factor of 50 or more. Hence, to preserve all available information, the accepted and rejected events have to be stored in separate halves of the analyser memory.

The topic of Compton suppression spectrometers is treated at length in Refs. 18 and 19. Among recent developments[20-22], multipurpose arrangements consisting of two large volume Ge(Li) detectors placed in an anticoincidence shield should be mentioned. The radioanalytical applications of such a system have been discussed by Cooper[21].

Compton suppression techniques can achieve a significant reduction of the interferences of the Compton continuum (by factors up to 50). They show their greatest potential in the analysis of biological samples, where the influence of the strongly interfering activated elements, Na, Cl, and Br, may be reduced considerably, enabling the trace analysis of over 30 other elements[23].

Although such anticoincidence arrangements were already built in the mid-1960s, they still remain a curiosity in typical activation laboratories. The reasons are obvious: they are very expensive and the maintenance of the complex electronics calls for highly qualified technical manpower[24]. To date, Compton suppression spectrometers have been more favoured in non-activation γ-ray spectrometry, such as low-level environmental radionuclide analysis[25-28] and nuclear spectroscopy.

1.1.3 Optimisation of irradiation and counting conditions

The ultimate possibilities of instrumental activation analysis are only reached when various experimental conditions are properly selected. In most research reactors the neutron energy distribution may vary considerably over the different irradiation sites. Some data for the Thetis swimming-pool-type reactor of the authors' Institute are summarised in Table 1.1 as an example.

Table 1.1 Neutron flux specifications at 250 kW operation in some of the irradiation sites (pneumatic tube facilities) of the swimming-pool reactor Thetis of Ghent University*

Site	$\Phi_{th} \times 10^{-11}$ †	$\Phi_{epi} \times 10^{-9}$	$\Phi_{fast} \times 10^{-10}$	Φ_{th}/Φ_{epi}	Φ_{th}/Φ_{fast}
1	18.4	72.5	29.7	25.4	6.19
2	15.4	47.3	13.2	32.5	11.7
3	16.3	65.6	26.3	24.9	6.21
14	10.8	22.3	5.83	48.3	18.5
15	5.10	5.50	0.991	92.7	51.5
16	2.82	1.88	0.312	150.0	90.4
8	1.80	0.93	0.170	193.5	105.9
9	26.7	111.3	64.8	24.0	4.12
17‡	43.0	293.0	307.0	14.7	1.40

* Data for thermal and epithermal neutron flux from Van der Linden[29], data for fast neutron flux from Maenhaut[30]

† Neutron fluxes expressed in neutrons cm^{-2} s^{-1}

‡ 'In-core' tube facility

It is clear that the sensitivity is partly determined by the choice of irradiation site. One has the impression, however, that insufficient attention is paid to this aspect. The generalised equation for the (n,γ) reaction rate:

$$R(\mathrm{n},\gamma) = \Phi_{\mathrm{th}}\sigma_0 + \Phi_{\mathrm{epi}}I = \Phi_{\mathrm{th}}\sigma_0\left(1 + \frac{\Phi_{\mathrm{epi}}}{\Phi_{\mathrm{th}}}\cdot\frac{I}{\sigma_0}\right)$$

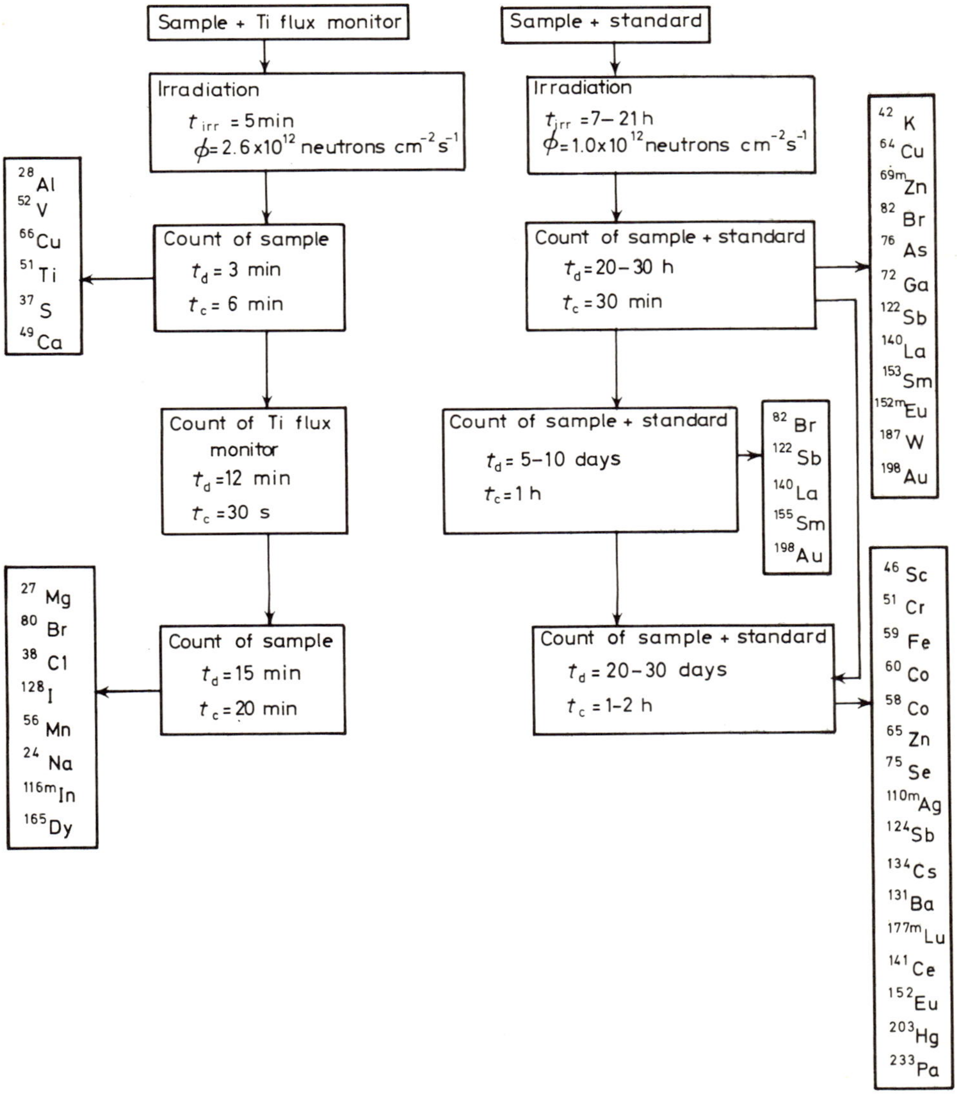

Figure 1.5 Irradiation and counting scheme for the analysis of air particulates collected on filter paper. ϕ = thermal neutron flux; t_{irr} = irradiation time; t_d = time between end of irradiation and start of the counting; t_c = counting time ('live-time') (Adapted from Dams *et al.*[44], by courtesy of The American Chemical Society)

where Φ_{th} and Φ_{epi} are the thermal and epithermal neutron flux, respectively, and σ_0 and I are the thermal neutron cross-section and resonance integral, respectively, shows that, for nuclides with a high I/σ_0 value, the ratio Φ_{epi}/Φ_{th} may be quite an important parameter. The ratio Φ_{fast}/Φ_{th}, on the other hand, is extremely important when (n,p) and (n,α) threshold reactions are involved. When dealing with very short-lived radionuclides, the neutron energy distribution is, of course, not the only criterion. In this case the absolute magnitude of the neutron flux and the sample transfer times are equally important.

'Epithermal activation'[31-33] in a nearly pure epithermal plus a fast neutron spectrum is achieved by suppressing the thermal neutron flux using cadmium or boron shieldings. It may appreciably enhance the sensitive determination of nuclides having high I/σ_0 ratios. A typical example occurs in the trace analysis of silicate rocks by Brunfelt and Steinnes[34, 35].

Advantages have been taken of the relatively low I/σ_0 value of the nuclides which give rise to major interfering activities on the one hand, and of the high I/σ_0 ratios of many trace elements on the other. Cadmium neutron filters are most commonly employed but, in general, no attempt has been made to optimise the filter itself, as stated by Rossitto *et al.*[36]. The latter authors derived approximate theoretical expressions for the optimum choice of Cd, B and B + Cd filters. Their essential conclusion is that a Cd filter is the most appropriate when resonances are located at low energy (less than about 2 eV); at or above 2 eV, B + Cd is more advantageous than Cd.

There are unfortunately many limitations to the epithermal activation technique, such as the relatively low epithermal flux of many small reactors, serious self-shielding effects in the sample in cases of overlapping resonance peaks, etc. These limitations are treated at length by Steinnes[35].

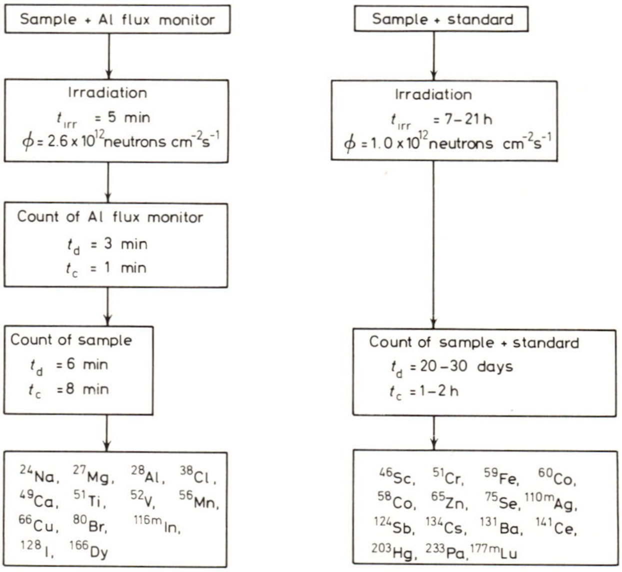

Figure 1.6 Shortened modification of the scheme displayed in Figure 1.5

The next step in optimisation is the choice of appropriate irradiation, cooling and counting times, so as to achieve maximal sensitivity for the elements of interest. Several authors[37-43] have developed calculation procedures which can guide the analyst in choosing optimal conditions for a given sample composition. In the case of multi-element analyses it is, of course, not possible to comply with these theoretical optimal conditions for every element, and it becomes necessary to develop for every type of sample a suitable irradiation and counting scheme using a single or limited number of irradiations and measurements. The measurement time of the different countings is mostly limited to obtain throughput of a reasonable number of samples. Examples of such schemes are given in Figures 1.5 and 1.6 and apply to the analysis of air particulates collected on Whatman 41 filter paper[44, 45]. Compared with the first procedure, the second one is shorter and the amount of spectra to be processed is smaller. This gain in counting and data-processing time per sample is, however, achieved at the expense of the precision and of the number of elements determined.

Another way of optimisation is the use of small very-high-resolution semiconductor detectors for a more selective assessment of low-energy radiation. There seems to be some growing interest in this field and it will be discussed later in detail.

1.1.4 Low-energy photon spectrometry

The application of low-energy peaks obtained from large Ge(Li) detectors is limited because the peaks are normally superimposed upon high Compton or Bremsstrahlung continua. This often results in rather poor counting statistics and occasionally introduces errors in the peak-area determinations.

The Si(Li) detectors now available, and especially the Ge(Li) small 'windowless' detectors, have acceptable detection efficiencies for energies up to 60 keV and 300 keV, respectively (Figure 1.7). In spectra obtained with such detectors, several factors enhance the peak-to-background ratio appreciably. First, there is the overall effect of peak-width narrowing caused by the better resolution at low energy. The thin beryllium window and the absence of an insensitive layer also markedly lower the absorption of soft γ-rays. As is clear from Figure 1.7, the detection yield for high-energy γ-rays is very low. In the case of the Ge(Li) detector, one can calculate that 81% of a parallel beam of 500 keV γ-rays, perpendicular to the detector surface, is not detected at all with a 5 mm thick detector. For a 50 mm long double-open-ended coaxial detector, this figure drops to 12%. The same calculation for 1 MeV γ-rays gives values of 86% and 22%, respectively. Table 1.2 gives the peak-to-background ratio for the small and large Ge(Li) detectors and clearly shows the advantage of the small detector for the sensitive detection of low-energy radiation. Further improvements in the performance of such small detectors are to be expected from the 'guard-ring detector' design[48], which eliminates some unwanted edge effects mainly connected with charge collection.

Especially favoured in low-energy spectrometry are the nuclides giving rise to intense X-ray emission[49, 50] (decay by electron capture or internal conversion), and nuclides which have their most intense or sole γ-radiation in the

Table 1.2 Peak-to-background ratio for low energy photon detector (LEPD) and two large Ge(Li) detectors (35 cm³ useful volume)

*Peak counted**		*Peak-to-background ratio*		
		LEPD	*Ge(Li) I*	*Ge(Li) II†*
^{241}Am	59.5 keV	58.2	2.14	5.74
^{57}Co	121.9 keV	12.4	1.74	2.23

* (^{241}Am + ^{137}Cs) and (^{57}Co + ^{137}Cs) sources were measured
† A detector with excellent performances at lower energies

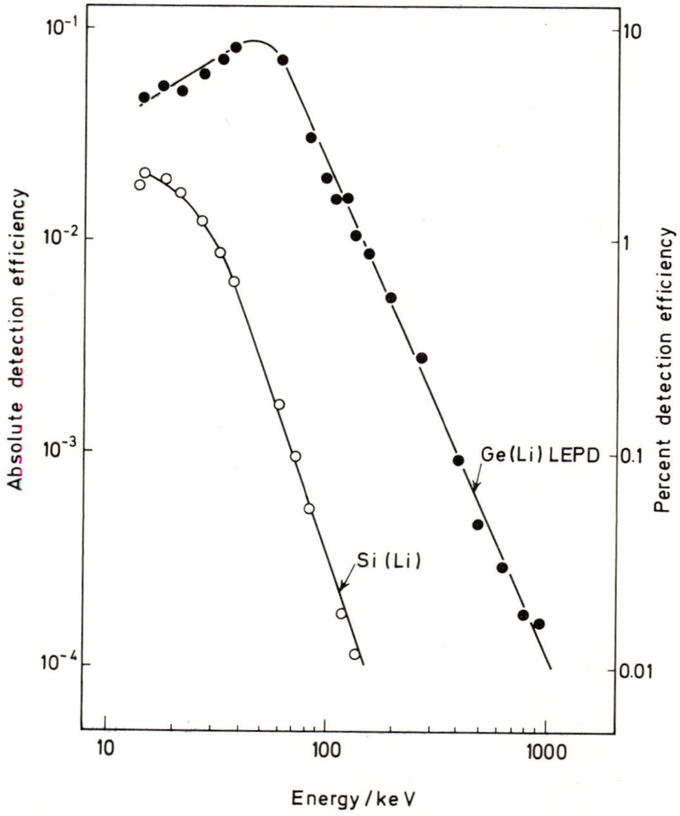

Figure 1.7 Full-energy peak efficiency curves of a Si(Li) and a small Ge(Li) detector for closest geometry ('point sources') [Si(Li):30 mm² × 4 mm; Ge(Li):78 mm² × 5 mm] (From De Neef[46] (unpublished))

50–200 keV energy range (e.g. most of the rare-earth elements, ^{182}Ta, ^{239}Np, etc). Owing to the larger energy range covered with Ge(Li) detectors, they are, in general, more suitable for activation analysis than Si(Li) detectors, except in cases where a more effective discrimination against high-energy radiation is required below the 50 keV region.

The usefulness of such detectors has been illustrated for several analytical problems[47,50-57]. The determination of trace elements in rock samples and the analysis of the platinum metals have been investigated. Preliminary results for rocks with a planar Ge(Li) detector were reported by Dran *et al.*[51]. Rosenberg and Wiik[52] determined 11 lanthanide elements in lunar samples. Some problems that are rather critical when using a small detector were considered by Hertogen and Gijbels[47]. They established its advantage over a large Ge(Li) detector for the quantitative analysis of a number of elements in silicate rocks. Contrary to the above cited authors, Zielinsky and Frey[53] first performed a chemical group separation of the lanthanides in volcanic rocks. De Bruin and Korthoven[54] discussed the determination of palladium in platinum, while Gijbels and Govaerts[55] showed the advantage of a small Ge(Li) detector over a NaI(Tl) wafer for the determination of Pd, Pt and Os in the presence of high-energy γ-emitters. Mantel and Amiel[57] investigated the feasibility of Si(Li) X-ray spectrometry in activation analysis.

It has to be emphasised that meaningful quantitative results with such small-size detectors are only obtained if geometry errors are avoided or properly corrected for. The same holds for the self-absorption phenomena of soft radiation. It has to be noted further that there are limits to the usefulness of X-rays: if many elements with close Z values are present in the sample, the $K\alpha$ and $K\beta$ X-rays may overlap to such an extent that accurate and precise peak-area integration becomes extremely difficult. A striking example is the X-ray spectrum of irradiated rock samples: from the large amount of X-ray peaks (some 40 in the energy region 25–70 keV), only a few appeared to be useful, owing to overlapping of X-rays of the elements from barium to tantalum[47].

1.1.5 Measurement problems in neutron activation spectrometry

In neutron activation spectrometry, a linear dependence between the measured full-energy peak activities and the actual source activity over as large an input-counting-rate range as possible is desired. Traditionally, in pulse-height analysis, deviations have been attributed to the dead-time from the slow process of analog-to-digital conversion. These losses have then virtually always been corrected with live-time clocks. In this way, however, pulse pile-up losses in the portions of the counting system that precede the multichannel analyser are not taken into account. Their relative contribution tends to increase with the development of increasingly fast converters, i.e. 100 MHz up to 200 MHz, and converters with a fixed short conversion time per pulse on one side, and amplifiers able to handle very high counting rates (several times 10^3 cps) with negligible losses in energy resolution on the other.

Although one has to deal simultaneously with both dead-time and pulse pile-up problems in many experiments, they are treated separately for the sake of clarity.

1.1.5.1 Dead-time problems in counting experiments

The following discussion is limited to dead-time effects with multichannel analysers, but the problem has to be tackled in single-channel countings as well. Appropriate information about mathematical approaches for single-channel analysis can be found in Refs. 58–60, while some electronic correction circuits are described in Refs. 61–65.

(a) *Live-timer correction at constant counting rate*—Most of the commercially available analysers are equipped with a live-timer circuitry for automatic dead-time correction[66]. The pulses lost because they reach the input of the ADC while it is busy processing a previous pulse are appropriately taken into account by extending the measurement interval by a time equal to the total time the instrument was busy (live-time mode of counting, with preset active time). Though the correct operation of the circuit is essential for accurate measurements, its actual performance is rarely checked. Crouch *et al.*[67], however, found inaccuracies ranging from 12% down to less than 0.5% for input counting rates up to 3×10^4 cps.

Extensive studies of the live-timer circuitry of one type of analyser are reported by Grimont *et al.*[68] and Huysmans *et al.*[69]. They observed that increasing overcompensation at higher counting rates[69] was consistent with the slowness of the electronic circuits used in Ref. 68.

There is an inherent shortcoming in the live-timer, already recognised in the original paper of Covell *et al.*[66], due to the finite width of the timing pulses. Its impact on the accuracy is treated in Refs. 69 and 70. To avoid these errors, Bartošek *et al.*[70] designed a circuit where the width is virtually infinitesimal.

It must be realised that even with an adequate operation of the live-timer, the active counting time is no longer free from statistical uncertainty. Indeed, the gating of the timing pulses is basically a statistical sampling process. The observed spread in the results is, therefore, always somewhat higher than the expected Poisson statistics in high-precision counting (Poisson statistics $<1\%$)[66, 68, 71], though the effect becomes appreciable at very short counting intervals only with the now available high-frequency clocks.

(b) *Range of validity of live-timer corrections for short-lived isotopes*—With a live-timer, lost *counts* are compensated by additional counting *time*. This is valid only when the time distribution of events is the same over the whole counting period, which is—at least theoretically—not the case for short-lived isotopes. Junod[72] calculated the errors introduced in the live-timer correction for one short-lived isotope, counted alone or in the presence of long-lived activities (Figure 1.8). From the figure it follows that, in general, one must keep the counting time, T_C, significantly smaller than the half-life, $T_{1/2}$, of the shortest-lived isotope of interest in a mixture. It can be seen further that the live-timer correction remains exact for the long-lived components. On the other hand, the acceptable range of $T_C/T_{\frac{1}{2}}$ is largely extended when using the clock-time mode of counting and applying the linear correction $A = A'T_C/T_L$

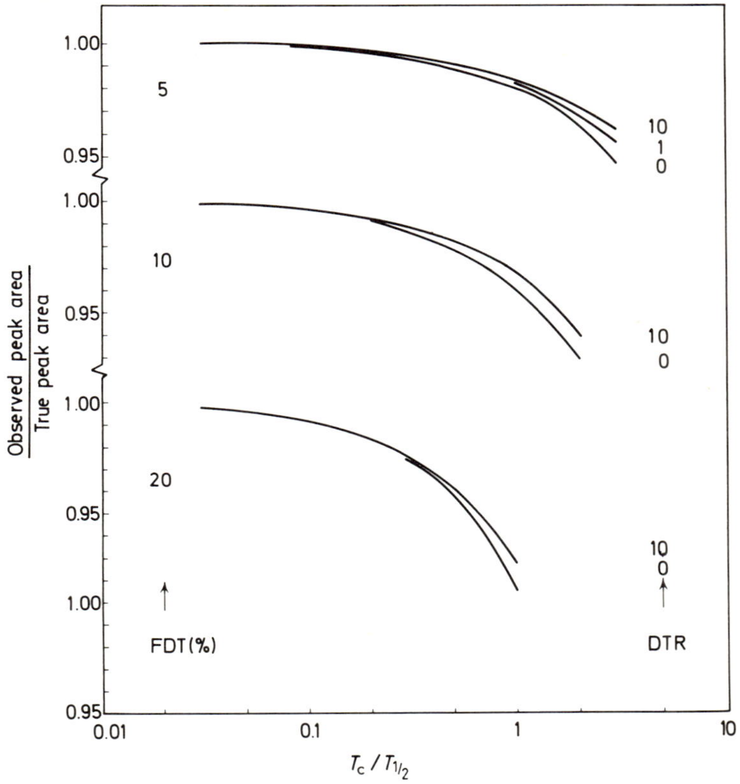

Figure 1.8 Calculated errors introduced in the automatic live-timer correction for one short-lived isotope, counted alone or in the presence of long-lived activities. FDT(%) = average fractional dead-time = total dead-time divided by the counting time (clock-time, T_C); DTR = dead-time ratio = total dead-time due to long-lived species divided by the total dead-time due to the short-lived isotope. The curves for DTR = 0 correspond to the case of a single radioisotope (Adapted from Junod[72], by courtesy of the Commissariat à l'Énergie Atomique)

(Figure 1.9). The corrected initial counting rate is then calculated following equation (1.2) (see Section 1.1.5.1(c)). A stands here for the true photopeak area, and A' for the observed area; T_C is the clock-time and the live-time T_L, is then measured by means of an external scaler, operated by the live-timer circuitry[72]. In most modern analysers, the live-time pulses are stored in channel zero and give directly T_L when used in the live-time mode of counting. A fixed counting time T_C can be applied by starting and stopping the measurement by means of an external timer. An inspection of Figures 1.8 and 1.9 shows that, for the same $T_C/T_{\frac{1}{2}}$ ratio, the errors are larger for higher average dead-time values.

(c) *Mathematical corrections for short-lived isotopes*—Nevertheless, the

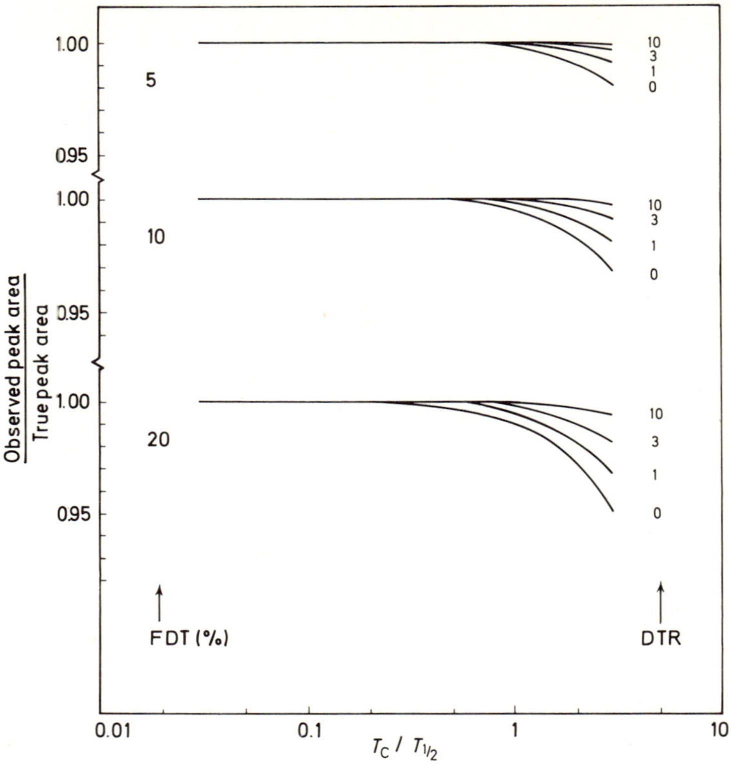

Figure 1.9 Calculated errors when using the clock-time mode of counting and applying the linear correction factor T_C/T_L (clock-time divided by the live-time). FDT(%) and DTR have the same meaning as in Figure 1.8 (Adapted from Junod[72], by courtesy of the Commissariat à l'Énergie Atomique)

serious restrictions on $T_C/T_{\frac{1}{2}}$, implied in the above outlined procedures, are seldom compatible with the requirements of sensitivity and good counting statistics. Therefore, many authors have tried to calculate dead-time corrections in more adverse situations. In view of the growing trend towards the use of short-lived radioisotopes for quantitative work, a critical assessment of the proposed solutions should be interesting. Some of them were previously reviewed by Wiernik[73].

The basic equation relating the observed peak area, A'_i, to the true initial photopeak counting rate, $R_i(0)$, is:

$$A'_i = R_i(0) \int_0^{T_C} \exp\,[-\lambda_i t][1 - DT(t)]\,\mathrm{d}t \tag{1.1}$$

where the suffix i stands for a given radionuclide. When $R_i(0)$ can be calculated from this equation, the corrected peak area is obtained from

$$A_i = R_i(0) \int_0^{T_C} \exp[-\lambda_i t]\,dt = \frac{1}{\lambda_i} R_i(0)[1 - \exp(-\lambda_i T_C)] \quad (1.2)$$

The main problem is to find a suitable analytical expression for the instantaneous dead-time $DT(t)$.

For *a single short-lived isotope*, the exact expression is[74]:

$$DT(t) = \frac{DT(0)\exp(-\lambda t)}{1 - DT(0)[1 - \exp(-\lambda t)]} \quad (1.3)$$

where $DT(0)$ is the dead-time at the start of the counting. Substituting equation (1.3) in (1.1) and solving the integral gives[75,76]:

$$R(0) = -\frac{A'\lambda DT(0)}{1 - DT(0)} \cdot \frac{1}{\ln\{1 - DT(0)[1 - \exp(-\lambda T_C)]\}} \quad (1.4)$$

When T_C and T_L are known, the foregoing formulae may be rearranged as[72,77]:

$$R(0) = \frac{A'}{T_D} \cdot \frac{\exp(\lambda T_D) - 1}{1 - \exp(-\lambda T_L)} \quad (1.5)$$

where $T_D = T_C - T_L$.

The link between equation (1.4) and equation (1.5) is

$$T_D = \int_0^{T_C} DT(t)\,dt = -\frac{1}{\lambda} \ln\left\{1 - DT(0)[1 - \exp(-\lambda T_C)]\right\} \quad (1.6)$$

which leads to an expression for $DT(0)$ as a function of T_C and T_D.

For *one short-lived isotope in the presence of long-lived activities*, Op de Beeck[75] derived the following expression:

$$DT(t) = \frac{DT_l(0) + DT_s(0)\exp(-\lambda_s t)}{1 - DT_s(0) + DT_s(0)\exp(-\lambda_s t)} \quad (1.7)$$

The parameters $DT_s(0)$ and $DT_l(0)$, the instantaneous dead-time at the start of the counting, due to the short- and long-lived components, respectively, are not directly observable and have to be calculated from a quite complex system of two equations. To solve the latter, one needs to measure $T_D = T_C - T_L$ and $DT(\infty)$, i.e. the dead-time at a moment the short-lived component has decayed. A simpler procedure[78] consists of measuring $DT(0)$ and $DT(\infty)$. It can then be shown that

$$DT(t) = \frac{(y - z)\exp(-\lambda_s t) + z}{(y - z)\exp(-\lambda_s t) + z + 1} \quad (1.8)$$

where $y = \dfrac{DT(0)}{1 - DT(0)}$ and $z = \dfrac{DT(\infty)}{1 - DT(\infty)}$. An analytical solution of equation (1.1) can be obtained after substitution of equation (1.7) or equation (1.8) for $DT(t)$.

The solution derived for this case in Ref. 79 is not theoretically correct, as it starts from the wrong assumption, i.e.

$$DT(t) = [DT(0) - DT(\infty)]\exp(-\lambda_s t) + DT(\infty) \quad (1.9)$$

The quite cumbersome procedure outlined by Junod[72] requires a least-squares analysis of the entire spectrum, which restricts its usefulness to NaI(Tl) spectrometry.

Nikolaenko *et al.*[80] analysed *the general problem of any mixture of isotopes*; they expressed $DT(t)$ as follows:

$$DT(t) = \frac{\sum_{i=1}^{n} DT_i(0) \exp(-\lambda_i t)}{1 - DT(0) + \sum_{i=1}^{n} DT_i(0) \exp(-\lambda_i t)} \tag{1.10}$$

This is in fact a generalisation of equation (1.7); n is the number of components present in the mixture; various long-lived components are formally considered as one component. The measurement of $DT(t)$ at n different moments in the course of the counting [including $DT(0)$] allows a system of n equations to be set up from which the n unknown parameters $DT_i(0)$ are calculated. In the case of a single short-lived isotope, one obtains the exact solution [equation (1.4)], whereas for two short-lived isotopes the authors found only an approximate analytical solution (error $<5\%$). Sterlinski[81], however, showed that after some rearrangement of intermediate formulae, an exact analytical solution for this case is still possible. In this way, any case can be solved in principle by numerical integration with a computer. Nevertheless, it would be difficult to estimate the precision of such a procedure.

Schonfeld[82] suggested the use of many successive readings from the dead-time indicator of the analyser, and solution of the problem by numerical integration using equation (1.1). An improved version of this approach was reported by De Wispelaere *et al.*[83]. They sampled the instantaneous dead-time values over many short periods by means of a high-precision external digital live-timer[70]. Furthermore, they avoided numerical integration by setting:

$$DT(t) = At^4 + Bt^3 + Ct^2 + Dt + E \tag{1.11}$$

In order to obtain the value of the coefficients, the sampled dead-time values are least-square fitted to a fourth-order polynomial. It was shown that the fourth-order polynomial was a sound approximation, even for rather complex mixtures. In the final experimental set-up, continuous sampling of dead-time was effected by feeding the live-time pulses to an auxiliary analyser, operating in the multiscaling mode. In this way, sampling periods down to a few milliseconds are possible and isotopes with half-lives in the seconds range may be processed if the live-timer clock frequency is properly increased. Unfortunately, this procedure requires a fair amount of additional equipment and it calls for extensive computer calculations. On the other hand, it seems an attractive procedure for incorporation in dedicated computer systems.

The solutions proposed by Gavron[84] and by Filippone and Munno[85] are typical for the least-squares resolution of scintillation spectra.

In an evaluation of these mathematical approaches, some final remarks must be made. The success of the corrections depends on the accuracy and precision of the adopted half-lives and of the measurement of total and instantaneous dead-times, T_D and $DT(t)$, respectively, In general, one can measure T_D quite easily by means of an external clock. The *a posteriori* calculation of T_D from the spectral data, as is done in many least-squares analysis programmes, is sometimes highly questionable. Indeed, pulses rejected at the input discriminator, or accepted pulses which fall outside the selected memory sub-group, may significantly contribute to the dead-time without being recorded in the spectrum. Precise instantaneous $DT(t)$ values are not so directly accessible, and one needs external circuitry[70, 79]. In any case, the precision and accuracy of readings from the dead-time indicator of most analysers is rather doubtful. To cope with possible smaller inaccuracies in these mathematical approaches, it is perhaps a good philosophy to choose a standard that is similar to the analysed samples.

(d) *Instrumental approaches*—The mixing of pulser signals with the detector signals in order to measure the overall counting losses has been recommended by several authors[86-88]. Abstraction is made here of pulse pile-up losses, as they will be treated below.

For adequate correction, the fractional losses, *FL*, suffered in the pulser peak must be equal to those suffered in the photopeaks of interest. The latter may be formulated as:

$$FL_i = \frac{A_i - A_i'}{A_i} = 1 - \frac{\int_0^{T_C} R_i(t)[1 - DT(t)]\,\mathrm{d}t}{\int_0^{T_C} R_i(t)\,\mathrm{d}t} \tag{1.12}$$

In order that $FL_p = FL_i$, two conditions must be fulfilled (the suffix i stands for a given isotope, p for the pulser):

(i) $R_p(t) = \alpha_p R_i(t) = \alpha_p \alpha_i N_i(t)$

i.e. the pulser frequency must be proportional to the total counting rate, $N_i(t)$, of the isotope of interest (α_p and α_i are proportionality constants).

(ii) The instantaneous dead-time losses of the pulser signals and the detector signals must be the same.

The first condition imposes the severest restrictions on the usefulness of the pulser technique. Pulsers with a constant frequency (whether random or not) can obviously correct only in cases of constant counting rate[86, 87], whereas for a single short-lived isotope the pulser must be triggered proportionally to the total detector counting rate[88]. For more complex cases, the pulser method is, at best, approximate.

The implications of the second condition were recognised by Bolotin *et al.*[88], and were fully treated by Wiernik[89]. When using a 'normal' pulser (i.e. with pulses equally spaced in time) or the externally triggered pulser in Bolotin's set-up, the pulser signals cannot arrive within the dead-time period owing to the processing of a previous pulser signal. As a consequence, the losses suffered by the pulser signals will correspond to the dead-time produced by the detector signals only, whereas the detector signals will be lost at a rate determined by the combined dead-time produced by the detector and pulser together[89]. The pulser frequency must therefore be kept low relative to the rate of the detector signals[88, 89] (50–100 cps.). When counting times are short, however, a low frequency introduces a rather high statistical uncertainty. As a matter of fact, the feeding of pulser signals into the spectral distribution is a sampling process, similar to that of the live-timer circuitry (see Refs. 68

and 71). Additional errors may be due to a poor estimation of the background continuum under the pulser peak. In very unfavourable conditions, this type of error can only be decreased by means of the rather complex 'leaky discriminator set-up', proposed by Anders[86]. Wiernik[90] showed via Monte Carlo calculations that the above-mentioned errors are largely eliminated when using a 'random' pulser, operated at much higher frequencies.

Görner *et al.*[91,92] have developed a 'corrector' circuit that performs an electronic integration of the integral

$$\int_0^{T_c} \exp(-\lambda_i t)[1 - DT(t)]\,dt$$

[see equation (1.1)]. Owing to the occurrence of the specific parameter λ_i, one obviously needs k-parallel correctors for the simultaneous assessment of k short-lived isotopes of interest. Two versions of the system were published: in the first[91] the exponential function is generated in an analogue way; in the second[92] in a digital way. A pulser frequency proportional to $[1 - DT(t)]$ is usually available at the output of the live-timer circuitry. Some comments on the digital version are found in Ref. 93. The device is not so suitable for multi-element determinations, but may be useful in some specific applications[94] and in routine 14 MeV activation analysis.

An elegant system to correct for dead-time losses is the 'dead-time stabiliser' proposed by Bartošek *et al.*[95]. The method does not depend on any assumption about the composition of the source and the half-life of the isotopes present. The method is based on the automatic creation of additional dead-time at the end of very short live-time measuring intervals (from 5 to 10 ms depending on the apparatus setting), so as to keep the fractional dead-time constant at a preset value during the length of the counting. The counts lost during the short measuring intervals are compensated by the normal live-timer correction, but so rapidly after the occurrence of the individual pulses that decay or change of the pulse rate do not affect the accuracy.

The dead-time correction system proposed by Harms[96] consists of storing a compensating pulse together with the next registered pulse in the same channel, each time the loss of a pulse has been observed at the input of the ADC. Harms showed by argument that the procedure yields correct results for spectra which do not change during the course of the measurement. In order to check the validity of the Harms dead-time correction for spectra which vary during the measurement, Monte Carlo calculations were performed by Masters and East[97]. The results are convincing and the system has been incorporated in several new multichannel analysers. The Harms correction, being entirely different from the live-timer method, has more in common with buffer memories, also called 'derandomisers', proposed to deal with counting losses in nuclear physics experiments involving pulse-burst analysis, etc.[98]. When appropriately designed, these devices are not only able to store the number of lost counts but also their energy information. The usefulness of such rather complex units for activation analysis is, however, seldom evaluated. A drawback of the Harms system, and of such derandomisers, is the difficulty in taking into account pulse pile-up effects in the amplifier part of the measuring chain.

1.1.5.2 Problems in high-counting-rate spectrometry

Neutron-irradiated samples sometimes must be measured in conditions of decreased solid angle to reduce the counting rate to a level which is acceptable for the spectrometer. The maximum acceptable counting rate is determined by the effect of pulse pile-up, which is responsible for such effects as counting losses, degradation of resolution and peak symmetry and, in extreme cases, of complete distortion of spectra. The effects of rate-dependent random coincidences were recognised in earlier scintillation spectrometry involving a least-squares analysis of the spectrum[99-102]. In fact, owing to pile-up a spectrum could no longer be regarded as a linear combination of single spectra. However, the problem was somewhat neglected in the first years of Ge(Li) spectrometry.

Analysis of the pulses lost by random pulse pile-up shows that there are two distinct effects. The first type of loss (peak pile-up) is the result of two pulses occurring within a specific time interval. These pulses are seen by the converter as a single pulse and thus produce a background up to the sum of the amplitudes of both pulses. The second type (tail pile-up) results from pulses falling on a non-negligible tail of a previous pulse. The ADC can detect these pulses as individual events, but photopeak tailing and broadening are introduced. Reduction of pile-up is essentially controlled by the pulse duration, i.e. the time interval during which the pulse has a non-negligible amplitude. Pulse-shaping should, therefore, produce unipolar pulses of the shortest possible duration. On the other hand, the optimum energy resolution for semiconductor detectors is frequently obtained at shaping time constants of up to 15 μs. Nevertheless, advances made in high-counting-rate spectrometry are probably more spectacular than in any other area of semiconductor γ-ray spectrometers and it is interesting to review the topic of high counting-rate γ-ray spectrometry in greater detail.

To a first approximation, fractional losses in counting rate under a photopeak can be expressed as[13, 103]:

$$\frac{R - R'}{R} \approx 2N\tau$$

where N is the total detector rate and τ the system's resolving time. Starting from this equation, Wyttenbach[103] developed a simple first-order correction based on the measurement of both the live-time and the clock-time. A slightly modified version of this method was reported by Verheijke[104].

The mixing of pulser signals with the detector signals at the pre-amplifier may be an appropriate method when dealing with constant input rates. A possible—but usually negligible—error is that a pulser signal does not pile-up with another pulser signal[105], even in the case of some pseudo-random pulsers[89]. The main limitations of the pulser technique, however, have been treated in Section 1.1.5.1(d).

Drawbacks of the above methods are:

(i) To a certain extent they can correct for the lost counts, but they cannot prevent spectrum distortion and resolution losses.

(ii) They cannot account for the complex interaction of dead-time and pile-up effects in the measurement of short-lived activities. Unfortunately, it is in this area especially that high counting rates frequently occur.

To achieve a higher precision in high-counting-rate spectrometry, circuits have been built to detect the pile-up effect and to split the input events into two categories: undistorted pulses are accepted whereas distorted ones are rejected. Peak pile-up rejection is generally accomplished with the aid of a fast anticoincidence and pulse-width discriminator operating on the pre-amplifier signals. Tail pile-up effects can be eliminated by an inspection circuit requiring that the reference base-line be identically zero before a new pulse can be accepted, or by processing only pulses that are preceded by a fixed time period in which no other event has occurred. The pile-up rejection circuit can only reduce distortions and not totally avoid them, so that there is always a resolving time within which the rejector cannot recognise two discrete interactions in the detector from a single one. Nevertheless, pile-up is reduced by one or two orders of magnitude. One drawback to the use of these circuits is that the counting losses introduced do not lend themselves easily to correction, and at moderate-to-high counting rates quantitative results are inaccurate due to counting losses. Several commercially available pile-up rejectors are equipped with a live-time correction system which accounts for counting losses introduced by pulse pile-up. The working principle has not been disclosed by the manufacturers. To cope with these errors, Bartošek *et al.*[106] modified a commercially available pulse pile-up rejector so that it could be interfaced properly with the analyser's live-timer circuit. Correction for the lost pulses is obtained by extending the measurement interval by stopping the live-time clock from the moment of occurrence of a peak pile-up event, excluding tail pile-up, until the time of arrival of the next unaffected pulse. When there is a delay between the acceptance of a pulse in the rejector and the occurrence of its output, it is also necessary to eliminate the processing of pulses during the dead-time period of the ADC. Obviously, a pulser is not required with this system. In the case of time-dependent counting rates as encountered when measuring short-lived radionuclides, this modified rejector was interfaced with the 'dead-time stabiliser', already mentioned in Section 1.1.5.1(d), yielding an almost all-round correction circuitry[107].

1.2 NEUTRON ACTIVATION ANALYSIS OF ATMOSPHERIC SAMPLES

1.2.1 Introduction

During the past decade, increasing attention has been focused on pollution of the natural environment. Air pollution is a growing hazard to public health, soil, water, vegetation, materials, animals, weather, visibility and personal comfort.

The importance of measuring concentrations of air pollutants as an inherent part of pollution abatement programmes is being recognised by both governmental and industrial agencies. Until a few years ago the main effort was concentrated on the determination of some gaseous components such as SO_2, CO, NO_x, O_3 and hydrocarbons. Airborne dust should, however, also be regarded as one of the main air pollution materials. No valid judgment as to the relative potential threat of this dust, to man's health or to his environment, is possible without a precise knowledge of its composition. The airborne dust may originate from the surface of land and ocean by the passage of winds or by any other natural effect such as eruptions of volcanoes etc., although technological and industrial development, coupled with vast urbanisation of the population, are probably the main sources of this kind of pollution. Stationary fuel combustion, solid waste disposal, transportation and industrial processes introduce a large variety of trace elements into the atmosphere. Thus, in order to trace the pollution sources of aerosols, and to determine how much the anthropogenic contribution is, it is necessary to determine the amount and elemental composition of atmospheric particulates and precipitation. For studies of sources, characteristics, atmospheric transport processes and removal rates, sensitive and multi-elemental analyses are required. Accurate analyses of airborne dust, dry deposition material and rain water is difficult because of the complex nature of the samples and the low concentrations involved. These studies generally require large numbers of analytical results. Therefore, the technique should not only be precise and sensitive but speed and ease are also desirable. Analytical techniques proposed for this purpose, such as colorimetry, emission spectrometry, atomic absorption, flame photometry and polarography may or may not involve a dissolution step of the sample and subsequent chemical and instrumental steps. Since atmospheric particulate material is often to a great extent carbonaceous[108], it is largely insoluble in water. Previous ashing, in order to obtain complete dissolution, without losses of volatile elements is very difficult. Also the sensitivity and selectivity of most of these techniques is inadequate to cope with the complexity of environmental samples. Within the past few years, instrumental neutron activation analysis using Ge(Li) γ-spectrometry and computer-assisted data reduction has been shown by a variety of authors to be a very powerful tool for the analysis of airborne particulate material. The sensitivity with which one may observe minor and trace elements and, therefore, also the number of components determined, has greatly increased during the past years.

Reproducibility and accuracy can be adequate even for trace elements, provided that a number of sources of error are avoided. Airborne particulates, collected on a filter or on an impaction surface, can often be analysed without any further treatment provided that the filter or surface is pure enough to allow a reasonable signal-to-blank ratio. The often-heard claim that neutron activation analysis is an expensive method is not tenable when many samples can be treated instrumentally and simultaneously and when more than 30 elements can be determined. Another advantage of the technique in its instrumental form is the availability of the sample for further analysis by other techniques. Obviously no single analytical technique can meet all the requirements set for air pollution measurements. Also, activation analysis has some shortcomings and drawbacks. Some important elements cannot be determined easily and there is no information on the chemical states of the elements. Thus, if one wants to ascertain the chemical form, which may be of vital importance in learning about the atmospheric reactions and in judging toxicological effects, pre-irradiation chemical separations must be carried out.

1.2.2 Sampling atmospheric pollutants for activation analysis

The major sampling techniques of atmospheric particulates are absorption by liquid impinger, electrostatic precipitation, filtration and impaction[109].

The first technique is generally very incomplete for dry particles. Electrostatic precipitation is a very effective technique, but requires very long sampling times, owing to the low obtainable flow rates. Therefore, aerosols are mostly sampled by passing air through fibrous or membrane-type filters. The important physical properties of the commonly used filter materials have been summarised by Lockhart *et al.*[110]. The principal requirements for a filter material include the collection efficiency, which is a function of particle size and vertical velocity, the pressure drop across the filter and the reduction of the flow rate through dust-loading. Of secondary importance are homogeneity, tensile strength, thickness, weight and hygroscopicity. Glass fibre filters have often been used for the sampling and determination of total suspended particulate, organics, radioactivity, sulphate and nitrate[111]. Unfortunately, these filters cannot be used for neutron activation analysis of inorganic trace elements, owing to their inorganic nature, unless the accumulated dust is separated from the filter, with the unavoidable risks of contaminations and losses. In order to select a filter material with low impurity levels of the elements to be determined, several authors have investigated by neutron activation analysis a number of organic filter materials whose physical properties are suitable for particulate sampling. Parkinson and Grant[112], Keane and Fisher[113], Brar *et al.*[114] and Spurny and Fiser[115], using short thermal neutron activations and NaI(Tl) γ-ray spectrometry, measured the concentrations of Na, Mg, Al, Cl, Mn and Br. On this basis, they chose Whatman 41, Esparto grass and Dexter X-1215. After the introduction of Ge(Li) detectors, more extensive studies of the blank levels of potential materials were performed. A number of investigators decided to use different types of millipore membrane filters, such as AAWP[116], RA[117], EH[118, 119] and WS[119].

Other groups of researchers chose different kinds of cellulose Whatman filters, such as No. 1[120], No. 40[121], No. 41[122-124] and 541[125]. The most extensive study of filter materials for non-destructive activation analysis of aerosols was probably made by Dams *et al.*[124]. Ten filter materials, ranging from membrane types to synthetic fibre and cellulose fibre filters, were analysed for the elements usually detected in airborne particulates. Impurity levels were given for the following 24 elements: Na, Mg, Al, S, Cl, K, Ca, Sc, Ti, Cr, V, Mn, Fe, Ni, Co, Cu, Zn, Br, Ag, Sb, Ba, Ce, La and Hg. The cleanest filter appeared to be Whatman No. 41. This was also found to allow a high flow rate, but it had a somewhat lower retentivity than the glass fibres, polystyrenes and membranes. At low linear velocity (5 cm s^{-1}), Cawse and Pierson[121] found a mean filtration efficiency for actual atmospheric aerosols of 91% for Whatman No. 41 and 97% for Whatman No. 40. At higher velocities (100 cm s^{-1}), both retentivities still exceed 98%, even for the elements associated with the smaller particles[110, 123].

Dams *et al.*[124] calculated a figure of merit, as the sample/blank ratio after 1 h sampling in a typical rural atmosphere at a flow rate of 13 l min^{-1} cm^{-2}. With Whatman No. 41 this figure of merit is less than 2 for only three elements, namely Cl, Na and Co. After 24 h sampling, the lowest figure of merit is obtained for Cl, namely 12. Also the silicon and lead concentrations were found to be negligible after a few hours sampling[125, 126].

The impurity levels of membrane filters, though not prohibitively high, are the principal limitation of detectable concentrations for several elements, because these filters show very high flow resistances and, therefore, reduce the obtainable flow rates strongly. Sampling for 24 h with membrane filters is therefore restricted to polluted areas. Since they have extremely high collection efficiencies, the 75% greater flow rate of a 0.8 μm pore size type represents a real sampling advantage over the 0.45 μm pore size. Polystyrene Microsorban (Delbag) filters generally have high Cl, Br, Cu and Zn values, precluding sensitive determinations of these elements. In addition, in neutron activation analysis the induced Cl and Br activities raise the detection limits of at least 15 other elements by an order of magnitude[124].

In attempting to identify sources of pollutants to study the atmospheric behaviour of aerosols and to learn about toxicological hazards, one can gain considerable advantage by measuring the mass *versus* size distribution of the particles containing the different elements. Therefore, a growing number of atmospheric samples are being taken with inertial impactors. In most cascade impactors, particles are segregated in a number of fractions, when they impact on to metal or glass plates[127]. When these surfaces are covered with impaction sheets consisting of a sufficiently pure material, these sheets can be analysed instrumentally by neutron activation and no separation of the dust is required. Several common organic materials such as polyethylene, Mylar, Teflon[124] and polycarbonate[116, 128] have been investigated for impurity levels by neutron activation analysis. Mylar contains high Mn and Sb concentrations. Owing to the high neutron cross-sections of these elements, their presence interferes significantly with the determination of several elements. The other materials can be obtained sufficiently pure. The impurities seem to be associated with the bulk of the material, rather than the surface, making thinnest sheeting the best.

Little information is available on the composition of the very fine particles (<0.1 μm diameter) that are not retained by impactors or filter materials, and on possible gaseous compounds of some elements. Pillay *et al.*[129, 130] sampled Se on filters and in a liquid trap beyond the filters. On average, in urban air, 50% of the Se was retained in the lead acetate solution and 50% on the Dexter X-1215 cellulose filters. It should, however, be recalled that Brar *et al.*[131] found in an earlier experiment that these Dexter filters retained only *ca.* 80% by weight of the particles, compared with a glass fibre filter. Moyers[132] collected Br on membrane filters and simultaneously, past an electrostatic precipitator, gaseous Br on clean activated charcoal. The authors concluded that between 20 and 70% of the total bromine produced by combustion of leaded gasoline may be released in the gaseous phase. Other experiments indicate that more than 80% of the bromine may be gaseous in an unpolluted marine atmosphere in Hawaii.

Almost no information is available on sampling dry deposition for subsequent non-destructive neutron activation analysis. Cawse and Pierson[121] sample-dry deposition for periods of one month on Whatman No. 541, mounted horizontally in a Perspex frame. Shelter from rain is provided by a sheet of Perspex placed 12 cm above the frame. Although the filter paper does not correspond to any natural surface such as soil, grass or other vegetation, the authors claim that this collector provides a useful measure of the relative dry deposition behaviour. Dry deposition for analysis by neutron activation can also be sampled in a container where the bottom is covered with one of the earlier discussed impaction surfaces. Routinely, wet and dry deposition are sampled simultaneously with standard deposit gauges[133, 134], consisting of a polyethylene funnel and a polystyrene bottle. After one month, soluble and insoluble matter are separated by filtration on Whatman No. 42[121] or membrane[135] filters. The solution can then be evaporated and ashed or carefully dried on Whatman No. 41 paper for subsequent neutron irradiation.

Snow or ice samples[136-139] can be sectioned and immediately trimmed back in order to remove contamination. Blocks are then sealed in polyethylene bags or tubing.

1.2.3 Preparation of atmospheric samples for neutron activation analysis

It is of prime importance to reduce manipulation of the samples before irradiation in order to minimise risks of contamination. Precautions in handling the filters or impaction surfaces are basically those ordinarily exercised during any quantitative trace analysis.

Routinely, a fraction of the filter is cut before irradiation and placed in pre-cleaned polyethylene vials or sealed in polyethylene bags or between two sheets of clean polyethylene. Membrane filters are quite brittle and are easily broken, which may result in loss of particulate matter. These problems were avoided by exposing the filter to acetone, so that it can be folded into a small volume and sealed in precleaned plastic film[116, 128]. Cellulose filters and polyethylene impaction surfaces can also be wrapped in an equal amount of the same clean material and pressed into a pellet (1.25 cm $\times$ 2 or 3 mm). To obtain a firm and stable pellet from the polyethylene, it should be heated to 60 °C. This pelletising procedure assures reproducible irradiation and counting geometry of the sample as well as convenience in transferring from irradiation to counting vials[140, 141].

When rain water, snow, ice or 'in-water' collected fallout is to be analysed, a major advantage of activation analysis, namely that no blank is introduced during analysis, is lost. Possible use of reagents and contact with the container material used for sampling, storage and irradiation favours contamination of the sample by material leaching from the container and leads to losses of trace elements due to adsorption on to the container walls. These problems were studied in detail by Robertson[142, 143]. To avoid problems associated with irradiation of liquids such as build-up of pressure due to radiolysis and loss of activities in volatile compounds, the samples can be irradiated in quartz vials in the frozen form[144]. These problems depend largely on the irradiation conditions. Comar[145] claims that it is possible to irradiate liquid samples in Teflon or polyethylene tubes if thermal neutrons are used at a flux not exceeding 5×10^{12} neutrons cm^{-2} s^{-1}, and if the irradiation time is not more than a day. For longer irradiation times, at higher fluxes, it is advisable to use thick-walled quartz containers and irradiate only freeze-dried samples.

1.2.4 Neutron activation analysis involving radiochemical separations

Neutron activation analysis of atmospheric samples has recently only occasionally been applied in connection with radiochemical separations. If the concentration from a single or from a limited number of major components is obtained after a radiochemical separation, other single-element techniques may offer more accurate or cheaper results. Radiochemical separations may be classified in two groups: individual and group separation.

1.2.4.1 Individual separations

In order to compare selenium/sulphur ratios in aerosols, in snow and rain, with those of fossil fuels, in igneous and sedementary rocks and in sulphide ores, several authors[146-148] have determined very low concentrations of Se by activation analysis. Selenium was separated by extractive distillation, precipitated and the ^{75}Se activity counted by NaI γ-spectrometry. The element In is sometimes used as a tracer for diffusion of emissions by stacks[149] and to study scavenging efficiencies of storms[150-152], because of its large neutron cross-section, its convenient half-life and its relative rarity as an atmospheric contaminant. A non-destructive analysis of In in rain water, even when using Ge(Li) γ-spectrometry, is generally not possible. After a few fast and simple separation steps based on liquid extraction, the short-lived isotope ^{116m}In can be measured by scintillation γ-spectrometry. When only small samples are available the counting statistics may be the limiting factors, and, therefore, β-counting is sometimes superior. Bhatki and Dingle[150] could, in this way, assess precisely the indium concentration in parts per trillion in rain water. Winchester *et al.*[153-157] have extensively investigated the atmospheric behaviour of the halogens in continental marine aerosols, in rain, in snow and in fresh water. After a short irradiation, the isotopes ^{38}Cl, ^{80m}Br and ^{128}I were separated using selective oxidation and reduction, and CCl_4 extraction. The β-activity was counted from silver halide sources. With the growing concern for Hg as an environmental pollutant, neutron activation analysis is being used on an increasing scale in the search for elevated levels in biological materials, in soils and in water samples and for establishing 'base-line' levels. Most of the procedures rely on a radiochemical separation involving a distillation and/or a precipitation step[158-170]. The absolute detection limit for Hg can be as low as 0.1 mg [159]. Owing to the high enrichment of Hg in the atmosphere, generally no radiochemical separation is performed for the detection of atmospheric samples, when using the long-lived ^{203}Hg isotope. A few investigators have determined small amounts of Na, Cl and S in Greenland and Antarctic snow by neutron activation analysis with radiochemical separations[136, 171, 172], and Hg, Se, Mn, Cu, As, Cd and Sb in glacial ice[210, 211].

1.2.4.2 Group separations

Group separations of the radionuclides into one or a few groups, leaving the job of final discrimination to the detector, are often of vital importance in the multi-elemental analysis of marine samples. For the analysis of marine aerosols, Dudey *et al.*[173] developed a simple separation scheme to remove the large Na activities and separated the other radioactive nuclides into two almost equally important fractions by passing the dissolved particulates over an anion resin in a Cl^- medium and a Chelex resin in an alkaline medium (pH 9). Hoffman *et al.*[174] avoided the Na and Cl interferences by a chloroform extraction of Cu, V and Al from a cupferron solution. Rancitelli *et al.*[175-177] use a group separation scheme appropriate for the multi-elemental analysis of rain, snow and aerosol samples. After irradiation, the frozen sample is melted, weighed and carriers of Cr, Co, Cu, Ag, Fe, Hg, Zn, Au, Mn, Sb, Se, Sc and La are added. From a small fraction of the sample the Na activity is measured. The solution is then transferred to a hot-water jacketed mixed-bed precipitation column, prepared from one layer each of OH^-, S^{2-} and CO_3^{2-} substituted Dowex X1–X2 resin. Fifteen elements including Sc, Mn, Co, Fe, Cr, Cu, Zn, Ag, Au, Hg, Se, Sb, Br, As and U are removed from the sample by precipitation or exchange reactions with the mixed resin. The resin is quantitatively transferred to a Ge(Li) detector and to an anticoincidence-shielded γ–γ-coincidence multidimensional analyser[178]. From the column effluent, the elements K, Cs, Rb and several alkaline earths are precipitated with sodium tetraphenylboron and counted with both detector systems. Also the US Geological Survey[158] has performed a few simple group separations for the analysis of fresh-water samples. Mercury is precipitated with $SnCl_2$, and by means of a sulphide precipitation As, Sb, Cd, Cr, Zn, Mo and Co are separated. Schramel *et al.*[179] determined 12 environmentally significant elements after distillation from a H_2O_2–HBr medium and a fast ion-exchange separation. Other automated ion-exchange systems routinely used for the analysis of geological, biological and water samples can also be applied. A critical discussion of existing procedures for separation is given by Adams *et al.*[180].

1.2.5 Multi-elemental instrumental neutron activation analysis

The development of high-resolution Ge(Li) detectors and the introduction of computer processing of γ-ray spectra has remarkably increased the possibilities of purely instrumental neutron activation analysis. Such a non-destructive technique is generally limited to those types of sample where the matrix does not become prohibitively radioactive. Most environmental samples satisfy this requirement. Airborne particulates collected on a clean organic filter are especially well-suited for non-destructive activation analysis. The major components of such a sample (filter + particulates), being organics, carbon, water, sulphate, nitrate and ammonia, do not produce important γ-activities. In the inorganic fraction, the elemental concentrations are balanced so that no single major activity is produced that can mask other elements. The most abundant elements Si, Fe, Ca, Mg and Cl have indeed relatively low cross-sections or small abundances of their γ-transitions. Only when marine aerosols are sampled does the large concentration of Na produce troublesome ^{24}Na ($T_{\frac{1}{2}}$ = 15 h) activities. In urban aerosols, the ^{82}Br ($T_{\frac{1}{2}}$ = 36 h) activity is predominant between 1 and 7 days after the end of the irradiation, while the

1779 keV photopeak of ^{28}Al ($T_{\frac{1}{2}} = 2.3$ min) shows the highest intensity a few minutes after a short neutron irradiation of continental airborne particulates. These interferences are, however, not so important as to make non-destructive analysis impossible.

The first attempts to analyse airborne particulate matter non-destructively date from 5 to 10 years ago[112-114, 181]. Short neutron irradiations and NaI γ-spectrometry resulted in the determination of Al, Br, Cl, Mn, Na, V and occasionally I, Ba and Mg. This simultaneous detection of six or more isotopes was possible by least-squares fitting and comparison with library spectra from irradiated individual standards. Later on, several authors started using non-destructive activation analysis with NaI γ-spectrometry for the detection of the longer-lived isotopes. Brar *et al.*[182] could detect the longer-lived isotopes of the elements Fe, Zn, Cr, Sb, Hg, Se, Ce, Ag, Co, La, Sc, Cs and Eu in Chicago surface air, by counting at intervals of 1, 2 and 9 months after irradiation. Although a few authors continued using scintillation detectors[183-186] for the detection of a limited number of elements, a major breakthrough occurred in 1970 when Zoller and Gordon[118] and Dams *et al.*[187] started utilising the excellent energy resolution of Ge(Li) detectors for instrumental neutron activation analysis of aerosols. When the parameters of analysis, which include irradiation, waiting and counting times, amount of sample and counting geometry, are optimised, one can generally observe more than 30 elements in airborne particulates. These parameters are dependent on the composition of the sample, the activation cross-sections of the elements present and the half-lives, γ-ray energies and intensities of the radionuclides, from not only the matrix elements but also from the trace or minor elements of interest. Because automatic reduction of Ge(Li) data can be markedly simpler than reduction of NaI data, large numbers of similar samples can be processed routinely at fairly rapid rates by relatively unskilled personnel if only the most intense radioisotopes are measured. The recent literature now contains many papers on the non-destructive activation of airborne particulates. Fields of investigation include urban[120, 122, 188, 189], industrial[117, 140, 190-194], marine[195, 196], rural[121, 197] and background aerosols[197].

1.2.5.1 Procedure

Although each investigating group has its own irradiation counting scheme, most are quite similar to the one described by Dams *et al.*[187]. This procedure consists of two irradiations and four to five γ-spectrometric measurements using a 20–50 cm^3 Ge(Li) detector with a resolution of approximately 2.0–2.5 keV for the ^{60}Co 1332.4 keV photopeak. For the analysis of the elements giving rise to short-lived isotopes, an approximate fraction of each filter is irradiated for 5 min. After a cooling time of 3 min, the isotopes ^{28}Al, ^{37}S, ^{49}Ca, ^{51}Ti, ^{52}V and ^{66}Cu can be detected. Some 15–20 min after the end of the irradiation, when the predominant ^{28}Al and ^{52}V activities have decayed, a second count is begun. It generally allows the detection of ^{24}Na, ^{27}Mg, ^{38}Cl, ^{56}Mn, ^{80m}Br, ^{116m}In, ^{128}I (Table 1.3) and in some samples ^{165}Dy and ^{125m}Sn. The same or another portion of the same filter or impaction surface is then irradiated for a longer period. After a 1 day cooling period, the samples are

Table 1.3 Nuclear data of isotopes measured in aerosols after 5 min irradiations*

Element	*Isotope*	*Half-life*	*γ-ray used* /keV	*24 h urban sample*/ng m^{-3}	
				Detection limit	*Typical concentration*
Al	^{28}Al	2.31 m	1778.9	8	3000
S	^{37}S	5.05 m	3102.4	5000	5000
Ca	^{49}Ca	8.8 m	3083	200	4000
Ti	^{51}Ti	5.79 m	320.0	40	200
V	^{52}V	3.76 m	1434.0	0.2	70
Cu	^{66}Cu	5.1 m	1039.0	20	100
Na	^{24}Na	15 h	1368.4	40	1200
			2753.6		
Mg	^{27}Mg	9.45 m	1014.1	600	1000
Cl	^{38}Cl	37.3 m	1642.0	100	5000
			2166.8		
Mn	^{56}Mn	2.58 h	846.9	0.6	200
			1810.7		
Br	^{80m}Br	17.6 m	617.0	4	200
In	^{116m}In	54 m	417.0	0.04	1
			1097.1		
I	^{128}I	25 m	442.7	1	15

* Detection limits after 24 h sampling in typical urban air with a high volume sampler, as obtained by the irradiation counting scheme described in Section 1.2.5.1. Typical concentrations in urban air (Ghent, Belgium)[140, 187]

counted for *ca.* 10 isotopes, namely ^{24}Na, ^{42}K, ^{64}Cu, ^{69}Zn, ^{72}Ga, ^{76}As, ^{82}Br, ^{122}Sb, ^{140}La and ^{153}Sm. Several other isotopes such as ^{99}Mo, ^{115}Cd, ^{152m}Eu, ^{175}Yb, ^{186}Re, ^{187}W and ^{198}Au can only be detected under favourable conditions, when the ^{24}Na and ^{82}Br activities are not too high. In order to avoid

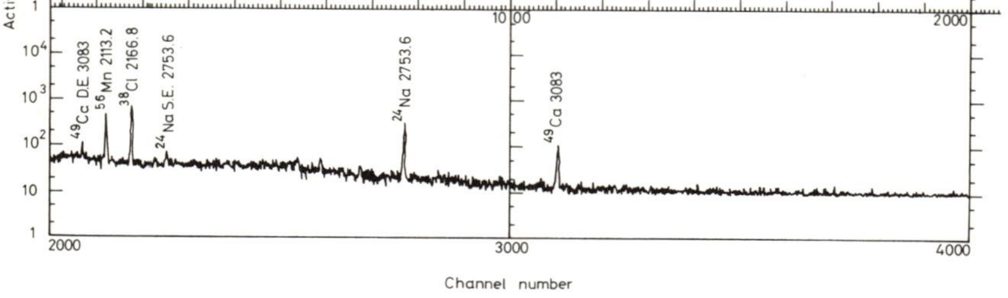

Figure 1.10 γ-Ray spectrum, recorded with 35 cm^3 Ge(Li) detector, of aerosol samples, t_{irr} = 5 min, t_{dec} = 3 min, t_{count} = 6 min

Table 1.4 Nuclear data of isotopes measured in aerosols after several hours irradiation*

Element	*Isotope*	*Half-life*	*γ-ray used* /keV	*24 h urban sample*/ng m^{-3} *Detection limit*	*Typical concentration*
K	^{42}K	12.5 h	1524.7	7.5	1700
Cu	^{64}Cu	12.5 h	511.0	5	100
Zn	^{69m}Zn	13.8 h	438.7	20	1000
As	^{76}As	26.3 h	657.0; 1215.8	4	50
Ga	^{72}Ga	14.3 h	834.1	1	2
Eu	^{152m}Eu	9.3 h	963.5	0.02	0.05
W	^{187}W	24.0 h	685.7	0.1	1
Br	^{82}Br	35.9 h	776.6; 619.0	0.5	200
Cd	^{115}Cd	53 h	492.5; 527.7	5	10
	^{115m}Cd	4.5 h	336.6		
Sb	^{122}Sb	66 h	564.0	1	30
La	^{140}La	40.3 h	1595.4	0.1	2
Sm	^{153}Sm	47.1 h	103.2	0.002	0.3
Au	^{198}Au	65 h	411.8	0.02	0.05
Sc	^{46}Sc	83.9 d	889.4; 1120.5	0.004	0.7
Cr	^{51}Cr	27.8 d	320.0	0.2	30
Fe	^{59}Fe	45.1 d	1098.6; 1291.5	20	3700
Co	^{60}Co	5.2 y	1173.1; 1332.4	0.02	4
Ni	^{58}Co	71.3 d	810.3	20	20
Zn	^{65}Zn	245 d	1115.4	1	1000
Se	^{75}Se	121 d	264.6; 400.6	0.1	10
Ag	^{110m}Ag	253 d	657.8; 937.2	1	0.5
Sb	^{124}Sb	60.9 d	1690.7; 602.6	1	30
Cs	^{134}Cs	2.07 y	795.8	0.05	1
Ba	^{131}Ba	11.5 d	216.1	40	100
Ce	^{141}Ce	32.5 d	145.4	0.2	4
Lu	^{177m}Lu	155 d	208.4	0.05	0.04
Hg	^{203}Hg	46.9 d	279.1	0.1	2
Th	^{233}Pa	27.0 d	311.8	0.04	0.5

* Detection limits after 24 h sampling in typical urban air with a high volume sampler, as obtained by the irradiation counting scheme described in Section 1.2.5.1. Typical concentrations in air of Ghent, Belgium[140,187]

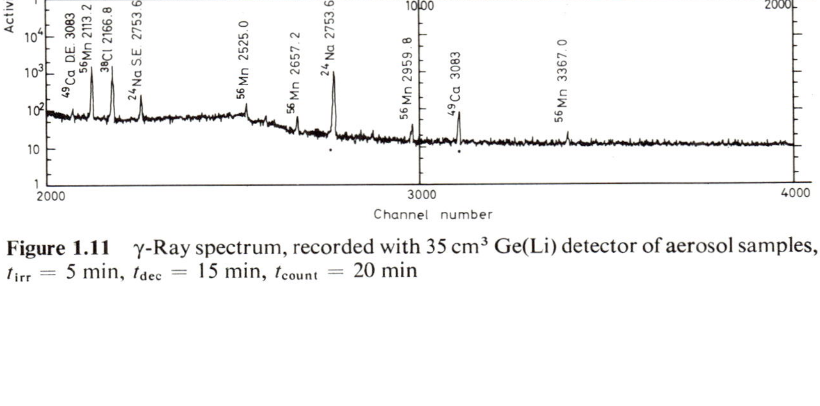

Figure 1.11 γ-Ray spectrum, recorded with 35 cm^3 Ge(Li) detector of aerosol samples, $t_{irr} = 5$ min, $t_{dec} = 15$ min, $t_{count} = 20$ min

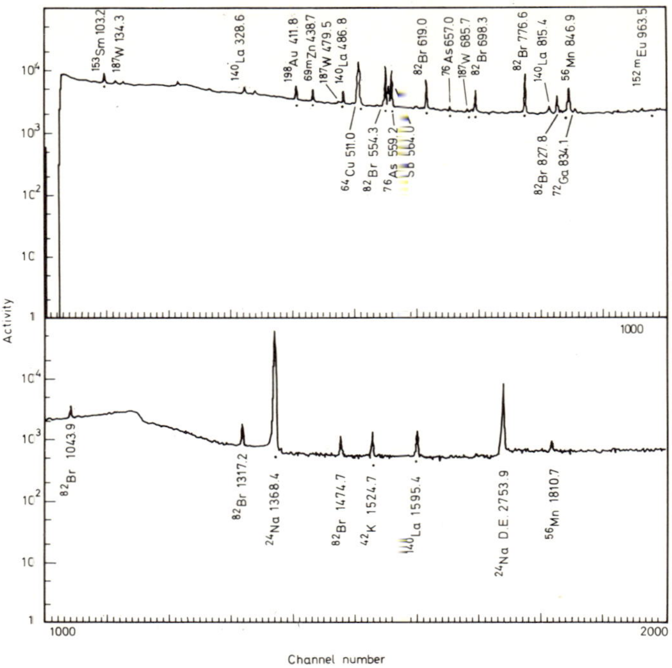

Figure 1.12 γ-Ray spectrum, recorded with 35 cm³ Ge(Li) detector of aerosol samples, $t_{irr} = 7$ h, $t_{dec} = 1$ d, $t_{count} = 20$ min

the interference of ^{24}Na, a measurement after *ca.* 5–10 days cooling is often helpful for the detection of the isotopes ^{99}Mo, ^{115}Cd, ^{122}Sb, ^{140}La, ^{153}Sm, ^{175}Yb and ^{198}Au. Unfortunately, most information is only obtained after waiting for 20 days after the complete decay of ^{82}Br. The net peak areas of the isotopes ^{46}Sc, ^{51}Cr, ^{59}Fe, ^{60}Co, ^{65}Zn, ^{75}Se, ^{124}Sb, ^{134}Cs, ^{141}Ce, ^{203}Hg and ^{233}Pa (from Th) can readily be determined, and often several other isotopes such as ^{58}Co (from Ni), ^{86}Rb, ^{110m}Ag, ^{131}Ba, ^{152}Eu, ^{177}Lu, ^{181}Hf, ^{182}Ta, ^{192}Ir can be detected. Tables 1.3 and 1.4 list the nuclear reactions, the half-lives and the main γ-transitions of the generally detectable isotopes. Figures 1.10–1.13 give γ-spectra recorded after 3 min, 15 min, 1 day and 20 days cooling time of a typical air particulate sample collected in Milan by the Commission of the European Communities[12].

A non-destructive technique can also be applied for the analysis of dry or wet deposited material since no major interfering activities are produced on neutron irradiation. Only very few results of instrumental neutron activation analysis have been published to date. Cawse and Pierson[121] determined concentrations in monthly rain-water samples of the following 27 elements: Na, Al, Cl, Ca, Sc, V, Cr, Mn, Fe, Co, Ni, Cu, Zn, As, Se, Br, Rb, In, Sb, I, Cs, La, Ce, W, Au, Hg and Th. The same authors[121] give reliable data on 19

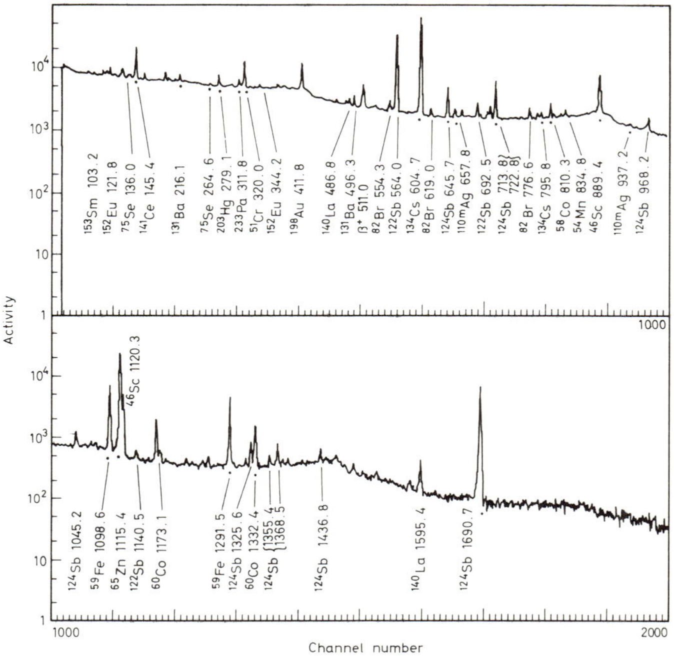

Figure 1.13 γ-Ray spectrum, recorded with 35 cm^3 Ge(Li) detector of aerosol samples, $t_{irr} = 7$ h, $t_{dec} = 20$ d, $t_{count} = 2$ h

elements in fall-out dust. For eight elements (Ni, Rb, In, I, La, W, Au and Hg) the concentrations are lower than the detection limit of the technique.

1.2.5.2 Detection limits

Tables 1.3 and 1.4 also give estimated detection limits in 24 h urban aerosol samples and list some typical concentrations.

The sensitivities are not immutable because they depend to a certain extent on the irradiation counting scheme used, and strongly on the size and the composition of the sample. Most air pollution studies of daily concentration variations and area-wide trace element distributions involve 24 h sampling of several hundred m^3 of air. However, studies of diurnal variation cycles, involving short-time samples, are also of considerable interest. When only a limited number of m^3 is being sampled, as in the case of such short samplings or with low-flow-rate pumps, the detection limits will be higher unless longer irradiation and counting periods are applied. For some elements, filter impurities may limit the sensitivity of such analyses. Rahn *et al.*[198] demonstrated that it is possible to determine 18 trace elements in 4 m^3 samples taken over 90 min in duration in a rural area of the USA, and 15 elements in 5 m^3 samples over 2 h in a residential area[199]. With sampling over 1–4 weeks at a low flow rate[121] or at a high flow rate in remote areas, extremely low concentrations of 30 elements can be determined. The levels of total suspended particulates occurring in these areas range between 2 and 20 μg m^{-3} compared with 40–400 μg m^{-3} in urban and industrial areas.

1.2.5.3 Accuracy

Interferences due to threshold reactions are very rare under normal irradiation conditions ($\Phi_{th}/\Phi_{fast} > 1$). Only the reaction $^{27}Al(n,p)^{27}Mg$ interferes strongly with the determination of magnesium through the reaction $^{26}Mg(n,\gamma)^{27}Mg$. Spectral interferences make the prominent photopeaks of a few isotopes unusable. Examples are 844 keV of ^{27}Mg (846 keV ^{56}Mn) and 121 keV of ^{75}Se (122 keV ^{152}Eu). The mono-energetic ^{203}Hg (279.1 keV) is interfered with by ^{75}Se (279.6 keV), but a correction based on the spectrum of pure ^{75}Se can be applied because the interference is usually less than 20% of the ^{203}Hg activity. Also the measurement of ^{64}Cu based upon 511.0 keV is interfered with by external pair production of high-energy γ-rays. As a matter of fact, the accuracy depends strongly on the composition and measurement of standards. Two alternatives are possible, namely co-irradiation of standard samples or separate irradiations of standard samples possibly coupled with flux monitoring. For the calibration of the short-lived isotopes, only very few investigators co-irradiate a standard sample and measure it in between both short counts[117]. Corrections for decay are difficult in such procedures because the counting times are of the same order of magnitude as the half-lives of measured isotopes. Most authors irradiate mixtures of the elements to be determined, or individual standards, separately and monitor the flux during each irradiation. During the counting of short-lived isotopes, the shape of the spectrum as well as the overall dead-time changes. The use of mixed standards not only provides a spectrum with a similar shape to that of the sample, but also a spectrum giving a similar variation of analyser dead-time, thus minimising dead-time errors.

Several elements such as Ti [187], Al, Rb [201] and Mn [131], are used as flux monitors during the production of short-lived isotopes. Monitors from elements producing sufficiently short-lived isotopes such as ^{51}Ti and ^{28}Al can be used repeatedly for a large number of analyses. Also for the calibration of the longer-lived isotope intensities, standards of separate elements are sometimes irradiated with a flux monitor[182] such as Sc. Belot *et al.*[202] added a known amount of Rb and measured the induced ^{86}Rb activity as an internal standard. Since the variation of the detector efficiency as a function of geometry is not very energy-dependent, this procedure allows measurement in different geometries for the separate samples. In addition, counting losses due to coincidence do not induce errors. Most investigators, however, co-irradiate a standard mixture of the elements to be detected with a number of unknown samples. Mixtures of standards are generally prepared by depositing two or three solutions of balanced mixtures of the appropriate elements on a highly pure substrate (ashless filter paper) and allowing to dry. The relative weight ratios of the various elements are adjusted to be roughly similar to those in airborne dust. For some elements at the border of detection limits, the relative weight may be made larger in order to obtain a sufficiently large peak area. Standard and sample sizes are adjusted to make the counting rates comparable in order to reduce errors due to coincidence summing and photopeak broadening. Another option is, of course, the use of a pile-up rejector.

1.2.5.4 Automatic data reduction

In order that instrumental neutron activation analysis should be applicable to large numbers of atmospheric samples, such as in routine monitoring, some sort of automatic data reduction is necessary. The different techniques described in Chapter 3 are chiefly used[185, 187, 203, 204]. Accuracy and speed is increased by computer-assisted data reduction, but human judgment should be retained in the examination of data. A computer program for the data reduction should include the following steps. (1) Qualitative determination of the presence of the isotopes searched. (2) Calculation of net peak areas. (3) Correction for decay and spectral interferences. (4) Conversion of peak areas into weights of elements. (5) Subtraction of analytical blanks due to filter materials, containers, etc. (6) Calculation of concentrations of trace elements in sampled air, rain, snow or fall-out.

1.2.6 Single or oligo-elemental instrumental neutron activation analyses

The detection of lead in atmospheric samples is obviously of prime importance. Unfortunately no γ-emitting isotope with a reasonable half-life and cross-section can be produced by neutron capture. Pillay *et al.*[129, 130] measured the extremely short-lived ^{207m}Pb ($T_{\frac{1}{2}} = 0.8$ s) after a 10–30 ms irradiation at a very high neutron flux (10^{14}–10^{15} neutrons cm^{-2} s^{-1}) and a rapid transfer (100 ms) to the detector. Standard, sample and flux monitor are irradiated simultaneously. A dual-parameter 1024 channel pulse-height analyser, incorporating two 100 MHz analog-to-digital converters with the added feature of sequential multispectra scaling, are used to count the samples simultaneously with two NaI(Tl) detectors. Four or eight spectra are taken from each sample, and the characteristic Pb X-rays are used to measure the lead content. The variations of dead-time of both ADCs during the counting periods are recorded[205]. Fluorine is also of special environmental importance. Recently a short irradiation counting scheme, using the short-lived ^{20}F ($T_{\frac{1}{2}} = 11.2$ s) isotope was developed for the determination in aerosols. In order to avoid errors due to variations in dead-time, a dead-time stabiliser is used. The interference of the threshold reaction ^{23}Na(n,α)^{20}F is prohibitively large in marine samples. Such samples are irradiated under Cd and the fluorine is determined from the threshold reaction ^{19}F(n,γ)^{19}O ($T_{\frac{1}{2}} = 27$ s). Interference of the (n,γ) reaction on oxygen from the filter paper is important. Se, Sc, Ag, and Hf are determined simultaneously with F, using very short-lived isotopes[200].

The possibilities of very thin Ge(Li) detectors, with extremely high resolution in the low-energy region, but very low efficiency, were investigated by Zoller and Gordon[118]. From a feasibility study it was found that only the rare earths Yb, Lu, Eu, Ce and the elements Se and Hf can be detected more accurately and probably with a higher sensitivity than with the larger Ge(Li) detectors. In addition, the X-rays of Ba and the rare earths Sm, Gd, Tm and Lu were observed but not used for their determination. This limited additional information will generally not justify the greater expense of this equipment and the long counting times necessary.

A major component of airborne particulates is Si. This element can be determined by 14 MeV neutron activation analysis using the reaction $^{28}Si(n,p)^{28}Al$[125]. NaI detection and single-channel measurement of the 1.778 keV γ-transition allows the determination of amounts ranging from 10 down to 0.05 μg m^{-3} in airborne particulates, when a neutron flux of 10^9 neutrons cm^{-2} s^{-1} is obtained at the irradiation position. A complete irradiation counting cycle takes less than 5 min. Persiani[206] uses a Cockcroft–Walton accelerator as a 14 MeV neutron source for the determination of Si, Al and Cl in aerosols. The elements V, Br and Mn are determined after thermalisation of the neutrons with a water moderator tank in front of the drift tube. With a 14 MeV neutron flux of 9×10^9 neutrons cm^{-2} s^{-1} and a thermal neutron flux of 6×10^8 neutrons cm^{-2} s^{-1}, sensitivities between 10 and 200 μg are obtained. This is *ca.* 10^3–10^4 times higher than for reactor thermal neutron activation analysis, but generally sufficient for the detection of all six elements in 24 h high-volume samples.

By the fission track method, using polycarbonate detectors, U levels can also be determined in ambient air. An irradiation of 24 h at a limited neutron flux of 5×10^{11} neutrons cm^{-2} s^{-1} is sufficient to determine 10^{-14} g U on a 1 cm^2 filter section[207]. The sensitivity of U determinations by the delayed neutron method, as used by the same author, is only 10^{-8} g of U, necessitating a larger dust sample. The agreement between both methods and with mass-spectrometric results is good[207, 208]. Concentrations range from 0.1 to 1.5 ng m^{-3} in rural locations of New York State to 0.02 ng m^{-3} in Sutton, England, 0.004 ng m^{-3} over the Atlantic and 0.003 in Antartica. These levels are attributed primarily to naturally occurring sources and non-nuclear industrial activity. Determination of the uranium content of airborne particulates may be particularly important in an area where the nuclear industry is situated or plans to enter.

1.2.7 Conclusions

Within the past 5 years, the application of neutron activation analysis to the study of atmospheric samples has greatly increased our ability to observe minor and trace elements from natural and anthropogenic sources. In pollution monitoring and abatement, where speed and reliability are essential, activation analysis could play a very important role, although it suffers from some important drawbacks. The results for the most important elements only become available after long irradiation, cooling and counting times. The costs per sample are quite high and some key elements such as Pb and Be cannot easily be determined. Neutron activation is, however, extremely valuable for checking results obtained by other analytical techniques, for

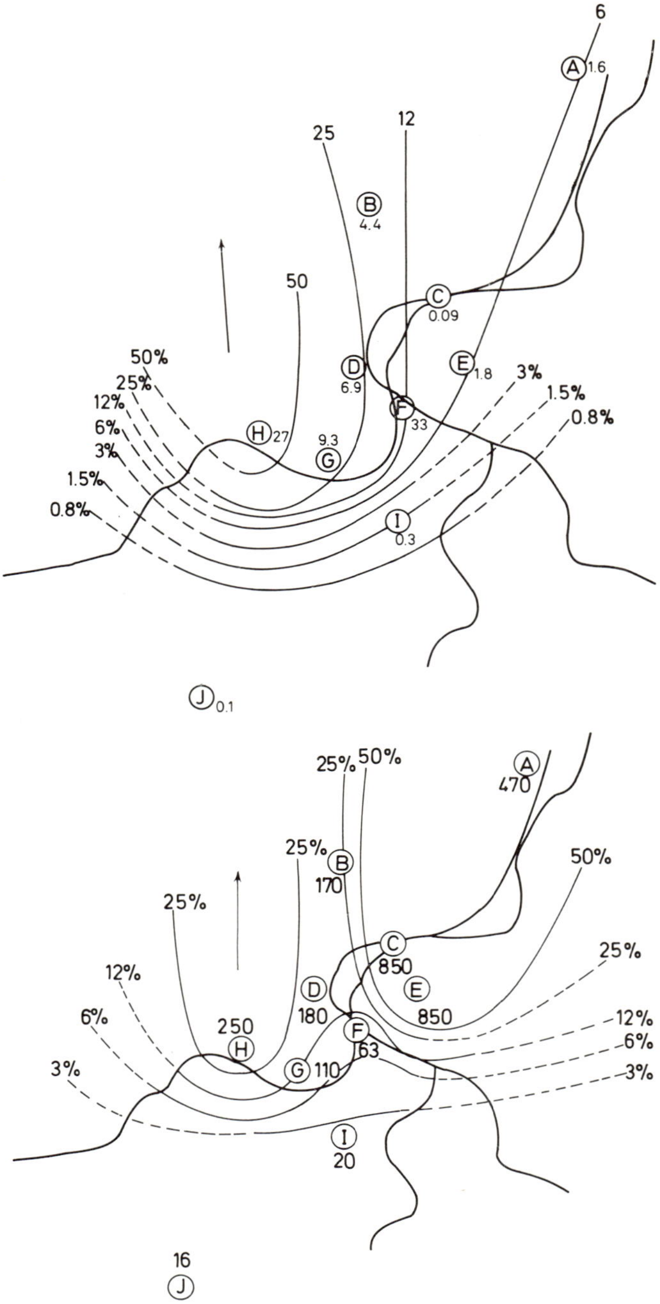

Figure 1.14 (a) Concentration maps of Fe in air particulates, in Liège, Belgium. (b) Concentration maps of Zn in air particulates, in Liège, Belgium

long-term observation and for fundamental research. This fundamental research should include studies of chemical composition, size distributions and atmospheric reactions of airborne emissions from various types of source, investigations on the harmful effects of elements or compounds, and experiments to gain a better knowledge of natural 'background' levels and compositions. Since no other analytical technique can give such an extensive picture of inorganic composition, activation analysis is ideally suited to do area-wide studies, individual source studies and chemical fingerprinting.

Finally, in order to illustrate the applicability in source location problems, a few results are given of a study recently[209] performed in Liège, Belgium. A set of ten 24 h high-volume air particulate samples, collected simultaneously in an industrial urban area, has been analysed by instrumental multi-elemental neutron activation analysis. From the geographical distribution patterns of the elements, it is obvious that several elements show large concentration variations over the area indicative of important local sources. In Figure 1.14(a) and 1.14(b), the concentration maps of Fe ($\mu g\ m^{-3}$) and Zn ($ng\ m^{-3}$) are plotted as examples. Concentration variations with two orders of magnitude occur within a distance of a few kilometres. As an aid to visualising the patterns, concentration isopleths were drawn in a geometric series at 50% of the maximum concentration, 25%, 12%, 6%, 3%, 1.5% and 0.8%. The element Fe, like most elements, had its maximum concentration at station H, slightly downwind of an important steel plant. The element Zn, together with some other metals such as Cu, Cr, Se, Sb and Hg, shows a second and even more pronounced maximum at station C, which is located downwind of several non-ferrous metal producing plants.

1.3 APPLICATIONS TO THE ANALYSIS OF HIGH-PURITY MATERIALS

1.3.1 Introduction

For high-purity materials, only those methods will be discussed which deal with the determination of trace constituents by means of reactor or 14 MeV neutron activation analysis, published from January 1968 onwards and which are referred to by the *NBS Bibliography* up to 1971[213], by *Analytical Abstracts* up to Vol. 24, No. 5 (May 1973), by *Chemical Titles* up to No. 26 (December 25, 1972), by *Chemical Abstracts* up to Vol. 78, No. 9 (March 5, 1973) or by *Nuclear Science Abstracts* up to Vol. 27, No. 8 (April 30, 1973). Also the papers presented at the International Colloquium, held from 2–6 October, 1972 at Saclay, France, have been taken into account.

Rigorous criteria were followed in selecting the articles. Papers mentioning no results or sensitivities were disregarded. The same was done for papers which dealt with methods that are too insensitive. The detection limits for the elements determined had to be $\leqslant 1$ p.p.m. Finally, the choice was restricted to activation analysis of high-purity elements and compounds, which rarely contain an overall impurity concentration of more than 10 p.p.m. For a number of reports and articles, only an abstract was available to us. From this it was sometimes hard to decide whether the above criteria were fulfilled and some excellent papers are probably omitted.

When surveying the methods described, two features were especially striking: first of all the fact that only some 20% of the methods involve the use of semiconductor detectors, and secondly that only some 30% are non-destructive. On the face of it this is probably astonishing in this age of automation, but is easily understandable if one considers that sensitivity is a very stringent requirement in the determination of trace impurities. To obtain a high sensitivity for a given element, it has indeed to be separated from other induced radioactivities. Even when use is made of high-resolution Ge(Li) detectors, separations are often necessary, especially when a high activity is present from a high-energy γ-ray emitter, because the Compton continuum hinders precise measurement of γ-ray emitters with lower energy.

1.3.2 Silicon and compounds

Since silicon compounds are among the most important materials for the semiconductor industry, much effort has been devoted to establishing appropriate methods for impurity analysis. Reactor neutron activation analysis is very suitable for the determination of trace elements in this matrix. By (n,γ) reaction on silicon only the 2.62 h ^{31}Si radioisotope is formed, which is nearly a pure β^- emitter, while by (n,α) and (n,p) reactions with the fast neutrons of the reactor spectrum only short-lived ($T_{\frac{1}{2}} < 10$ min) aluminium and magnesium radioisotopes are produced. Non-destructive analysis of the impurities in silicon can easily be performed after a 1 day cooling time provided that NaI(Tl) or preferably high-resolution Ge(Li) γ-ray spectrometry is applied. By non-destructive Ge(Li) γ-ray spectrometry, Keenan and Larrabee[214] determined Na, Cu, As, Sb, Au, Hf and Zr in semiconductor silicon, while Kotas and Kukula[215] determined W, Cu, Na, La, Fe, Cr, K and Au. If shorter-lived radioactivities have to be measured, it is necessary to eliminate the ^{31}Si matrix activity. Distillation of SiF_4 as described by De Corte *et al.*[216, 217], Lambrev *et al.*[218] and Kotas and Kukula[215, 219] is a very useful technique. Kotas and Kukula[215, 219] measured the distillation residue without further chemical separation and were able to determine Na, Cu and As in thin layers of SiO_2 and Si_3N_4.

If one is interested in determining particular elements, specific separations, followed by NaI(Tl) or Geiger–Müller counting, is most advisable. De Corte *et al.*[216] gave a detailed separation scheme for 21 elements. Extraction, ion exchange and precipitation as well as distillation were applied.

1.3.3 Germanium and compounds

By reactor irradiation of germanium, a more complex matrix activity is formed than for silicon. Non-destructive Ge(Li) γ-ray spectrometry is, however, possible after a cooling period of 100 days. In this way, Adams *et al.*[220] could determine 13 trace constituents with extremely high sensitivity (from 1 p.p.b. down to 10^{-5} p.p.b.). For selective elimination of the germanium matrix activity, distillation of $GeCl_4$ seems well suited. When determining copper in germanium, care has to be taken to eliminate contamination. Nagy and Giber[221] investigated different pre- and post-irradiation etching techniques and concluded that a mixture of HNO_3 and HCl is best suited. For washing, water 'de-coppered with germanium', which is prepared by shaking water with germanium, should be used.

The determination of arsenic in germanium is interfered with by ^{76}As and ^{77}As activities formed from $^{74}Ge(n,\gamma)^{75}Ge \xrightarrow{\beta^-} {}^{75}As(n,\gamma)^{76}As$ and $^{76}Ge(n,\gamma)^{77}Ge \xrightarrow{\beta^-} {}^{77}As$ respectively. Lima and Silva[222] combined isotope dilution and activation analysis to dispose of this problem. In their method, arsenic is separated from germanium before irradiation and the yield of the chemical separation is determined by isotope dilution.

1.3.4 Gallium, indium, thallium, phosphorus, arsenic, antimony and compounds

This group of elements and compounds, of which gallium and gallium arsenide especially have been investigated, presents rather a challenge to activation analysis. Often by reactor irradiation, a high matrix activity is induced, and moreover self-shielding can be important, although this is not always considered by authors. To avoid these problems, the matrix can be separated from the trace elements before irradiation, as reported by Nagy *et al.*[223-225]. These authors separated gallium before irradiation by distillation as $GaCl_3$. By such a pre-irradiation treatment, however, the great advantage of activation analysis over other analytical techniques is lost, namely that no impurities are introduced from the chemicals.

Because of the high matrix activity, which is induced in most cases, non-destructive analysis of a large number of trace constituents in one sample is not usually applied. Non-destructive techniques have only been developed for the determination of single elements or a maximum of two. The irradiation conditions, cooling time and counting techniques are then specially chosen to present minimal interference from the matrix. In this way, Rausch and Salamon[226] could determine selenium and tellurium in gallium, in arsenic and in gallium arsenide by non-destructive NaI(Tl) γ-ray spectrometry. Orvini[227] used Ge(Li) γ-ray spectrometry to determine iodine in Ga_2Se_3 and Ga_2S_3, Daiev *et al.*[228] determined indium in thallium and in thallium compounds and Wölfle *et al.*[229, 230] used an improved selective γ–γ coincidence arrangement for the non-destructive determination of sub-micro traces of copper in high-purity thallium.

Amongst destructive methods developed for the determination of trace impurities in Ga, In, Tl, As, Sb, P and their compounds, only two are suitable for the determination of more than 10 elements. Gil'bert and Pronin[231] described a method, involving extraction, for the determination of 20 elements in thallium, while Neeb *et al.*[232] used precipitation, distillation, extraction and ion exchange to determine 21 elements in gallium and in gallium arsenide.

In matrices which only give rise to short-lived radioisotopes, such as gallium, gallium arsenide and arsenic, non-destructive analysis of a large number of traces by means of long-lived radioisotopes should be possible when the matrix activity has decayed. Although this method has not yet been applied, Neeb *et al.*[232] state that, by Ge(Li) γ-ray spectrometry, Ag, Cd, Co, Cr, Fe, Hg, Ni, Zn and other elements can be determined non-destructively in gallium and gallium arsenide.

1.3.5 Aluminium and compounds

The (n,γ) reaction on aluminium yields only the short-lived ^{28}Al radioisotope ($T_{\frac{1}{2}} = 2.240$ min). Non-destructive analysis of trace impurities in aluminium after a 1 h cooling time is, however, seriously hindered by a high ^{24}Na activity, formed from $^{27}Al(n,\alpha)^{24}Na$. Therefore, in a systematic neutron activation analysis of aluminium and compounds, chemical separations are often applied. Lesbats[233] described a method for the determination of about 30 elements. The method involves crystallisation of aluminium as $AlCl_3 \cdot 6H_2O$, precipitation, ion exchange and extraction.

When the ^{24}Na matrix activity has decayed or after irradiation in a pure thermal flux, non-destructive Ge(Li) γ-ray spectrometry can easily be performed. Blouri *et al.*[234] developed a very sensitive method for the determination of 30 elements. The detection limits range from 5×10^{-2} down to 5×10^{-6} p.p.b.

In the determination of oxygen in aluminium by 14 MeV neutron activation analysis, special care has to be taken to prevent positive errors. Dugain *et al.*[235] investigated the apparent oxygen content as a function of different etching and irradiation modes and reached a lower limit of *ca.* 1 p.p.m.

1.3.6 Lead, bismuth, tin, zinc and compounds

Little attention has been paid to the activation analysis of trace constituents in lead and bismuth. As far as zinc is concerned, a detailed separation and analysis scheme for the determination of 13 elements has been described by Lanfranco and Bianchini[236], whereas Cuypers *et al.*[237] developed a method involving ion exchange and separation on inorganic exchangers for the determination of 25 elements.

A method for the determination of 16 elements in tin has been described by Lanfranco[238]. This method involves many chemical separations and is rather time-consuming, but requires only simple counting apparatus, namely a Geiger–Müller tube and a NaI(Tl) detector coupled to a multichannel analyser. Maenhaut *et al.*[239, 240] developed a method for the determination of 18 elements in high-purity tin, in which chemical separations are restricted to a minimum. The tin and ^{125}Sb matrix activities are distilled as bromides, after which the non-volatile trace constituents are determined by Ge(Li) γ-ray spectrometry of the residue. After some additional separations on the distillate, arsenic, selenium and antimony are determined. Nuclear interferences on the tin matrix, giving rise to apparent cadmium, indium and antimony contents, are taken into account.

1.3.7 Cadmium, mercury and compounds

The thermal neutron absorption cross-section of cadmium being very high (2440 barns), irradiation of cadmium in a thermal neutron flux easily leads to very important self-shielding. Special techniques have therefore to be used when neutron activation analysis is applied for the determination of trace elements in cadmium and compounds. Mizuike *et al.*[241] separated gold from cadmium before irradiation by means of electrolysis with vitreous carbon electrodes and was able to determine 6 p.p.b. gold in 99.999% cadmium. Orvini[227] applied epi-cadmium activation to eliminate the matrix self-shielding, whereas Filatova *et al.*[242] irradiated very thin films of CdS on glass supports.

1.3.8 Sulphur, selenium, tellurium and tellurium compounds

Although reactor irradiation of selenium yields a high and long-lived ^{75}Se activity, neutron activation analysis has often been applied to determine trace constituents in this element. Nearly all methods developed involve chemical separations. Very useful methods for a systematic analysis of more than 10 elements have been described by Kiss[243], Gil'bert and Pronin[244], Ballaux *et al.*[245] and Mignonsin[246]. Gil'bert and Pronin[244] have combined group separation and NaI(Tl) γ-ray spectrometry for the determination of 18 elements. Ballaux *et al.*[245] applied distillation of the matrix isotopes, after which 15 non-volatile elements can be determined by simple Ge(Li) γ-ray spectrometry. This method is very elegant since the germanium, arsenic and bromine activities also formed from the selenium matrix are distilled.

Only a few methods have been described for the neutron activation analysis of trace elements in sulphur, tellurium and tellurium compounds. Practically all are non-destructive, but no more than three elements are determined in any one method.

1.3.9 Iron, copper, molybdenum, titanium and titanium compounds

The neutron activation analysis of trace constituents in high-purity iron has been thoroughly investigated by Albert[247]. Methods have been described for the determination of 55 elements. Most of these are separated individually so that numerous chemical separations are necessary. More recently, Lesbats and Tardy[248] have developed methods, involving extraction, ion exchange and precipitation, which allow the determination of 15 trace elements in high-purity iron. Special techniques for the determination of a few impurities are described by Samadi and May[249] and by Steinnes[250]. Samadi and May[249] have applied irradiation with fast neutrons from a high-flux reactor in order to determine titanium and lead by means of the isotopes ^{47}Sc and ^{203}Pb, which are formed via threshold reactions on ^{47}Ti and ^{204}Pb, respectively. Steinnes[250] has separated manganese from iron before irradiation so that it can be determined without any interference from the ^{56}Fe(n,p)^{56}Mn reaction.

Although non-destructive neutron activation analysis of traces in copper should be possible after a 1 week cooling time when the ^{64}Cu activity has decayed, it has not been widely used for systematic analysis. Even the destructive techniques described are only suitable for the determination of a maximum of two elements. More research has been applied to the systematic analysis of molybdenum. Fedoroff[251] has presented a method which can be applied for the determination of 40 elements in molybdenum with a sensitivity better than 0.1 p.p.m.

As far as the analysis of high-purity titanium and compounds is concerned, a useful and sensitive method involving group separations and Ge(Li) γ-ray spectrometry has been developed by Neirinckx *et al.*[252] for the determination of some 21 elements in single crystals of TiO_2.

1.3.10 Graphite

Because neutron irradiation of graphite does not yield any matrix activity, non-destructive Ge(Li) γ-ray spectrometry of the irradiated sample can easily be applied. By this method, Cojocaru and Filby[253] have determined more than 20 elements, whereas Reed and Hilton[254] identified 13 elements. However, if some dominant radioactivities are formed from impurities which have high activation cross-sections, destructive techniques can be useful. May and Pinte[255, 256] extracted trace elements from the graphite sample with fuming nitric acid, after which precipitation, extraction and ion exchange were applied for further separation of different elements.

1.3.11 Reagents

Ultra-pure reagents, which are used in the fabrication and analysis of semiconductor materials, are now routinely produced. The purity control is usually performed by spectroscopic methods[257]. However, neutron activation can also be applied for the analysis of some impurities. A non-destructive technique for the determination of Al, Cl, Cu, Na and Mn in nitric acid has been described by Filippova *et al.*[258]. In this technique, 2 g of nitric acid are irradiated for 5–30 min in a flux of 10^{12} neutrons cm^{-2} s^{-1}. Kosta and Ravnik[259] have developed a method wherein liquid reagents are evaporated in irradiation capsules before irradiation. After irradiation, specific and fast separations are carried out to determine Cl, As, Cu, Mn and Na. Some authors have applied pre-separation techniques for the analysis of trace metals in liquids. Byrne *et al.*[260] separated mercury from hydrochloric acid before irradiation by adsorption on an anion-exchange resin, whereas Lima and Atalla[261] used co-precipitation for the pre-separation of copper from different solutions.

1.3.12 Miscellaneous

Neutron activation analysis has not been applied very frequently to the analysis of other elements and compounds except those discussed above.

Hence, this survey of the application of neutron activation analysis to high-purity materials will be concluded by mentioning some typical papers. Artiukhin *et al.*[262] have developed a method which allows the determination of 24 elements in silver. The method involves extraction, extraction chromatography and precipitation. Lesbats[263] has developed a systematic analysis of zone-refined magnesium allowing the determination of 26 elements. A non-destructive method involving Ge(Li) γ-ray spectrometry for the determination of 17 trace elements in niobium has been described by Sukhov and Firsov[264]. Even high-purity gold has been analysed by neutron activation analysis. May *et al.*[265] have described a method in which self-shielding is taken into account.

1.3.13 Survey table

Matrix	*Elements determined*	*Sample weight* /g	*Chemical separations involved**	*Counting mode*	*Concentrations* /p.p.m.	*Detection limits* /p.p.m.	*Ref.*
Ag	24 elements	0.2–0.3	electrol., extr., extr. chromat., pptn.	NaI(Tl), G.M.(Bi,Te)	3×10^{-4}(Ir)–8.4(Cr)	—	262
Al	S	7	pptn., ion exch., co-pptn.	G.M.	2×10^{-2}	10^{-2}	266
	P, S	7	pptn., ion exch., co-pptn.	G.M.	up to 0.2 (P)	10^{-3}(P) 10^{-2}(S)	
	33 elements	1–1.5	non-destructive; ion exch. (Zn, Zr, Fe)	Ge(Li)	3.8×10^{-4}(Sc)–0.8(Fe)	2×10^{-6}(Sc)–0.4(Ge)	234
	O	12.8	non-destructive	Na(I)Tl	down to 2	—	235
	Cd	—	subst. extr.	—	0.52–40	—	267
	30 elements	1	pptn., ion exch., extr.	γ-spectr. β-spectr.(Ca)	up to 5	10^{-5}(Au, As)–10^{-2} (Ca, Sr, Na, Ni)	233
	Ge	0.5	distill., extr., ion exch., pptn.	liq. scint.	4×10^{-3}–5.8×10^{-2}	2×10^{-3}	268
	Ti	1–2	pptn., extr., ion exch.	NaI(Tl)	0.010–2.03	5×10^{-3}	249
	O	2.5	^{18}F separation	—	—	1	269
	Sc, Au, Cr, Sb, U	0.5	non-destructive	NaI(Tl); Ge(Li)	1.8(U)	—	270
Al nitride	12 elements	—	non-destructive	—	—	10^{-2}–10^{-1}	271
Al oxide	22 elements	—	pptn., ion exch., electrol., extr., non-destructive (Mn, Ga, Na, Dy, Eu, La)	NaI(Tl)	6×10^{-3}(Co)–16(Ga)	~10^{-2} μg	272
	~50 elements	0.4–1	non-destructive	Ge(Li)	6×10^{-4}(Mn)–8.3(Sr) Cr: up to 317; Ba: up to 265	10^{-4}(Au, In)–100(Zr)	273

1.3.13 Survey table—*continued*

Matrix	*Elements determined*	*Sample weight /g*	*Chemical separations involved**	*Counting mode*	*Concentrations /p.p.m.*	*Detection limits /p.p.m.*	*Ref.*
Al oxide—*continued*	Cu, Zn, Mn	0.1	extr.	NaI(Tl)	—	3×10^{-2}(Cu); 2(Zn); 3×10^{-3}(Mn)	274
As	Se, Te	0.2–0.5	extr.	NaI(Tl), G.M. (low Te content)	—	2×10^{-2}	226
			non-destructive	NaI(Tl)	—	2	
Au	29 elements	1	electrol., pptn.	Ge(Li)	5×10^{-3}(Co)–530(Pd)	10^{-6}(Cu)–1(Fe)	265
B	Ta, W	—	non-destructive	γ-spectr.	—	0.1–1 μg(Ta) 10^{-2}–0.1 μg(W)	275
Be	Cu	0.41–0.44	non-destructive	NaI(Tl)[γ–γ coinc.]	8.9; 15.1	6×10^{-6}	229
	Cu	0.44	non-destructive	NaI(Tl)[γ–γ coinc.]	8.9	10^{-4}	230
Be oxide	Mn, Na, W, Mo	1	—	NaI(Tl), Ge(Li)	0.35(Mn), 4(Na) 1(W), 50(Mo)	—	276
Bi	Cu	0.57–1.55	non-destructive	NaI(Tl)[γ–γ coinc.]	2.7–2.73	6×10^{-6}	229
	Cu	0.57	non-destructive	NaI(Tl)[γ–γ coinc.]	2.73	10^{-4}	230
C	P, Cu, As, W	—	—	NaI(Tl); β-spectr. (P)	5.2×10^{-3}(Cu)–9.4(W)	—	277
	Co, Sc, Zn	0.2	—	Ge(Li)	2×10^{-3}(Co), 2×10^{-4}(Sc), 10^{-2}(Zn)	—	276
	>20 elements	—	non-destructive	Ge(Li)	—	10^{-4}–10^{-5} μg	253
	Mg, Al, V, S, La, Dy, Mn	0.1–0.2	non-destructive	γ-spectr.	10^{-2}(Dy)–54(Mg)	5×10^{-3}(Dy)–50(Mg)	278
	20 elements	10	pptn., ion exch., extr.	NaI(Tl), G.M.	1.1×10^{-4}(In)–27(K)	—	255
	rare earths	10	pptn., ion exch. (pre-sepn.)	NaI(Tl)	3×10^{-6}(Nd)–0.112(Ce)	—	256
	13 elements	—	non-destructive	Ge(Li)	2×10^{-4}(Sc)–9(Mo)	—	254
	Au, Sb	—	non-destructive	γ-spectr.	—	8×10^{-4} μg(Au)–5×10^{-3} μg(Sb)	279

	Dy	0.2–0.4	non-destructive	NaI(Tl)	0.04–1.5	—	280
$CaCO_3$	Cu, Mn	—	subst., extr.	—	0.22(Cu), 0.19(Mn)	—	267
Cd	Au	—	electrodepos. (pre-sepn.), extr.	NaI(Tl)	6×10^{-3}	—	241
$Cd_xHg_{1-x}Te$	Au, Cu, Zn	0.2	pptn., ion exch., extr.	γ-spectr.	—	4.3×10^{-4}(Au); 1.7×10^{-3}(Cu); 2.3×10^{-2}(Zn)	281
CdS	Co, Cu, Fe, Ga, Zn, Au	10^{-3}	ion exch., extr.	NaI(Tl)	—	10(Co); 5(Cu, Ga); 100(Fe); 50(Zn); 0.1(Au)	242
	I	6×10^{-3}–0.15	extr., non-destructive	NaI(Tl) Ge(Li)	0.05–900	5×10^{-3} μg	227
CdTe	I	6×10^{-3}–0.15	extr., non-destructive	NaI(Tl) Ge(Li)	0.05–900	5×10^{-3} μg	227
Cu	S	0.5–1	electrol., co-pptn., pptn., ion exch.	—	5–10; 0.2	—	282
	P	0.5–1	electrol., pptn., ion exch.	—	0.01–0.03; 0.07	—	
	Hg	2	non-destructive	Ge(Li)	0.03–0.05	—	283
	Au	—	subst. extr.	—	0.03–0.04	—	267
	Se, Te	—	ion exch., chromat., inorg. exch.	NaI(Tl)	0.3–0.94(Se); 0.24–0.72(Te)	5×10^{-4}(Se) 5×10^{-3}(Te)	284
	O	—	non-destructive	NaI(Tl)	4	2	285
	S	10	—	plastic scintil.	1.5–2	—	286
Fe	55 elements	1	extr., pptn., ion exch., electrol., distill.; non-destructive (V, Al)	γ-spectr. β-spectr. (Si, S, P, Ni)	16 elements found in zone-refined iron 10^{-3}(Au, Sb)–48(Al)	for 39 elements: sum of detection limits = 1 p.p.m.	247
	S	1	extr., pptn., co-pptn.	β-spectr.	0.24	—	287
	P, S	1	extr., isot. exch., pptn., co-pptn.	β-spectr.	0.3(P); 7–8(S)	—	
	Cu, Ni	—	ion exch., extr., pptn.	—	—	5×10^{-3} μg(Cu); 5×10^{-2} μg(Ni)	288
	O	8.5	—	—	1 p.p.m. level	—	289
	Au	—	subst. extr.	Ge(Li)	0.04–0.2	—	290

1.3.13 Survey table—*continued*

Matrix	*Elements determined*	*Sample weight /g*	*Chemical separations involved**	*Counting mode*	*Concentrations /p.p.m.*	*Detection limits /p.p.m.*	*Ref.*
Fe—*continued*	Ni, Na, Mn, Cu, Fe, Ga, Au, Sb, As	—	extr., ion exch., pptn.	—	5×10^{-5}(Au)–1.8(Ga)	—	248
	W, Mo	—	ion exch.	—	0.08(Mo)	—	
	Ag, Cr, Co, Zn	—	extr., ion exch.	—	0.05(Cr); 0.03(Co)	0.05(Zn)	
	W	—	extr.	—	20–1000	—	
	Se	—	ion exch.	—	10–70	—	
	Te	—	distill.	—	10–50	—	
	Ge	0.5	distill., extr., ion exch., pptn.	liq. scintill.	3.8×10^{-2}–1.41	2×10^{-3}	268
	Ti	1–2	extr., pptn., ion exch.	NaI(Tl)	0.010–4.6	5×10^{-3}	249
	Mn	0.2	ion exch. (pre-sepn.); distill.	NaI(Tl)	5.66	—	250
Ga	In, Au, Mn	5×10^{-2}	pptn., extr.	—	6.4×10^{-3}(Au), 0.12(In), 0.22(Mn)	1.2×10^{-3}(Au), 4.8×10^{-3}(In), 5.2×10^{-3}(Mn)	291
	Pt, Au, Hg, Sc, Zn, Sn	0.2–0.3	pptn., extr., ion exch.	NaI(Tl)	0.17(Au)–98(Pb)	—	292
	Hg	0.2–0.3	subst. extr.	NaI(Tl)	0.39	10^{-2} μg	293
	Cu, Au	0.1	$GaCl_3$ distill. (pre-sepn), ion exch., pptn.	γ-spectr.	0.63, 2.1(Cu); 3.7×10^{-3}, 4.3×10^{-2}(Au)	—	223, 224
	Cu, Au	0.1–0.2	$GaCl_3$ distill. (pre-sepn.), ion exch.	—	0.1–1(Cu); 10^{-3}–10^{-2}(Au)	—	225
	21 elements	0.5	pptn., distill., ion exch., extr.	NaI(Tl)	—	2×10^{-6}(Au)–2×10^{-3}(Te, Ge, Ni)	232

	Se, Te	0.2–0.5	extr.	NaI(Tl); G.M. (low Te content)	—	10^{-2} μg	226
			non-destructive	NaI(Tl)	—	1 μg	
GaAs	Mn, Ni, Cu, Te, Zn, Fe, Cr, Co, Ni	1	extr., ion exch., pptn.	—	—	10^{-4}–0.1	294
	Mn, Cu, Zn, Cd	0.24–0.54	distill., extr., subst. extr.	NaI(Tl); G.M.	3×10^{-3}(Mn)–4.2×10^{-2}(Cd)	—	295
	Ni, Au, Ag	0.1–0.5	subst. extr.	NaI(Tl); G.M.	~0.06(Ni); ~7×10^{-3}(Au); ~4×10^{-3}(Ag)	—	296
	Mn, Cu, Ni, Ag, Au, Zn, Cd	—	distill., extr. subst., extr.	γ-spectr.	6×10^{-5}(Au)–0.096(Cu)	—	297
	Mn, Ni, Cu, Zn, Ag, Cd, Au	—	subst. extr.	—	2×10^{-3}(Ag)–0.08(Ni)	—	267
	Se	—	—	γ-spectr.	—	0.5	298
	Te, Cu, Zn, Au, In	0.1	distill., extr., pptn.	γ-spectr.	4.6(Te); 5×10^{-2}(Cu); 3(Zn); 5×10^{-2}(Au); 2.4(In)	—	299
	21 elements	0.5	pptn., distill., ion exch., extr.	NaI(Tl)	—	2×10^{-6}(Au)–2×10^{-3}(Te, Ge, Ni)	232
	Se, Te	0.2–0.5	extr.	NaI(Tl); G.M. (low Te content)	—	10^{-2} μg	226
			non-destructive	NaI(Tl)	—	1 μg	
	Mn	0.01–0.4	extr.	NaI(Tl)	7×10^{-4}–5.1×10^{-3}	10^{-4} μg	300
	Mn, Cu, Ni, Cr, Zn, Co	0.1	extr. chromat.	NaI(Tl)	10^{-2}(Mn)–5(Cr)	5×10^{-2}(Mn)–10(Ni)	301
	Si, In, Au, Ge, Zn, P, Cr	—	non-destructive	—	0.37–4.1(Si); 15.8–25.7(Cr)	0.61(Zn)–1.84(Au)	302
GaAsP	Mn	0.01–0.4	extr.	NaI(Tl)	7×10^{-4}–5.1×10^{-3}	10^{-4} μg	300
Ge	13 elements	0.5	non-destructive	Ge(Li)	1.5×10^{-6}(Co); 1.7×10^{-4}(Zn); 3.6×10^{-4}(U);	2×10^{-8}(Ir)–7×10^{-4}(Ni)	220

1.3.13 Survey table—*continued*

Matrix	*Elements determined*	*Sample weight /g*	*Chemical separations involved**	*Counting mode*	*Concentrations /p.p.m.*	*Detection limits /p.p.m.*	*Ref.*
Ge—*continued*							
	P, Cu, W, As	—	—	—	$\leqslant 5 \times 10^{-6}$(Sb) 2×10^{-4}(As, Cu)– 3×10^{-2}(W)	—	278
	Sn, Hg, Cr, Sb, Sc, Zn, Co	0.1	non-destructive	NaI(Tl)	—	10^{-4}(Sc)–0.2(Sn)	303
	Au	2.4	distill., pptn.	Ge(Li)	0.024–0.037	10^{-3}	304
	As	0.22–0.42	co-pptn. (pre-sepn.), Gutzeit method	NaI(Tl)	0.3	—	222
	Cu	—	distill., pptn., ion exch., subst. extr.	NaI(Tl)[γ–spectr. or γ–γ coinc.]	10^{-3}–0.364	—	221
	Cu, Au, Sb, As, Cd, Zn, Hg, Fe	0.2	pptn., ion exch.	γ-spectr.	5×10^{-4}(Cu, Au)–5(Tl)	—	305
	Au, Cu, Ni	0.1	extr., pptn.	Ge(Li); G.M. (^{65}Ni)	10^{-4}(Au)–11(Ni)	1(Ni)	306
GeO_2	Mo	0.5	subst. extr.	scintill. spectr.	0.1–0.2	—	307
	Sn, Hg, Cr, Sb, Sc, Zn, Co	0.1	non-destructive	NaI(Tl)	—	10^{-4}(Sc)–0.2(Sn)	303
HgTe	Au, Cu, Zn	0.2	pptn., ion exch., extr.	γ-spectr.	—	4.3×10^{-4}(Au); 1.7×10^{-3}(Cu); 2.3×10^{-2}(Zn)	288
InAs	Au, Se, Te, Ge, Sb, Cu, Zn, Cd	0.05–1	pptn., distill., extr., ion exch.	NaI(Tl); G.M.(^{71}Ge)	up to 3×10^{-3}(Ge)– up to 50(Zn, Cd)	5×10^{-4}(Au)–0.5(Se, Cd)	308
InSb	Cu, As, Au, Co, Cd	0.1	pptn., extr.	NaI(Tl)	0.1(Cu); 0.5(As); 0.89(Au); 0.46(Co); 4.8(Cd)	3×10^{-1}(Cu); 10^{-1}(As); 1.2×10^{-3}(Au); 1.4×10^{-1}(Co); 1.6(Cd)	309
	Au, Te, Ge, As, Cu, Zn, Cd	0.05–1	extr., chromat., distill., pptn.	NaI(Tl)	down to 10^{-3}(Au)– down to 5(Cd)	5×10^{-4}(Au)–0.5(Cd)	310
KCl	Co	0.05	non-destructive	NaI(Tl)	—	2×10^{-1}–2	311
LiF	I, Cl, Br	0.5	distill., pptn., extr.	NaI(Tl)	—	1.4×10^{-3}(I); 5×10^{-3}(Cl);	312

						2×10^{-3}(Br)	
Mg	S	—	pptn., extr.	G.M.	2×10^{-3}–4×10^{-3}	2×10^{-3}	266
	P, S	—	pptn., extr.	G.M.	up to 10^{-2}	—	
	26 elements	0.1–1	pptn., electr., extr., ion exch.	—	Na down to 0.02	—	263
Mo	Al	0.05–0.5	ion exch.	NaI(Tl)	0.6–40	0.1	313
	Cu, Mn, Na, Ni, W	0.05–0.8	ion exch., pptn.	NaI(Tl); β-counter (^{65}Ni)	0.03(Mn)–28(Ni); W: 38–220	10^{-4} μg(Mn)–7×10^{-3} μg(Ni)	314
	Pd	0.05–0.5	ion exch., pptn.	G.M., NaI(Tl) [X-ray]	<1	6.5×10^{-5} μg (β-counting); 2×10^{-3} μg (X-ray counting)	315
	S	0.05–0.8	ion exch., pptn.	G.M.	0.08–0.1	—	316
	40 elements	—	ion exch., extr., pptn. Nd: pre-sepn.	—	total impurities <1 (except W)	0.1	251
	Cu, Zn	0.02	pptn. extr.	—	0.51(Cu); 3.3×10^{-2}(Zn)	—	317
	W, Hf	—	—	γ-spectr.	—	6(W); 0.1(Hf)	298
	Mn, Zn, Cu, Cd, Ag, Hg	0.5	subst. extr., extr. chromat.	NaI(Tl) β-counter (Cd)	—	10^{-2} or higher	318
Na salts	Co, Fe, Cd	0.07–0.11	inorg. chromat., ion exch.	—	down to 5×10^{-2}	—	319
Na_2CO_3	Cu	—	subst. extr.	—	0.64–0.66	—	267
Nb	17 elements	2	non-destructive	Ge(Li)	2×10^{-3}(Au)–2(Mo)	10^{-4}(Mn)–2.5(Fe)	264
Nb_2O_5	Ta	0.1–0.3	subst. extr.	—	0.36	—	320
Ni	S	—	pptn., co-pptn., isot. exch.	G.M.	0.3	—	287
	P, S	—	pptn., co-pptn., extr., isot. exch.	G.M.	6×10^{-3}(P); 1(S)	—	
P	Cu, As, Sb	0.2	pptn., ion exch.	gas flow counter NaI(Tl)	—	3.5×10^{-3}(Cu); 2.5×10^{-3}(As); 4.5×10^{-3}(Sb)	321
Pb	Bi	0.5–1.5	extr., spont. depos.	ZnS(Ag); Si-surface barrier	9.73×10^{-2} for homogenised sample	—	322
	W, Sb	0.1	pptn.	NaI(Tl)	—	10^{-2}	317
	Sb	—	non-destructive	NaI(Tl)	—	1	

1.3.13 Survey table—*continued*

Matrix	*Elements determined*	*Sample weight /g*	*Chemical separations involved**	*Counting mode*	*Concentrations /p.p.m.*	*Detection limits /p.p.m.*	*Ref.*
Pb—*continued*	Cu	1.5–2.2	non-destructive	NaI(Tl)[γ–γ coinc.]	4.38×10^{-2}–4.91×10^{-2}	6×10^{-6}	229
	Cu	2.12	non-destructive	NaI(Tl)[γ–γ coinc.]	4.38×10^{-2}	10^{-4}	230
PbO	Ag	0.1	subst. extr.	Ge(Li); NaI(Tl)	13–15	1	323
Reagents							
filter paper; dithizon; thioacetamide; L-cysteine–HCl; ion-exch. resin; HCl; HNO_3; H_2SO_4; cotton wool; polythene; PVC	Hg	—	Two methods: {1. Volatilisation {2. Extr., ion exch. for HCl: ion exch. (pre-sepn.)	γ-spectr.	10^{-4}–6×10^{-2} for dithizon: 0.3–193	2×10^{-4} µg	260
HNO_3	Cl, Cu, Na, Mn, Al	2	non-destructive	NaI(Tl)	4×10^{-3}(Mn)–2(Al)	2.5×10^{-4}(Al)–4×10^{-2}(Cu, Na)	258
Mineral acids; NH_4OH; H_2O_2;	As, Cu, Cl, Mn, Na	—	pptn., extr. non-destructive (Na)	NaI(Tl) Ge(Li)	7×10^{-5}(Mn)–6(Na)	—	259
$(NH_4)_2S$; NH_4OH; HCl; HNO_3	Cu	—	co-pptn. (pre-sepn.)	NaI(Tl)	$\sim 10^{-2}$	—	261
S	Se, Te	—	—	γ-spectr.	—	0.07 µg	324
	Se, Te	0.1	non-destructive	NaI(Tl)	—	0.7	325
	Se	0.13	—	NaI(Tl) [γ-spectr. or coinc.]	—	0.3	326
	As, Se, Te	0.1–0.3	non-destructive; low content: pptn., distill.	Ge(Li)	0.15–6.8(As) 0.5–15(Se) 15(Te)	0.01 µg(As) 0.05 µg(Se) 0.15 µg(Te)	327

Se	Te	0.1	distill., extr., pptn.	plastic scintill. + NaI(Tl): β–γ coinc.	1.5×10^{-2}–2.5×10^{-2}	—	328
	Cl	2	distill., pptn.	NaI(Tl)	0.4–1	—	329
	P, S	0.05	distill., pptn.	G.M.	0.065–0.52(P); 1.5–4.6(S)	—	
	Cl, Br, I	2	distill., pptn.	Ge(Li)	0.44(Cl); 1–2(Br); 0.7(I)	0.14(I); 0.6(Br)	330
	15 elements	0.1	distill. pptn. (Tl)	Ge(Li); G.M. (^{204}Tl)	6.3×10^{-4}(Sc)–3.07(Cr)	—	245
	Sb	1	pptn., distill.	Ge(Li)	2.1×10^{-2}–2.9×10^{-2}	5×10^{-3}	331
	18 elements	0.1–0.2	extr.	NaI(Tl)	—	2×10^{-5} μg(Ir, Au)–0.3 μg(Sn)	244
	18 elements	—	alkaline reduction, extr., reduction, hydrol., pptn.	Ge(Li)	8×10^{-5}(Au)–23(Ge)	1.5×10^{-6}(Au)–3.8×10^{-2}(Fe)	243
	Mn, Ni, Cu	2	distill., extr., subst. extr.	γ-spectr.	10^{-3}(Mn)–0.18(Ni)	—	297
	Zn, Au, Ag, Cd	2	distill., subst. extr.	γ-spectr.	$\sim 4 \times 10^{-2}$(Zn)	3×10^{-6}(Au)	
	11 elements	—	ion exch.	—	7×10^{-3}(Ga)–3.99(S)	—	246
	Te	5×10^{-3}–3×10^{-2}	distill., pptn.	Ge(Li)	6.7–72	0.05–2	332
	Te, Cl, Tl, Ni	10^{-3}–10^{-2}	distill., extr., pptn.	NaI(Tl); flow proport. counter (^{35}S, ^{204}Tl)	—	2×10^{-3} μg(Te); 5×10^{-3} μg(Ce); 5×10^{-3} μg(Tl); 5×10^{-2} μg(Ni)	333
	Br	1	distill., absorption on $AgNO_3$ loaded porous glass powder column	NaI(Tl)	0.021–10.5	—	334
	Cu	0.16–0.18	non-destructive	NaI(Tl)[γ–γ coinc.]	9.78×10^{-2}–0.11	6×10^{-6}	229
	Cu	0.18	non-destructive	NaI(Tl)[γ–γ coinc.]	0.11	10^{-4}	230
Si	P	1	pptn.	G.M., liq. scintill.	down to 1.55	—	335
	21 elements	1	distill., extr., ion exch., pptn.	NaI(Tl), Ge(Li), plastic scintill. (Tl)	1.7×10^{-7}(Au)–2.62×10^{-3}(Cu)	10^{-7}(Au)–2×10^{-2}(Fe)	216, 217

1.3.13 Survey table—*continued*

Matrix	*Elements determined*	*Sample weight /g*	*Chemical separations involved**	*Counting mode*	*Concentrations /p.p.m.*	*Detection limits /p.p.m.*	*Ref.*
Si—*continued*	Na, Cu, As, Sb, Au, Hf, Zr	—	non-destructive	Ge(Li)	6.2×10^{-5}(Sb)–1.6×10^{-3}(As)	—	214
	Na	1	non-destructive	NaI(Tl)	3–200	0.5	336
	W, Cu, Na, La, Fe, Cr, K, Au	—	non-destructive	Ge(Li)	2×10^{-3}(Au)–14(Fe)	—	215
	68 elements	—	non-destructive	Ge(Li), NaI(Tl)	1.4×10^{-4}(Au); 2.7×10^{-3}(Cu) 1.1×10^{-3}(Na)	1.8×10^{-5}(Au)–150(Y)	337
	As, Cu, P, S	—	—	NaI(Tl)	10^{-4}(As, Cu)–5(S)	—	338
	Cu, Au, Ag	—	—	NaI(Tl)	7.8×10^{-5}(Au); 2.7×10^{-3}(Cu) 4.8×10^{-2}(Ag)	—	221
	Sb, As, P	—	—	NaI(Tl); β-counter (P)	down to $\sim 5 \times 10^{-2}$	2×10^{-5} μg(As, Sb); 2×10^{-4} μg(P)	339
	P	—	subst. extr.	4π-counter	down to 0.2 ng	8.8×10^{-4}	340
	P, As, Sb, Au, Cu, Na	—	non-destructive	NaI(Tl), Ge(Li)	—	1.4×10^{-3}(Au)–0.18(P)	341
	Ta	0.1	subst. extr.	γ-spectr., scintill. counter	0.1	—	342
	As, Au, Cu, Ga, Mo, Sb, Zn	—	extr., pptn.	NaI(Tl)	—	10^{-5} μg(Au)–10^{-2} μg(Zn)	343
	Ag	1	isot. exch. (AgCl)	NaI(Tl)	1×10^{-3}–1.7×10^{-2}	10^{-3}	344
$SiCl_4$	Cu, Zn, Mn, Sb, As	—	distill., subst. extr.	NaI(Tl)	1.5×10^{-5}(Sb)–7.7×10^{-3}(Zn)	—	218
Si_3N_4 layers	Na, Cu, As	5×10^{-5}–10^{-4}	SiF_4 evaporation	Ge(Li)	2–4.5(Na); 0.4–0.7(Cu); 2–3(As)	—	219

	W, Cu, Na, La, Fe, Cr, K, Au	—	SiF_4 evaporation	Ge(Li)	0.4–0.7(Na); 0.4–0.7(Cu)	—	215
SiO_2 layers	Na, Cu, As	5×10^{-5}–10^{-4}	SiF_4 evaporation	Ge(Li)	2–4.5(Na); 0.4–0.7(Cu); 2–3(As)	—	219
	W, Cu, Na, La, Fe, Cr, K, Au	—	SiF_4 evaporation	Ge(Li)	0.4–0.7(Na); 0.4–0.7(Cu)	—	215
	Na, Ga, Cu, Fe, Ag	—	non-destructive	γ-spectr.	—	10^{-2}–1	345
	Na	—	—	γ-spectr.	—	0.5×10^{-3} μg–10^{-3} μg	346
SiO_2	Mn, Zn	1	subst. extr.	—	0.062(Mn); 0.25(Zn)	—	267
Sn	16 elements	2	distill., ion exch., pptn., extr., co-pptn.	NaI(Tl); G.M. (^{210}Bi, ^{71}Ge)	10^{-4}(Au)–0.5(As)	—	238
	As, Se, Sb	1	distill., pptn., spont. depos.	Ge(Li)	10^{-3}(Sc)–0.6(Sb) —	—	239
	In, Mn	0.5–2	ion exch. or non-destructive	Ge(Li); NaI(Tl)	—	10^{-3}(In); 10^{-2}(Mn)	347
	19 elements	0.1–2	distill., pptn., metallic depos., ion exch. (In, Mn) or non-destructive (In, Mn)	NaI(Tl); Ge(Li)	2.3×10^{-4}(Au)–0.5(Sb)	2×10^{-5}(Sc)–0.2(Fe)	240
	Cu	1.8–2.4	non-destructive	NaI(Tl)[γ–γ coinc.]	0.124–0.162	6×10^{-6}	229
	Cu	2.22	non-destructive	NaI(Tl)[γ–γ coinc.]	0.162	10^{-4}	230
	Sb	0.1	extr.	NaI(Tl)	2.78; 20.4	0.2	348
Te	Sb	0.01	—	NaI(Tl)	—	5×10^{-3}	349
	I	0.01	non-destructive	Ge(Li)	1–100	1	350
Te compounds [K tellurite; K tellurate]	I	0.01	non-destructive	Ge(Li)	1–100	1	350
$Th(NO_3)_4 5H_2O$	rare earths	2–5	extr. (pre-sepn.), ion exch., pptn.	Ge(Li); CH_4 counter (low activities)	2×10^{-3}(Cu)–5.5(Ce)	—	351
Ti	Hf, V, Al, Cu, Mn, Na, As, Sb	—	non-destructive	NaI(Tl)	—	10^{-3}(Mn)–2(Cu)	352
	Cr	—	sepn.	NaI(Tl)	—	2×10^{-2}	

Matrix	*Elements determined*	*Sample weight /g*	*Chemical separations involved**	*Counting mode*	*Concentrations /p.p.m.*	*Detection limits /p.p.m.*	*Ref.*
Ti—*continued*	Au, Cu, Zn	0.01–0.1	pptn., extr.	NaI(Tl)	1.7×10^{-3}(Au); 170(Cu); 0.28(Zn)	10^{-2} μg(Au); 10^{-3} μg(Cu); 0.1 μg(Zn)	318
	N	—	pptn.	$\beta + \gamma, \gamma$ coinc.	—	0.3	353
	Hf	5×10^{-3}–1.5×10^{-2}	—	NaI(Tl)	—	0.1	354
	V	0.05–1	pre-sepn., pptn., ion exch. or scavenging	Ge(Li)	4.3–7.8	—	355
	Al	0.05–1	non-destructive	NaI(Tl)	up to 58	⩾0.15	
	Mn, In, U	0.05	non-destructive	Ge(Li)	up to 61(Mn)	10^{-2}(In); 0.5(Mn); 5(U)	356
			ion exch., extr.	Ge(Li)	0.93–8.4(Mn); 7.2(U)	10^{-3}(In); 0.2(Mn); 0.2(U)	
$TiCl_4$	Au, Mn, Ni, Fe, Cu, Co, Cr	0.5	extr., ion exch.	γ-spectr.	—	5×10^{-5}(Mn)–0.5(Fe)	357
TiO_2	Hf, V, Al, Cu, Mn, Na, As, Sb	—	non-destructive	NaI(Tl)	—	10^{-3}(Mn)–2(Cu)	352
	Cr	—	sepn.	NaI(Tl)	—	2×10^{-2}	
	Hf	5×10^{-3}–1.5×10^{-2}	—	NaI(Tl)	—	0.1	354
	21 elements	0.05	ion exch., extr.	Ge(Li)	1.7×10^{-4}(Au)–1.32(Cu)	2×10^{-5}(Ir)–2(Ba)	252
TiO_2	Mn, In, U	0.05	non-destructive	Ge(Li)	down to 4.6×10^{-2} (Mn)	2×10^{-3}(In); 5×10^{-2}(Mn); 0.5(U)	356
			ion exch., extr.	Ge(Li)	—	2×10^{-4}(In); 5×10^{-3}(Mn); 5×10^{-2}(U)	
Tl	In	—	non-destructive	NaI(Tl)	5×10^{-2}–55	5×10^{-6} μg(^{116m}In); 1 μg(^{114m}In)	228
	20 elements	up to 0.1	extr.	NaI(Tl)	1.2×10^{-3}(Ag)–42(Cu)	—	231

	Cu	0.89–1.31	non-destructive	NaI(Tl)[γ–γ coinc.]	9.45×10^{-3}–1.05	6×10^{-6}	229
	Cu	1.31	non-destructive	NaI(Tl)[γ–γ coinc.]	1.05	10^{-4}	230
Tl compounds (nitrate, oxide, sulphate, chloride)	In	—	non-destructive	NaI(Tl)	5×10^{-2}–55	3×10^{-6} μg(^{116m}In) 1 μg(^{114m}In)	228
U	Yb, Ho, Dy, Gd, Eu, Sm	1	ion exch. (pre-sepn.), co-pptn., ion exch.	NaI(Tl), G.M.	—	10^{-6}(Eu)–5×10^{-3}(Ho)	358
	Eu, Sm, Dy, Yb, Ho, La, Gd	1	ion exch. (pre-sepn.), ion exch., pptn.	NaI(Tl), β-counter (^{153}Sm)	—	10^{-6}(Eu)–5×10^{-4}(Gd)	359
$UO_2(NO_3)_2 \cdot 6H_2O$	rare earths	—	ion exch., extr. (pre-sepn.), ion exch., pptn.	Ge(Li), CH_4 counter (low activ.)	3×10^{-4}(Tb)–0.225(Ce)	—	351
U oxide	Cd	0.5	ion exch. (pre-sepn.), pptn.	NaI(Tl)	5–10	10^{-2}	360
V_2O_5	Mo	0.1	distill.	scintill. spectr.	0.02–14	~10^{-3} μg	361
	Au, Mn, Co	0.02	non-destructive	NaI(Tl)	1.8(Au); 25(Mn); 47(Co)	5×10^{-1}–5	362
	As, Sb, W, Mo	0.02	distill., pptn., extr.	β-spectr.	0.16(As); 0.2(Sb); 20(W); 2×10^{-2}(Mo)	5×10^{-2}–0.5	
W	P, Co, Cr, Fe, Zr, Hf, Ta, Th	0.1	pptn., ion exch., extr.	—	0.02(Ta)–9.0(Fe)	0.01(Co, Hf, Ta)–0.5(Fe, Zr)	363
	Ga	0.5–1	extr.	Ge(Li)	2×10^{-4}–18	10^{-3}	364
Zn	Cu, Fe, Co, Ni	—	pptn., co-pptn., paper chromat.	—	—	0.05(Cu, Ni); 0.02(Co); 4(Fe)	365
	25 elements	—	ion exch., inorg. chromat.	—	p.p.m. range or less	p.p.b. range	237
	Cu, In, Cd	0.5–1	subst. extr.	—	~5(Cu); 3.4×10^{-4}(In); 0.33–14.3(Cd)	—	268
	13 elements	2–5	carrier-element technique	—	—	10^{-6}–0.1 μg	236
	Ag, Co	0.2–1.5	ion exch.	NaI(Tl)	1.3–1.88(Ag); 0.11–0.16(Co)	—	366

1.3.13 Survey table—*continued*

Matrix	*Elements determined*	*Sample weight /g*	*Chemical separations involved**	*Counting mode*	*Concentrations /p.p.m.*	*Detection limits /p.p.m.*	*Ref.*
Zn	Ge	0.5–2	ion exch., pptn.	NaI(Tl)–[X-ray]	down to 4.9×10^{-2}	4×10^{-3}	367
	Cu	0.1	subst. spont. depos.	—	0.18	—	368
ZnS	Cu	0.05–0.1	pre-sepn.: (1) pptn. exch., extr., ion exch. (2) paper chromat.	γ-spectr.	—	5×10^{-3} μg–10^{-2} μg	369
$ZnSO_4$	Cu, Fe, Co, Ni	—	pptn., co-pptn., paper chromat.	—	—	0.05(Cu, Ni); 0.02(Co); 4(Fe)	365
	Ag	0.1	subst. extr.	NaI(Tl); Ge(Li)	0.13–0.19	1[Ge(Li)]	323
Zr	Hf	0.015–0.045	non-destructive	NaI(Tl)	0.88	0.5 μg	370
ZrO_2	Hf	7×10^{-3}–5.4×10^{-2}	non-destructive	NaI(Tl)	0.29–5.3	0.05	371

* The following abbreviations are used: electrol. = electrolysis; extr. = extraction; subst. = sub-stoichiometric; chromat. = chromatography; pptn. = precipitation; exch. = exchange; distill. = distillation; sepn. = separation; inorg. = inorganic; isot. = isotopic; spont. = spontaneous; depos. = deposition; hydrol. = hydrolysis

References

1. Meinke, W. W. (1972). *Proc. Conf. Saclay, Modern Trends in Activation Analysis*; (1973). *J. Radioanal. Chem.*, **15,** 419
2. Bertolini, G., Cappellani, F. and Restelli, G. (1973). *Nucl. Instr. Meth.*, **112,** 219
3. Palms, J. M., Venugopala Rao, P. and Wood, R. E. (1968). *Nucl. Instr. Meth.*, **64,** 310
4. Johnson, L. O. and Heath, R. L. (1970). *IEEE Trans. Nucl. Sci.*, **NS-17(1),** 276
5. Kandiah, K. (1971). *Nucl. Instr. Meth.*, **95,** 289
6. Dunn, D. R., Kaifer, R. C. and McQuaid, J. H. (1972). *IEEE Trans. Nucl. Sci.*, **NS-19(1),** 461
7. Goulding, F. S., Walton, J. and Malone, D. F. (1969). *Nucl. Instr. Meth.*, **71,** 273
8. Goulding, F. S., Walton, J. and Pehl, R. H. (1970). *IEEE Trans. Nucl. Sci.*, **NS-17(1),** 218
9. Landis, D. A., Goulding, F. S., Pehl, R. H. and Walton, J. T. (1971). *IEEE Trans. Nucl. Sci.*, **NS-18(1),** 115
10. Adams, F. (1970). *Proc. Symp. München, Semiconductor Detectors for Nuclear Radiation*, 167 (P. Eichenes and S. Keil, editors)
11. Fowler, I. L. and Toone, R. J. (1972). *Nucl. Instr. Meth.*, **98,** 193
12. Wagner, S., Trammel, R. and Walter, F. J. (1972). *IEEE Trans. Nucl. Sci.*, **NS-19(3),** 380
13. Cline, J. E. (1968). *IEEE Trans. Nucl. Sci.*, **NS-15(3),** 198
14. McKenzie, J. M., Donovan, P. F. and Gynn, A. C. (1967). *Nucl. Instr. Meth.*, **54,** 147
15. Cooper, J. A. (1971). *Nucl. Instr. Meth.*, **94,** 289
16. Palms, J. M., Wood, R. E. and Puckett, O. H. (1968). *IEEE Trans. Nucl. Sci.*, **NS-15(3),** 397
17. Grenier, G. and Poussier, C. (1970). *Nucl. Instr. Meth.*, **89,** 199
18. Adams, F. and Dams, R. (1970). *Applied Gamma-ray Spectrometry*, 150 (Oxford: Pergamon Press)
19. Camp, D. C. (1972). *Proc. Conf. Vanderbilt University, Radioactivity in Nuclear Spectroscopy*, Vol. 1, 135 (J. H. Hamilton and J. C. Manthuruthil, editors) (New York: Gordon and Breach)
20. Konijn, J., Goudsmit, P. F. A. and Lingeman, E. W. A. (1973). *Nucl. Instr. Meth.*, **109,** 83
21. Cooper, J. A. (1971). *Anal. Chem.*, **43,** 838
22. Brauer, F. B., Kaye, J. H., Fager, J. E., Mitzlaff, W. A. and Rieck, M. G. (1972). *Nucl. Instr. Meth.*, **102,** 223
23. Cooper, R. D., Linekin, D. M. and Brownell, G. L. (1967). *Proc. Symp. Amsterdam, Nuclear Activation Techniques in the Life Sciences*, 65 (Vienna: IAEA 1967)
24. Lyon, W. S. (1969), *Proc. Conf. Gaithersburg, Modern Trends in Activation Analysis*, Vol. 2, 1033 (J. R. De Voe, editor) (Washington: NBS Spec. Publ. 312 (1969))
25. Zimmer, W. H. (1972). *IEEE Trans. Nucl. Sci.*, **NS-19(1),** 151
26. Ohlwiller, R. W. (1972). *IEEE Trans. Nucl. Sci.*, **NS-19(1),** 166
27. Lather, N. and Sardersi, C. G. (1972). *IEEE Trans. Nucl. Sci.*, **NS-19(1),** 141
28. Markwell, F. R., Bolch, W. E., Roessler, C. E. and Ellis, W. H. (1968). *Trans. Amer. Nucl. Soc.*, **NS-15(2),** 639
29. Van der Linden, R. (1974). *Ph.D. Thesis*, State University of Ghent
30. Maenhaut, W. (1973). Unpublished results
31. Brune, D. and Jirlow, K. (1964). *Nukleonik*, **6,** 272
32. Högdahl, O. T. (1965). *Proc. Symp. Salzburg, Radiochemical Methods of Analysis*, Vol. 1, 23 (Vienna: IAEA 1965)
33. Prouza, Z. and Rakovic, M. (1967). *Isotopenpraxis*, **3,** 389
34. Brunfelt, A. O. and Steinnes, E. (1969). *Anal. Chim. Acta*, **48,** 13
35. Steinnes, E. (1971). *Proc. Symp. Kjeller, Activation Analysis in Geochemistry and Cosmochemistry*, 113 (A. O. Brunfelt and E. Steinnes, editors) (Oslo: Universitets-forlaget)
36. Rossitto, F., Terrani, M. and Terrani, S. (1972). *Nucl. Instr. Meth.*, **103,** 77
37. Okada, M. (1961). *Anal. Chim. Acta*, **24,** 410
38. Isenhour, T. L. and Morrison, G. H. (1964). *Anal. Chem.*, **36,** 1089

39. Isenhour, T. L., Evans, C. A. and Morrison, G. H. (1965). *Proc. Conf. Houston, Modern Trends in Activation Analysis*, 123
40. Isirlin, Yu., A., Gelfman, A. Ya. and Mokhir, E. P. (1966). *Zh. Anal. Khim.*, **21,** 1012
41. Quittner, P. (1967). *Acta Chim. Acad. Sci. Hung.*, **51,** 371
42. Fedoroff, M. (1971). *Nucl. Instr. Meth.*, **91,** 173
43. Fedoroff, M. (1972). *Proc. Conf. Saclay, Modern Trends in Activation Analysis*; (1973). *J. Radioanal. Chem.*, **15,** 435
44. Dams, R., Robbins, J. A., Rahn, K. A. and Winchester, J. W. (1970). *Anal. Chem.*, **42,** 861
45. Heindryckx, R. and Dams, R. (1974). *Radiochem. Radioanal. Lett.*, **16,** 209
46. De Neef, J. (1973). *Ph.D. Thesis*, State University of Ghent
47. Hertogen, J. and Gijbels, R. (1971). *Anal. Chim. Acta*, **56,** 61
48. Landis, D. A., Goulding, F. S. and Jarret, B. V. (1972). *Nucl. Instr. Meth.*, **101,** 127
49. Pillay, K. K. and Miller, W. W. (1969). *J. Radioanal. Chem.*, **2,** 97
50. Weaver, J. N. (1973). *International Laboratory*, 29
51. Dran, J. C., Gijbels, R. and Gordon, G. E. (1968). *MIT Laboratory for Nuclear Science, Chemistry Progress Report, MIT 905-133*, Section IV, D
52. Rosenberg, J. and Wiik, H. B. (1971). *Radiochem. Radioanal. Lett.*, **6,** 45
53. Zielinsky, R. A. and Frey, F. A. (1970). *Contrib. Mineral. Petrol.*, **29,** 242
54. De Bruin, M. and Korthoven, P. J. M. (1970). *Proc. Symp. München, Semiconductor Detectors for Nuclear Radiation*, 174 (P. Eichenes and S. Keil, editors)
55. Gijbels, R. and Govaerts, A. (1972). *Proc. Conf. Saclay, Modern Trends in Activation Analysis*; (1973). *J. Radioanal. Chem.*, **16,** 7
56. Zoller, W. H. and Gordon, G. E. (1970). *Anal. Chem.*, **42,** 257
57. Mantel, M. and Amiel, S. (1972). *Anal. Chem.*, **44,** 548
58. Axton, E. J. and Ryves, T. B. (1963). *Int. J. Appl. Radiat. Isotopes*, **14,** 159
59. Porges, K. G. A. and De Volpi, A. (1971). *Int. J. Appl. Radiat. Isotopes*, **22,** 581
60. Rossitto, F. and Terrani, M. (1968). *Int. J. Appl. Radiat. Isotopes*, **19,** 871
61. Rau, F. and Wolf, G. H. (1963). *Nucl. Instr. Meth.*, **27,** 321
62. Seufert, H. (1966). *Nucl. Instr. Meth.*, **44,** 335
63. Sěda, J. (1968). *Nucl. Instr. Meth.*, **59,** 179
64. Sěda, J. (1969). *Nucl. Instr. Meth.*, **70,** 121
65. Taylor, D., Blatt, S. J. and Henderson, I. A. (1970). *Nucl. Instr. Meth.*, **77,** 177
66. Covell, D. F., Sandomire, M. M. and Eichen, M. S. (1960). *Anal. Chem.*, **32,** 1086
67. Crouch, D. F. and Heath, R. L. (1963). *USAEC Report IDO-16923*
68. Grimont, Marin, Richard and Sicard (1966). *Report CEA-R 2082*, France
69. Huysmans, K., Gijbels, R. and Hoste, J. (1974). *J. Radioanal. Chem.*, **20,** 51
70. Bartošek, J., Windels, G. and Hoste, J. (1972). *Nucl. Instr. Meth.*, **103,** 43
71. Deighton, M. O. (1961). *Nucl. Instr. Meth.*, **14,** 48
72. Junod, E. (1966). *Report CEA-R 2980*, France
73. Wiernik, M. (1971). *Nucl. Instr. Meth.*, **95,** 13
74. Löw, K. (1964). *Nucl. Instr. Meth.*, **26,** 216
75. Op de Beeck, J. (1966). Unpublished results
76. Das, H. A., Van Raaphorst, J. G. and Umans, H. J. (1969). *Proc. Conf. Gaithersburg, Modern Trends in Activation Analysis*, Vol. 1, 291 (J. R. De Voe, editor) (Washington: NBS Spec. Publ. 312 (1969))
77. Shanks, J. (1962). *TEES-2671-2*, 5/15
78. Hertogen, J. (1973). Unpublished results
79. Wiernik, M. and Amiel, S. (1969). *J. Radioanal. Chem.*, **3,** 245
80. Nikolaenko, O. K., Sidorov, A. V. and Shtan, A. S. (1970). *Radiation Technology*, Ed. 5, Atomizdat, 120
81. Sterlinski, S. (1972). Personal communication
82. Schonfeld, E. (1966). *Nucl. Instr. Meth.*, **42,** 213
83. De Wispelaere, C., Op de Beeck, J. P. and Hoste, J. (1973). *Anal. Chem.* **45,** 547
84. Gavron, A. (1969). *Nucl. Instr. Meth.*, **67,** 245
85. Filippone, W. and Munno, F. J. (1972). *Nucl. Sci. Eng.*, **47,** 150
86. Anders, O. U. (1969). *Nucl. Instr. Meth.*, **68,** 205
87. Strauss, M. G., Sifter, L. L., Lenkszus, F. R. and Brenner, R. (1968). *IEEE Trans. Nucl. Sci.*, **NS-15(3),** 518

88. Bolotin, H. H., Strauss, M. G. and McClure, D. A. (1970). *Nucl. Instr. Meth.*, **83,** 1
89. Wiernik, M. (1971). *Nucl. Instr. Meth.*, **96,** 325
90. Wiernik, M. (1970). *Israel AEC Report IA-1231*
91. Görner, W. and Höhnel, G. (1970). *Nucl. Instr. Meth.*, **98,** 371
92. Görner, W., Peters, D. and Zshau, J. (1972). *Nucl. Instr. Meth.*, **98,** 371
93. Rao, D. N., Glen, J. L. and Taylor, D. (1972). *Nucl. Instr. Meth.*, **103,** 175
94. Görner, W. and Förster, H. (1972). *J. Radioanal. Chem.*, **12,** 497
95. Bartošek, J., Adams, F. and Hoste, J. (1972). *Nucl. Instr. Meth.*, **103,** 45
96. Harms, J. (1967). *Nucl. Instr. Meth.*, **53,** 192
97. Masters, C. F. and East, L. V. (1971). *IEEE Trans. Nucl. Sci.*, **NS-18(1),** 383
98. Arque Almarez, H. (1965). *Nucl. Instr. Meth.*, **33,** 61
99. Máthé, G. (1963). *Nucl. Instr. Meth.*, **23,** 261
100. Kennett, T. J., Prestwich, W. V. and Keech, G. L. (1964). *Nucl. Instr. Meth.*, **29,** 325
101. Jasiński, A., Ludziejewski, J. and Bialkowski, J. (1964). *Nucl. Instr. Meth.*, **31,** 90
102. Quittner, P. (1972). *Gamma-ray Spectroscopy*, 100 (London: Adam Hilger)
103. Wyttenbach, A. (1971). *J. Radioanal. Chem.*, **8,** 335
104. Verheijke, M. L. (1973). *Philips Nat. Lab. Tech. Note 15/73*
105. Debertin, K. (1971). *Atomkernenergie*, **17,** 97
106. Bartošek, J., Mašek, J., Adams, F. and Hoste, J. (1972). *Nucl. Instr. Meth.*, **104,** 221
107. Adams, F., Hoste, J., Bartošek, J. and Mašek, J. (1972). *Proc. Conf. Saclay, Modern Trends in Activation Analysis*; (1973). *J. Radioanal. Chem.*, **15,** 479
108. Junge, C. E. (1963). *Air Chemistry and Radioactivity* (New York: Academic Press)
109. Giever, P. M. (1968). *Air Pollution*, Vol. 2, 249 (A. C. Stern, editor) (New York: Academic Press)
110. Lockhart, L. B., Patterson, R. L. and Anderson, W. L. (1964). *US Naval Research Laboratory Report 6054* (Washington, D.C.: US Naval Lab.)
111. Jutze, G. A. and Foster, K. E. (1967). Air Pollution Control Association Committee TR-2; *J. Air Pollut. Control. Assoc.*, **17,** 17
112. Parkinson, T. F. and Grant, L. G. (1963). *Nature* (*London*), **197,** 479
113. Keane, J. R. and Fisher, E. M. R. (1968). *Atmos. Environ.*, **2,** 603
114. Brar, S. S., Nelson, D. M., Kanabrocki, E. L., Moore, C. E., Burnham, C. D. and Hattori, D. M. (1968). *Proc. Symp. Gaithersburg, Modern Trends in Activation Analysis*, 43 (Washington D.C.: NBS (1969))
115. Spurny, K. and Fiser, J. (1970). *Staub*, **30,** 249
116. Gray, D., McKown, D. M., Kay, M., Eichor, M. and Vogt, J. R. (1972). *IEEE Trans. Nucl. Sci.*, **19,** 174
117. Mamuro, T., Matsuda, Y., Mizohata, A., Takeuchi, T. and Fujita, A. (1970). *Ann. Rep. Rad. Ctr. Osaka*, **11,** 1; **12,** 1; (1971). *Radioisotopes*, **20,** 111, 117
118. Zoller, W. H. and Gordon, G. E. (1970). *Anal. Chem.*, **42,** 257
119. Tuttle, R. F., Vogt, J. R. and Parkinson, T. F. (1970). *Proc. Symp. Salzburg, Nuclear Techniques in Environmental Pollution*, 119 (Vienna: IAEA (1971))
120. Hasan, O. and Spyrou, N. M. (1972). *Proc. Symp. Bled, Activation Analysis in the Life Sciences*, 263 (Vienna: IAEA (1972))
121. Cawse, P. A. and Pierson, D. H. (1972). *AERE Report R7134*, (An Analytical Study of Trace Elements in the Atmospheric Environment)
122. Belot, Y., Diop, B. and Marini, T. (1971). *Symp. Rome, Radioecology applied to the Protection of Man and his Environment*, paper 30
123. Lindeken, P. L., Morgin, R. L. and Petrock, K. F. (1963). *Health Phys.*, **9,** 305
124. Dams, R., Rahn, K. A. and Winchester, J. W. (1972). *Environ. Sci. Techn.*, **6,** 441
125. Van Grieken, R. and Dams, R. (1972). *Anal. Chim. Acta*, **63,** 369
126. Janssens, M. J. and Dams, R. (1973). *Anal. Chim. Acta*, **65,** 41
127. Andersen, A. A. (1966). *Ind. Hyg. Assoc.*, **27,** 320
128. McKown, D., Kay, M., Gray, D., Abu-Samra, A., Eichor, M. and Vogt, J. (1971). *Proc. ANS Meeting, Nuclear Methods in Environmental Research*, 150 (Columbia: Univ. Missouri)
129. Pillay, K. K. S. and Thomas, C. C., Jr. (1969). *Report W NY-046* (Buffalo, N.Y.: Western New York Nuclear Research Center)
130. Pillay, K. K. S. and Thomas, C. C., Jr. (1971). *J. Radioanal. Chem.*, **7,** 107

131. Brar, S. S., Nelson, D. M., Kanabrocki, E. L., Moore, C. E., Burnham, C. D. and Hattori, D. M. (1970). *Environ. Sci. Techn.*, **4,** 50
132. Moyers, J. L., Zoller, W. H., Duce, R. A. and Hoffman, G. L. (1972). *Environ. Sci. Techn.*, **6,** 68
133. American Society for Testing Materials, (1971). *Annual Book of ASTM Standards*, 425 (Philadelphia: ASTM)
134. Meetham, A. R. (1964). *Atmospheric Pollution*, Chap. 11 (London: Pergamon Press)
135. Murakami, Y., Bando, S., Kobayashi, M. and Yamane, Y. (1970). *Proc. Symp. Salzburg, Nuclear Techniques in Environmental Pollution*, 169 (Vienna: IAEA (1971))
136. Federer, B. (1970). *Pure Appl. Geophys.*, **79,** 120
137. Murozumi, M., Chow, T. J. and Patterson, C. C. (1969). *Geochim. Cosmochim. Acta*, **33,** 1247
138. Duce, R. A., Winchester, J. W. and Van Nahl, T. W. (1966). *Tellus*, **2,** 238
139. Winchester, J. W., Zoller, W. H., Duce, R. A. and Benson, C. S. (1967). *Atmos. Environ.*, **1,** 105
140. Heindryckx, R. and Dams, R. (1972). *Colloquium CNRS Saclay, Activation Analysis of Microquantities*; (1974). *J. Radioanal. Chem.*, **19,** 339
141. Dams, R., Temmerman. G. and Vanhoorne, M. (1972). *Proc. Symp. Bled, Activation Analysis in the Life Sciences*, 263 (Vienna: IAEA (1972))
142. Robertson, D. E. (1968). *Anal. Chem.*, **40,** 1067
143. Robertson, D. E. (1970). *Anal. Chem.*, **42,** 533
144. Brune, D. and Landström, O. (1966). *Radiochim. Acta*, **5,** 228
145. Comar, D. (1968). *Advances in Activation Analysis*, Vol. 1, 109 (J. M. A. Lenihan and S. J. Thompson, editors) (London: Academic Press)
146. Hashimoto, Y. and Winchester, J. W. (1967). *Environ. Sci. Techn.*, **1,** 338
147. Pillay, K. K. S., Thomas, C. C. and Sondel, J. A. (1971). *Environ. Sci. Techn.*, **5,** 74
148. Pillay, K. K. S., Thomas, C. C. and Kaminski, J. W. (1969). *Nucl. Appl. Techn.*, **7,** 478
149. Dahl, J. B., Steinnes, E. and Thomassen, J. (1970). *Proc. Symp. Salzburg, Nuclear Techniques in Environmental Pollution*, 283 (Vienna: IAEA (1971))
150. Bhatki, K. S. and Dingle, A. N. (1970). *J. Appl. Meteorol.*, **9,** 276
151. Bhatki, K. S. and Dingle, A. N. (1970). *Radiochem. Radioanal. Lett.*, **3,** 71
152. Gatz, D. F., Dingle, A. N. and Winchester, J. W. (1969). *J. Appl. Meteorol.*, **8,** 229
153. Duce, R. A. and Winchester, J. W. (1965). *Radiochim. Acta*, **4,** 100
154. Lininger, R. L., Duce, R. A., Winchester, J. W. and Matson, W. R. (1966). *J. Geophys. Res.*, **71,** 2457
155. Winchester, J. W. and Duce, R. A. (1967). *Proc. Symp. Amsterdam, Nuclear Activation Techniques in the Life Sciences*, 631 (Vienna: IAEA (1968))
156. Tiffany, M. A., Winchester, J. W. and Loucks, R. M. (1969). *J. Water Poll. Control.*, **41,** 1319
157. Loucks, R. H., Winchester, J. W., Matson, W. R. and Tiffany, M. A. (1968). *Proc. Symp. Gaithersburg, Modern Trends in Activation Analysis*, 1 (Washington, D.C.: NBS (1969))
158. Tatcher, L. L. and Johnson, J. O. (1970). *Proc. Symp. Salzburg, Nuclear Techniques in Environmental Pollution*, 323 (Vienna: IAEA (1971))
159. Lunggren, K., Sjöstrand, B., Johnels, A. G., Olsen, M., Otterbind, G. and Westermark, T. (1970). *Proc. Symp. Salzburg, Nuclear Techniques in Environmental Pollution*, 373 (Vienna: IAEA (1971))
160. Reiniger, P., Frissel, M. J., Poelstra, P. and Beek, H. (1970). *Proc. Symp. Salzburg, Nuclear Techniques in Environmental Pollution*, 407 (Vienna: IAEA (1971))
161. Byrne, A. R., Dermelj, M. and Kosta, L. (1970). *Proc. Symp. Salzburg, Nuclear Techniques in Environmental Pollution*, 415 (Vienna: IAEA (1971))
162. Läg, J. and Steinnes, E. (1970). *Proc. Symp. Salzburg, Nuclear Techniques in Environmental Pollution*, 429 (Vienna: IAEA (1971))
163. Strohal, P. and Muljev, D. (1970). *Proc. Symp. Salzburg, Nuclear Techniques in Environmental Pollution*, 439 (Vienna: IAEA (1971))
164. De Goeij, J. J. M., Houtman, J. P. W. and Tsjioe, P. S. (1972). *Proc. Symp. Bled, Nuclear Activation Techniques in the Life Sciences*, 295 (Vienna: IAEA (1972))
165. Das, M. A. and Van der Sloot, H. A. (1972). *Proc. Symp. Bled, Nuclear Activation Techniques in the Life Sciences*, 306 (Vienna: IAEA (1972))

166. Lyon, W. S., Bate, L. C. and Emmery, J. F. (1972). *Proc. Symp. Bled, Nuclear Activation Techniques in the Life Sciences*, 253 (Vienna: IAEA (1972))
167. Leddicotte, G. W. and McLain, M. E., Jr. (1971). *Proc. ANS Meeting, Nuclear Methods in Environmental Research*, 216 (Columbia: Univ. Missouri)
168. Kennedy, E. J., Ruch, R. R., Gluskoter, M. J. and Shimp, N. F. (1971). *Proc. ANS Meeting Nuclear Methods in Environmental Research*, 205 (Columbia: Univ. Missouri)
169. De Goeij, J. J. M. (1971). *Proc. ANS Meeting Nuclear Methods in Environmental Research*, 226 (Columbia: Univ. Missouri)
170. Weiss, H. V., Koide, M. and Goldberg, E. D. (1971). *Science*, **174**, 692
171. Murozumi, M., Nakamura, S. and Patterson, C. C. (1970). *Bunseki Kagaku*, **19**, 1057
172. Murozumi, M. and Patterson, C. C. (1970). *Bunseki Kagaku*, **19**, 1049
173. Dudey, N. D., Ross, L. E. and Noshkin, V. E. (1968). *Proc. Symp. Gaithersburg, Modern Trends in Activation Analysis*, 65 (Washington, D.C.: NBS (1969))
174. Hoffman, G. L., Duce, R. A. and Zoller, W. M. (1969). *Environ. Sci. Techn.*, **3**, 1207
175. Rancitelli, L. A., Tanner, T. M. and Dean, J. M. (1969). *USAEC Report B.N.W.L.-1051*, Pt. 2 (Brookhaven, N.Y.: Brookhaven National Lab.)
176. Rancitelli, L. A. and Perkins, R. W. (1970). *J. Geophys. Res.*, **75**, 3055
177. Rancitelli, L. A., Perkins, R. W., Tanner, T. M. and Thomas, T. M. (1970). *Proc. Conf. 700601, Richland, Precipitation Scavenging*, (R. J. Engelmann and W. G. N. Slinn, editors)
178. Perkins, R. W. (1964). *Nucl. Instr. Meth.*, **33**, 71
179. Schramel, P., Samsahl, K. and Pavly, J. (1972). *Colloquium CNRS Saclay, Activation Analysis of Microquantities*; (1974). *J. Radioanal. Chem.*, **19**, 329
180. Adams, F., Op de Beeck, J. P., Van den Winkel, P., Gijbels, R., Desoete, D. and Hoste, J. (1971). *Critical Reviews in Anal. Chem.*, **1**, 455
181. Gordon, C. M. and Larson, R. E. (1964). *J. Geophys. Res.*, **69**, 2881
182. Brar, S. S., Nelson, D. M., Kline, J. R., Gustafson, P. F., Kanobrocki, E. L., Moore, C. E. and Hattori, D. M. (1970). *J. Geophys. Res.*, **75**, 2939
183. Randerson, D. (1970). *Atmos. Environ.*, **4**, 249
184. Atkins, D. H. F., Fisher, E. M. R. and Salmon, L. (1972). *AERE Report R6724*, (Studies of the Analysis of Airborne Particulate Material by Radioactivation and Sodium Iodide Spectrometry)
185. Wainerdi, R. E., Fite, L. E. and Kuykendahl, W. E. (1970). *Proc. Symp. Salzburg, Nuclear Techniques in Environmental Pollution*, 459 (Vienna: IAEA (1971))
186. Bando, S., Yamane, Y. and Murakami, Y. (1970). *Proc. Symp. Salzburg, Nuclear Techniques in Environmental Pollution*, 169 (Vienna: IAEA (1971))
187. Dams, R., Robbins, J. A., Rahn, K. A. and Winchester, J. W. (1970). *Anal. Chem.*, **42**, 861
188. Pillay, K. K. S., Thomas, C. C. and Hycke, C. M. (1971). *Nucl. Techn.*, **10**, 226
189. Navarette, M., Galvez, L., Tzontlimatzin, E. and Ley, A. (1972). *Proc. Symp. Bled, Nuclear Activation Techniques in the Life Sciences*, 277 (Vienna: IAEA (1972))
190. Harrison, P. R., Rahn, K. A., Dams, R., Robbins, J. A., Winchester, J. W., Brar, S. S. and Nelson, D. M. (1971). *J. Air Poll. Control. Assoc.*, **21**, 563
191. Dams, R., Robbins, J. A., Rahn, K. A. and Winchester, J. W. (1970). *Proc. Symp. Salzburg, Nuclear Techniques in Environmental Pollution*, 139 (Vienna: IAEA (1971))
192. Nifong, G. D. and Winchester, J. W. (1970). *USAEC Report No. 600-1705-8* (Ann Arbor.: Univ. Michigan)
193. Dams, R., Rahn, K. A., Nifong, G. D., Robbins, J. A. and Winchester, J. W. (1971). *Proc. 2nd Clean Air Congress*, 509 (B. Englund, editor) (New York: Academic Press)
194. Cowan, G. A., Gregoru, T. G., Jr., Guthals, P. R., Sedlacek, W. A. and Smith, H. L. (1971). *ACS Meeting, Los Angeles*
195. Wilkniss, P. E. and Larson, R. E. (1970). *Proc. Symp. Salzburg, Nuclear Techniques in Environmental Pollution*, 159 (Vienna: IAEA (1971))
196. Bressan, D. J., Carr, R. A., Hannan, P. J. and Wilkniss, P. E. (1972). *Colloquium CNRS Saclay, Activation Analysis of Microquantities*; (1974). *J. Radioanal. Chem.*, **19**, 373
197. Rahn, K. A. (1970). *USAEC Report No. 600-1705-9* (Ann Arbor: Univ. Michigan)
198. Rahn, K. A., Dams, R., Robbins, J. A. and Winchester, J. W. (1971). *Atmos. Environ.*, **5**, 413
199. Rahn, K. A., Wesolowski, J. J., John, W. and Ralston, W. R. (1971). *J. Air Poll. Control. Assoc.*, **21**, 406

200. Dams, R., Billiett, J. and Hoste, J. (1975). *J. Environ. Anal. Chem.*, **4,** 141
201. Heindryckx, R. and Dams, R. (1974). *Radioanal. Radiochem. Lett.*, **16,** 209
202. Belot, Y., Diop, B., Marini, T., Caput, C. and Quinault, J. M. (1972). *Colloquium CNRS Saclay, Activation Analysis of Microquantities*; (1974). *J. Radioanal. Chem.*, **19,** 319
203. Salmon, L. and Creevy, M. G. (1970). *Proc. Symp. Salzburg, Nuclear Techniques in Environmental Pollution*, 47 (Vienna: IAEA (1971))
204. Edgington, D. N. and Lucas, H. F., Jr. (1970). *Proc. Symp. Salzburg, Nuclear Techniques in Environmental Pollution*, 311 (Vienna: IAEA (1971))
205. Wiernik, M. and Amiel, S. (1969). *J. Radioanal. Chem.*, **3,** 245, 403
206. Persiani, C. (1971). *J. Amer. Ind. Hyg. Assoc.*, **32,** 573
207. Hamilton, E. I. (1970). *Health Phys.*, **19,** 511
208. McEachern, P., Myers, W. G. and White, F. A. (1971). *Environ. Sci. Techn.*, **5,** 701
209. Hoste, J., Dams, R., Block, C., Demaynk, M. and Heindryckx, R. (1975). *Final Progress Report, Study of National Air Pollution by Combustion*, Inst. Nucl. Science, Univ. Ghent
210. Weiss, H. V., Koide, M. and Goldberg, E. (1971). *Science*, **172,** 261, 692
211. Weiss, H. V. and Bertine, K. K. (1973). *Anal. Chim. Acta*, **65,** 253
212. Finzi, S. (1973). *Preliminary Report on the Preparation and Characterisation of a Sample of Urban Air Particulates*, Docum. 1183/73 (Brussels: Euratom)
213. Lutz, G. J., Boreni, R. J., Maddock, R. S. and Wing, J. (1972). 'Activation Analysis: A Bibliography through 1971', *NBS Tech. Note* 467 (Washington, D.C.: NBS)
214. Keenan, J. A. and Larrabee, G. B. (1971). *Chem. Instr.*, **3,** 125
215. Kotas, P. and Kukula, F. (1972). *Radioisotopy*, **13,** 253
216. De Corte, F., Speecke, A. and Hoste, J. (1971). *J. Radioanal. Chem.*, **8,** 287
217. De Corte, F., Speecke, A. and Hoste, J. (1971). *J. Radioanal. Chem.*, **9,** 9
218. Lambrev, V. G., Vlasov, V. S. and Myakin'kova, T. V. (1970). *Zh. Anal. Khim.*, **25,** 1983
219. Kotas, P. and Kukula, F. (1970). *Radiochem. Radioanal. Lett.*, **5,** 137
220. Adams, F., Maenhaut, W. and Hoste, J. (1972). *Anal. Chim. Acta*, **61,** 127
221. Nagy, L. G. and Giber, J. (1968). *Periodica Polytech. Chem. Engng.*, **12,** 37
222. Lima, F. W. and Silva, C. N. (1970). *J. Radioanal. Chem.*, **4,** 197
223. Nagy, L. G., Torok, G., Giber, J. and Szokolyi, L. (1970). *Periodica Polytech. Chem. Engng.*, **14,** 149
224. Nagy, L. G., Torok, G., Szokolyi, L. and Giber, J. (1969). *Proc. Symp. Gaithersburg, Modern Trends in Activation Analysis*, 527 (Washington, D.C.: NBS (1969))
225. Nagy, L. G., Torok, G., Szokolyi, L. and Giber, J. (1969). *Magy. Kem. Foly*, **75,** 496
226. Rausch, H. and Salamon, A. (1968). *Talanta*, **15,** 975
227. Orvini, E. (1972). *Radiochem. Radioanal. Lett.*, **10,** 145
228. Daiev, Kh., Aleksandrov, S. and Daieva, L. (1968). *Z. Anal. Chem.*, **239,** 369
229. Wölfle, R., Herpers, U. and Herr, W. (1968). *Z. Anal. Chem.*, **233,** 241
230. Wölfle, R., Herpers, U. and Herr, W. (1969). *J. Radioanal. Chem.*, **2,** 171
231. Gil'bert, E. N. and Pronin, V. A. (1968). *J. Anal. Chem. USSR*, **23,** 1536
232. Neeb, K. H., Stoeckert, H., Braun, R. and Bleich, H. P. (1969). *Z. Anal. Chem.*, **245,** 233
233. Lesbats, A. (1968). *Ann. Chim.*, **3,** 293
234. Blouri, J., Fedoroff, M. and Revel, G. (1972). *Colloquium CNRS Saclay, Activation Analysis of Microquantities*; (1973). *J. Radioanal. Chem.*, **17,** 185
235. Dugain, F., André, M. and Speecke, A. (1970). *Radiochem. Radioanal. Lett.*, **4,** 35
236. Lanfranco, G. and Bianchini, A. (1968). *Chim. Ind. (Milan)*, **50,** 332
237. Cuypers, J., Girardi, F. and Mousty, F. (1972). *Colloquium CNRS Saclay, Activation Analysis of Microquantities*; (1973). *J. Radioanal. Chem.*, **17,** 115
238. Lanfranco, G. (1971). *Metallurgia Ital.*, **63,** 225
239. Maenhaut, W., Adams, F. and Hoste, J. (1970). *J. Radioanal. Chem.*, **6,** 83
240. Maenhaut, W., Adams, F. and Hoste, J. (1972). *Anal. Chim. Acta*, **59,** 209
241. Mizuike, A., Fukuda, K. and Mitsuya, N. (1968). *Radioisotopes*, **17,** 199
242. Filatova, T. P., Chernova, M. A. and Yarmarkin, V. K. (1971). *Aktiv. Anal., Mater. Vses. Soveshch., 2nd 1968*, 99
243. Kiss, L. G. (1971). *Periodica Polytech. Chem. Engng.*, **15,** 351
244. Gil'bert, E. N. and Pronin, V. A. (1969). *Izv. Sib. Otdel. Akad. Nauk SSSR, Ser. Khim. Nauk*, **2,** 168

245. Ballaux, C., Dams, R. and Hoste, J. (1969). *Anal. Chim. Acta*, **47,** 397
246. Mignonsin, E. P. (1972). *Colloquium CNRS Saclay, Activation Analysis of Microquantities*; (1974). *J. Radioanal. Chem.*, **19,** 33
247. Albert, P. (1968). *Mem. Sci. Rev. Met.*, **65,** Spec. No. 3
248. Lesbats, A. and Tardy, M. (1972). *Colloquium CNRS Saclay, Activation Analysis of Microquantities*; (1973). *J. Radioanal. Chem.*, **17,** 127
249. Samadi, A. A. and May, S. (1969). *Bull. Soc. Chim. Fr.*, 3776
250. Steinnes, E. (1972). *Radiochim. Acta*, **17,** 119
251. Fedoroff, M. (1971). *Ann. Chim.*, **6,** 159
252. Neirinckx, R., Adams, F. and Hoste, J. (1969). *Anal. Chim. Acta*, **48,** 1
253. Cojocaru, V. and Filby, R. H. (1970). *Inst. Fiz. Atom* (*Report*) IFA-NR-31
254. Reed, D. and Hilton, D. A. (1970). *Radiochem. Radioanal. Lett.*, **5,** 293
255. May, S. and Pinte, G. (1969). *Proc. Symp. Gaithersburg, Modern Trends in Activation Analysis*, 655 (Washington, D.C.: NBS (1969))
256. May, S. and Pinte, G. (1969). *J. Radioanal. Chem.*, **3,** 329
257. Alimarin, I. P. (1968). *Analysis of High-Purity Materials*, 553 (Israel Programme for Scientific Translations Ltd.)
258. Filippova, N. V., Smakhtin, L. A. and Miglina, N. V. (1970). *J. Anal. Chem. USSR*, **25,** 2001
259. Kosta, L. and Ravnik, V. (1971). *Radiochem. Radioanal. Lett.*, **7,** 295
260. Byrne, A. R., Dermelj, M. and Kosta, L. (1970). *J. Radioanal. Chem.*, **6,** 325
261. Lima, F. W. and Atalla, L. T. (1970). *Radiochem. Radioanal. Lett.*, **3,** 23
262. Artiukhin, P. I., Startseva, E. A. and Nikolaev, A. V. (1972). *Colloquium CNRS Saclay, Activation Analysis of Microquantities*; (1973). *J. Radioanal. Chem.*, **17,** 173
263. Lesbats, A. (1969). *Ann. Chim.*, **4,** 29
264. Sukhov, G. V. and Firsov, V. I. (1972). *Zav. Lab.*, **38,** 368
265. May, S., Piccot, D. and Pinte, G. (1972). *Colloquium CNRS Saclay, Activation Analysis of Microquantities*; (1973). *J. Radioanal. Chem.*, **17,** 153
266. Albert, P., Blouri, J., Cleyrergue, C., Deschamps, N. and Le Héricy, J. (1968). *J. Radioanal. Chem.*, **1,** 297
267. Kudo, K. and Suzuki, N. (1972). *Colloquium CNRS Saclay, Activation Analysis of Microquantities*; (1974). *J. Radioanal. Chem.*, **19,** 55
268. May, S. and Samadi, A. A. (1970). *Bull. Soc. Chim. Fr.*, 1628
269. Schramel, P. (1968). *Int. Leichmetalltagung Leoben*, 27
270. Van Reenen, T. J. and de Wet, W. J. (1968). *J. S. African Chem. Inst.*, **21,** 120
271. Khusnutdinov, R. I., Minagaliev, G. G. and Leushkina, G. V. (1971). *Aktiv. Anal., Mater. Vses. Soveshch., 2nd 1968*, 109
272. Dhalenne, G., Rouchaud, J. C. and Revel, G. (1971). *C.R. Acad. Sci.* (*Paris*), **272C,** 538
273. Thompson, B. A. and Miller, E. C. (1971). *J. Res. Nat. Bur. Stand.*, **75A,** 429
274. Veveris, O., Miculis, J. and Riekstina, D. (1971). *Izv. Akad. Nauk Latv. SSR, Ser. Fiz. Tekh. Nauk*, **1,** 16
275. Lobanov, E. M., Krasivina, L. E., Zverev, B. P. and Usmanova, M. M. (1968). *Aktivatsionnyi Analiz Chistykh Materialov*, 16 (B. P. Zverev, editor) (Tashkent: Izdatelstvo Fan Uzbekskoii SSR)
276. Borisov, G. I., Demidov, A. M. and Zakharov, E. A. (1969). *Atom. Energ.*, **26,** 16
277. Aveyanova, V. P., Ginzburg, M. I., Marunina, N. I. and Sukhov, G. V. (1970). *Soviet J. Non-fer. Metals*, **11,** 53
278. Lobanov, E. M., Valieva, M. G. and Usmanova, M. M. (1971). *Neitron-Aktiv. Anal.*, 75
279. Valieva, M. G., Lobanov, E. M. and Usmanova, M. M. (1971). *Aktiv. Anal., Mater. Vses. Soveshch., 2nd 1968*, 106
280. Zmijewska, W. and Sorantin, H. (1971). *J. Radioanal. Chem.*, **8,** 83
281. Bocharova, N. N., Razumova, G. N. and Shuba, I. D. (1972). *Radiokhimiya*, **14,** 158
282. Albert, P., Blouri, J., Cleyrergue, C., Deschamps, N. and Le Héricy, J. (1968). *J. Radioanal. Chem.*, **1,** 389
283. Gillette, R. K. (1971). *Anal. Lett.*, **4,** 563
284. Mignonsin, E. P. and Guillaume, M. (1972). *Colloquium CNRS Saclay, Modern Trends in Activation Analysis*; (1973). *J. Radioanal. Chem.*, **16,** 235
285. Taczanowski, S. and Janczyszyn, J. (1971). *Radiochem. Radioanal. Lett.*, **6,** 101

286. Vinitskaya, G. P., Koval'skaya, G. E., Kazachevskii, I. V., Levkovskii, V. N. and Sokol'skii, V. V. (1971). *Aktiv. Anal., Mater. Vses. Soveshch., 2nd 1968*, 191
287. Albert, P., Blouri, J., Cleyrergue, C., Deschamps, N. and Le Héricy, J. (1968). *J. Radioanal. Chem.*, **1**, 431
288. Borák, J., Křivánek, M., Kukula, F. and Vinš, V. (1968). *Chem. Zvesti*, **22**, 701
289. Furuya, K., Koizumi, M. and Kikuchi, T. (1970). *Jap. Anal.*, **19**, 775
290. Kukula, F. (1969). *Chem. Zvesti*, **23**, 521
291. Ganiev, A. G. and Khudaibergenov, U. (1968). *Aktivatsionnyii Analiz Chistykh Materialov*, 69 (B. P. Zverev, editor) (Tashkent: Izdatelstvo Fan Uzbekskoii SSR)
292. Křivánek, M., Kukula, F. and Vinš, V. (1968). *J. Radioanal. Chem.*, **1**, 219
293. Kukula, F. and Křivánek, M. (1968). *Isotopenpraxis*, **4**, 57
294. Dogadkin, N. N. and Yakovlev, Y. V. (1971). *Yadernofiz. Metody Anal. Veshchestva*, 29
295. Kudo, K. and Hishinuma, N. (1969). *J. Radioanal. Chem.*, **3**, 369
296. Kudo, K. and Hishinuma, N. (1970). *J. Radioanal. Chem.*, **5**, 331
297. Kudo, K., Shigematsu, T. and Iso, H. (1972). *Colloquium CNRS Saclay, Modern Trends in Activation Analysis*; (1973). *J. Radioanal. Chem.*, **16**, 139
298. Leushkina, G. V., Lobanov, E. M., Dutov, A. G. and Matveeva, N. P. (1971). *Aktiv. Anal., Mater. Vses. Soveshch., 2nd 1968*, 23
299. Lobanov, E. M., Ganiev, A. G., Silvanovich, Y. A. and Khudaibergenov, U. (1968). *Aktivatsionnyii Analiz Chistykh Materialov*, 33 (B. P. Zverev, editor) (Tashkent: Izdatelstvo Fan Uzbekskoii SSR)
300. Rausch, H. and Salamon, A. (1969). *J. Radioanal. Chem.*, **3**, 265
301. Shulepnikov, M. N., Schmanenkova, G. I., Yakovlev, Y. V. and Dogadkin, N. N. (1971). *Zh. Anal. Khim.*, **26**, 1043
302. Wolfstirn, K. B. (1970). *J. Phys. Chem. Solids*, **31**, 601
303. De Neve, R., De Soete, D. and Hoste, J. (1968). *Anal. Chim. Acta*, **40**, 379
304. Gijbels, R. (1973). *Talanta*, **20**, 346
305. Razumova, G. N., Shuba, I. D. and Vasilév, I. Ya. (1970). *Radiokhimiya*, **12**, 133
306. Simkova, M. and Kukula, F. (1970). *Radiochem. Radioanal. Lett.*, **3**, 295
307. Baishya, N. K. and Heslop, R. B. (1970). *Anal. Chim. Acta*, **50**, 209
308. Vasil'ev, I. Ya., Razumova, G. N. and Shuba, I. D. (1968). *Radiokhimiya*, **10**, 596
309. Ganiev, A. G., Khudaibergenov, U. and Mukhtarov, R. M. (1968). *Aktivatsionnyii Analiz Chistykh Materialov*, 55 (B. P. Zverev, editor) (Tashkent: Izdatelstvo Fan Uzbekskoii SSR)
310. Vasil'ev, I. Ya., Razumova, G. N. and Shuba, I. D. (1969). *Radiokhimiya*, **11**, 573
311. Lobanov, E. M., Balbakov, D., Shalpykov, A., Chanyshev, A. I. and Kozhogulov, O. (1969). *Izv. Akad. Nauk Kirg. SSR*, **5**, 28
312. Abashin, E. G., Smakhtin, L. A. and Gorshkov, V. V. (1971). *Aktiv. Anal., Mater. Vses. Soveshch., 2nd 1968*, 133
313. Fedoroff, M. (1968). *C. R. Acad. Sci. (Paris)*, **267C**, 1227
314. Fedoroff, M. (1970). *Bull. Soc. Chim. Fr.*, 1233
315. Fedoroff, M. (1970). *C. R. Acad. Sci. (Paris)*, **270C**, 486
316. Fedoroff, M. (1970). *C. R. Acad. Sci. (Paris)*, **271C**, 399
317. Ganiev, A. G., Nazmitdinov, M. K., Rybnov, V. V. and Mukhtarov, R. M. (1968). *Aktivatsionnyii Analiz Chistykh Materialov*, 61 (B. P. Zverev, editor) (Tashkent: Izdatelstvo Fan Uzbekskoii SSR)
318. Stepanets, O. V., Yakovlev, Y. V. and Alimarin, I. P. (1970). *J. Anal. Chem. USSR*, **25**, 1637
319. Avsaragov, Kh. B. and Mikhailov, B. N. (1971). *Metody Anal. Miner. Syr'ya*, 103
320. Bilimovich, G. N., Alimarin, I. P. and Tikhonova, T. V. (1971). *J. Anal. Chem. USSR*, **26**, 104
321. Bando, S. and Kishi, H. (1970). *Jap. Anal.*, **19**, 17
322. De Boeck, R., Adams, F. and Hoste, J. (1969). *J. Radioanal. Chem.*, **2**, 109
323. Kukula, F. (1970). *Isotopenpraxis*, **6**, 303
324. Arroyo, A. and Toro, G. (1968). *Rep. Inst. Asunt. Nucl.*, (IAN-Q-8)
325. Arroyo, A. and Toro, G. (1969). *Proc. Symp. Gaithersburg, Modern Trends in Activation Analysis*, 541 (Washington, D.C.: NBS (1969))
326. Glazov, V. M. and Korshunov, I. A. (1968). *Trudy Khim. Khim. Tekhnol. (Gorkii)*, **1**, 107

327. Verot, J. L. and Jaumier, J. J. (1972). *Colloquium CNRS Saclay, Activation Analysis of Microquantities*; (1973). *J. Radioanal. Chem.*, **17**, 237
328. Ballaux, C., Dams, R. and Hoste, J. (1968). *Anal. Chim. Acta*, **41**, 147
329. Ballaux, C., Dams, R. and Hoste, J. (1968). *Anal. Chim. Acta*, **43**, 1
330. Ballaux, C., Dams, R. and Hoste, J. (1969). *Anal. Chim. Acta*, **45**, 337
331. Gijbels, R. and Dams, R. (1972). *Anal. Chim. Acta*, **62**, 191
332. Neeb, K. H., Martin, J. and Franke, R. (1969). *Z. Anal. Chem.*, **247**, 27
333. Neeb, K. H., Franke, R. and Neidl, H. (1969). *Z. Anal. Chem.*, **247**, 225
334. Soyka, W., Havezov, I. K. and Erdtmann, G. (1972). *J. Radioanal. Chem.*, **11**, 23
335. De Corte, F., Speecke, A. and Hoste, J. (1971). *J. Radioanal. Chem.*, **8**, 277
336. Kenna, B. T. and Conrad, F. J. (1968). *Talanta*, **15**, 418
337. Larrabee, G. B. and Keenan, J. A. (1971). *J. Electrochem. Soc.*, **118**, 1351
338. Martin, J. (1969). *Proc. Symp. Gaithersburg, Modern Trends in Activation Analysis*, 517 (Washington, D.C.: NBS (1969))
339. Perezhogin, G. A. and Meshcheryakov, V. G. (1969). *Industrial Laboratory USSR*, **35**, 979
340. Perezhogin, G. A. and Sidorova, L. P. (1969). *Zh. Anal. Khim.*, **24**, 435
341. Rausch, H., Bereznai, T. and Bogáncs, J. (1972). *Colloquium CNRS Saclay, Activation Analysis of Microquantities*; (1974). *J. Radioanal. Chem.*, **19**, 77
342. Zmijewska, W. (1970). *Chem. Anal.*, **15**, 471
343. Zmijewska, W., (1970). *Chem. Anal.*, **15**, 1025
344. Zmijewska, W. (1971). *Radiochem. Radioanal. Lett.*, **8**, 285
345. Lobanov, E. M., Mamadzhanov, F. I. and Chubarov, L. B. (1969). *Dokl. Akad. Nauk Tadzh. SSR*, **12**, 37
346. Nagashima, N. (1969). *Bunseki Kiki*, **7**, 31
347. Maenhaut, W., Adams, F. and Hoste, J. (1971). *J. Radioanal. Chem.*, **9**, 27
348. Oka, Y., Kato, T. and Tsai, H. T. (1968). *Bull. Chem. Soc. Jap.*, **41**, 2427
349. Abrarov, O. A., Imamov, T. K., and Sabirov, S. S. (1968). *Aktivatsionnyii Analiz Chistykh Materialov*, 66 (B. P. Zverev, editor) (Tashkent: Izdatelstvo Fan Uzbekskoii SSR)
350. Tamai, T., Hashimoto, T., Matsushita, R. and Iwata, S. (1970). *Jap. Anal.*, **19**, 502
351. Bachmann, K. and Lieser, K. H. (1970). *Z. Anal. Chem.*, **250**, 172
352. Blinkov, D. I. and Lobanov, E. M. (1969). *Trudy Khim. Khim. Tekhnol.*, **3**, 108
353. Kudinov, B. S., Pronman, I. M. and Shchulepnikov, M. N. (1968). *Metody Opredeleniya i Issledovaniya Sostoyaniya Gazov v Metallakh Izdatelstvo Nauka, Moskwa*, 122
354. Lobanov, E. M., Blinkov, D. I., Dutov, A. G. and Zinovev, N. V. (1968). *Izv. Akad. Nauk Uzbek. SSR, Ser. Fiz.-Matem. Nauk, No. 2*, 94
355. Neirinckx, R., Adams, F. and Hoste, J. (1969). *Anal. Chim. Acta*, **47**, 173
356. Neirinckx, R., Adams, F. and Hoste, J. (1970). *Anal. Chim. Acta*, **50**, 31
357. Sukhanovskaya, A. I. and Shafran, I. G. (1971). *Yadernofiz. Metody Anal. Veshchestva*, 35
358. Alimarin, I. P., Miklishanskii, A. Z. and Yakovlev, Y. V. (1970). *J. Radioanal. Chem.*, **4**, 45
359. Miklishanskii, A. Z. and Yakovlev, Y. V. (1971). *Yadernofiz. Metody Anal. Veshchestva*, 23
360. Ruf, H. (1969). *Z. Anal. Chem.*, **247**, 297
361. Adams, F. and Hoste, J. (1970). *Radiochem. Radioanal. Lett.*, **3**, 31
362. Ganiev, A. G., Nazmitdinov, M. K. and Sabirov, S. (1968). *Izv. Akad. Nauk Uzbek. SSR, Ser. Fiz.-Matem. Nauk, No. 1*, 40
363. Döge, H. G. and Grosse-Ruyken, H. (1968). *Isotopenpraxis*, **4**, 262
364. Salamon, A. (1970). *J. Radioanal. Chem.*, **4**, 81
365. Bellobono, I. R. (1970). *Rend. Inst. Lomb. Sci. Lett. A*, **104**, 918
366. Machiroux, R. and Mousty, F. (1968). *Anal. Chim. Acta*, **42**, 371
367. Machiroux, R. and Mousty, F. (1969). *Anal. Chim. Acta*, **48**, 219
368. Miller, J. (1970). *J. Radioanal. Chem.*, **4**, 35
369. Hegedus, D., Noszticiusz, Z. and Erdey, L. (1971). *Magy. Kem. Lapia*, **26**, 344
370. Haerdi, W., Balsenc, L. and Monnier, D. (1968). *J. Radioanal. Chem.*, **1**, 51
371. Okada, M., Yonezawa, C. and Tamura, N. (1972). *Radiochem. Radioanal. Lett.*, **11**, 193
372. Bruggeman, A. (1973). Unpublished results.

2
Activation Analysis—Part 2

R. CORNELIS, J. HOSTE, A. SPEECKE, C. VANDECASTEELE and J. VERSIECK
Rijksuniversiteit Gent

and

R. GIJBELS
Universitaire Instelling Antwerpen

2.1 APPLICATIONS OF NEUTRON ACTIVATION ANALYSIS IN GEOCHEMISTRY AND COSMOCHEMISTRY

2.1.1 Introduction

It is the task of the geochemist to determine the abundance and the distribution of the chemical elements in the earth and in earth-related materials, including meteorites, tektites, lunar rocks and soil. For convenience, the chemical elements have been grouped into major (>1%), minor (0.01–0.1%) and trace elements (<0.01% or <100 p.p.m.). In the case of extremely low concentrations, such as for iridium in many terrestrial silicate rocks (0.0X p.p.b.), it is perhaps appropriate to talk about ultratrace elements. In most terrestrial rocks, tektites, meteoritic stones, lunar rocks and soils, the elements fall generally in the following classes: major: O, (Na), Mg, Al, Si, (K), Ca, (Ti), Fe; minor: H, C, (F), (Na), P, S, Cl, (K), (Ti), V, Cr, Mn, (Co), (Ni), Sr, Zr, Ba; and trace elements: the noble gases, Li, Be, B, N, (F), Sc, (Ni), (Co), Cu and all elements beyond Cu, i.e. the last two-thirds of the Periodic Table. Elements in parentheses may fall into two abundance classes; for example, Na and K, on the average, are >1% in rocks found in the earth's crust, and <1% in meteorites, lunar rocks and soil[1]. The above classification of the elements is, of course, not valid for ores and minerals.

Activation analysis has predominantly been used as an ultrasensitive method of trace analysis. The more recently developed instrumental neutron activation analysis (INAA) using semiconductor detectors and on- or off-line computers has made activation analysis a promising tool not only for chemists and specialised geochemists, but also for geologists in general. It is a fact that emission spectrography, X-ray fluorescence, atomic absorption and even mass spectrometry have gained greater acceptance and are more commonly used in geology and geochemistry departments and mining institutes than activation analysis. An increased emphasis on activation analysis as a tool in geochemistry in the years to come is, however, to be expected[2]. Indeed, in reviewing the results obtained by various groups, selected by NASA, on the powdered, reasonably homogeneous lunar fines from Apollo 11[3], Smales pointed out that activation analysis was clearly the most frequently used analytical method, not only for trace elements, but even for major elements[4].

Some recent conferences have revealed a tendency to use activation analysis for determining major elements, indeed, and to place emphasis more on precision and accuracy than on ultimate sensitivity. Fast, automated 'non-destructive' analyses are possible in a number of cases with a modest isotopic neutron source[5], which is very attractive for a variety of routine applications.

The literature on activation analysis applied to geological materials is already quite impressive[6]. It can be studied, for instance, on the basis of the application (e.g. terrestrial, meteoritic and lunar studies), or on the basis of the concentration range of the element determined (major, minor, trace). We consider it more convenient to discuss the analytical possibilities of various activation sources: isotopic neutron sources, neutron generators and nuclear reactors. Bremsstrahlung and charged particle sources will not be considered here. It is hoped that the general applicability of activation analysis will be realised through the discussion of a few specific applications and the presentation of some synoptic tables.

2.1.2 The choice of a neutron source

The choice of a neutron source depends on the problem to be solved. It is obvious that a research reactor or even a high-flux reactor is to be recommended when a very high sensitivity is required or desirable, e.g. for determining p.p.b. concentrations of rhenium in minerals or flue dusts. If oxygen must be determined, a 14 MeV neutron generator is most suitable because of the high threshold energy ($\sim$9 MeV) of the most useful nuclear reaction $^{16}O(n,p)^{16}N$. An unmoderated isotopic neutron source (e.g. $^{241}Am/Be$) is perhaps the best choice for determining the fluorine content of fluorine ores and concentrates via the reaction $^{19}F(n,\alpha)^{16}N$ (threshold energy $\sim$2 MeV) without interference from the $^{16}O(n,p)^{16}N$ reaction. For the precise instrumental determination of manganese in pyrolusite ores, via the sensitive reaction $^{55}Mn(n,\gamma)^{56}Mn$, a relatively modest isotopic source of moderated neutrons is ideally suited. Nuclear reactors, accelerators and isotopic neutron sources are so different in neutron production and neutron energy that they can be considered competitors only for some applications.

Other criteria to be considered are, of course, cost of the neutron source, possibility for transportation, dimension of source plus shielding, the necessity of having highly specialised personnel, etc. In mobile installations, for geophysical prospection, radioisotope sources with yields of less than 10^8 neutron s^{-1}, with or without moderator, have mostly been used, mainly in the USSR[7]. The neutron flux in the irradiation zone with the above source strength is usually not more than 10^5 neutron $cm^{-2}\ s^{-1}$, depending upon the design. Experiments in the USA and in the USSR have shown that deuteron accelerators (pumped and sealed tube) producing 14 MeV, 3 MeV or moderated neutrons are also transportable[7-14].

In stationary conditions, heavier shielding is only a cost not a transportation problem, hence more powerful neutron sources can be used than in field conditions (isotopic neutron sources of greater strength; laboratory neutron generators and so-called neutron multipliers).

Reactor neutron activation analysis, and γ-activation analysis (Bremsstrahlung of varying energy, 3–40 MeV, obtained from electrostatic, linear wave-guide and cyclic electron accelerators, betatron and synchrotons) is practically restricted to stationary applications. The main trends of the use of various neutron and Bremsstrahlung sources for the study of rocks and ores have been described by Savosin *et al.*[15].

2.1.3 Applications using isotopic neutron sources and neutron multipliers

The use of isotopic neutron sources and neutron multipliers in ore and rock analysis has been discussed in some detail by Gijbels[16]. One can use the original neutron spectrum with neutron energies up to 4 MeV, or thermalised neutrons, or a combination of both. The fast neutrons have frequently been used for determining fluorine by the $^{19}F(n,\alpha)^{16}N$ reaction; the thermalised neutrons have often been used for determining manganese, copper, and tungsten by (n,γ) reactions; silicon and aluminium can be determined by irradiating the sample subsequently with fast and moderated neutrons, the relevant nuclear reactions being $^{28}Si(n,p)^{28}Al$ and $^{27}Al(n,\gamma)^{28}Al$.

Russian workers have mostly used $^{210}Po/Be$ and $^{239}Pu/Be$ sources, up to 10 Ci. For Western countries, a number of papers deal with the application of $^{241}Am/Be$ (typically 5 Ci) and ^{252}Cf (typically 100 μg) in neutron activation analysis of geological materials. $^{228}Th/Be$ and $^{226}Ra/Be$ have not been utilised so often, because of their high γ-dose rate; the same is true for $^{124}Sb/Be$. A source that is quite promising both in investment and replenishment costs is $^{227}Ac/Be$ ($T_{\frac{1}{2}} = 21.8$ years).

Table 2.1 lists a number of typical fast, instrumental major element determinations in geological material by isotopic neutron source activation analysis. (This table is accompanied by a rather extensive list of references, since in most reviews on the use of activation analysis, no special attention has been drawn to this particular aspect.) In addition, the AECL (Canada) has investigated the feasibility of determining a number of economically interesting elements in various minerals, using a 6000 Ci $^{124}Sb/Be$ source[17]: Al in bauxite, Ag in friebergite, As in skutterudite, Au in auriferous quartz, Ba in barite, Ca in limestone, Co in skutterudite and in Cu–Ni ore, Cr in chromite, Cu in chalcopyrite, in Cu–Ni ore, in Cu–Zn–Pb ore and in malachite, Dy in apatite, F in fluorite, In in sphalerite, Mg in dolomite and in brucite, Mn in manganese ore, Mo in molybdenite, Ni in pentlandite, rare earths and Th in monazite, Sb in stibnite, Se in Cu–Zn ore, Sn in cassiterite, Ti in ilmenite, W in wolframite, U in uranitite and carnotite, V in carnotite, Zn in sphalerite and Zr in zircon. This mobile system is built around an $^{124}Sb/Be$ source and allows thermal fluxes of 10^8 neutron $cm^{-2}s^{-1}$. The whole can be placed on a 12 m long trailer[17].

Isotopic neutron source activation can also be applied to common rocks. Janghorbani *et al.*[18], for instance, used a 1.25 mg ^{252}Cf source for determining Mn, Al, Na and Ti in lunar rocks. The induced activities were, however, rather low in the 0.5–1 g samples available.

Usually the method is optimised for determining one single element of (economic) interest, although sometimes more elements are sought for, e.g. Al, Cr and Si in chrome ores. Countings are usually performed with NaI(Tl) detectors coupled to single-channel analysers. In special cases, such as

Table 2.1 Neutron activation analysis of ores using isotopic neutron sources

Element determined	*Activation reaction*	*Half-life*	*Type of ore analysed*	*Ref.*
(*a*) *Unmoderated neutrons*				
Aluminium	$^{27}Al(n,p)^{27}Mg$	9.46 m	chrome ores and concentrates	25–27
Chromium	$^{52}Cr(n,p)^{52}V$	3.75 m	chrome ores, concentrates and tailings	25–27
Fluorine	$^{19}F(n,\alpha)^{16}N$	7.14 s	fluorite ores and concentrates, apatite	28–36
Iron	$^{56}Fe(n,n'\gamma)$	inelastic neutron scatt.	iron ores, concentrates and filings, manganese ores	37
	$^{56}Fe(n,p)^{56}Mn$	155 m	taconite ore	38
Silicon†	$^{28}Si(n,p)^{28}Al$	2.24 m	chrome ores, concentrates and tailings, pyrite, pyrite smelter slag, nickel smelter slag, taconite ore	25–27, 38
Uranium*	$^{238}U(n,\gamma)^{239}U$	23.5 m	uranium ore (equilibrated and not)	39
(*b*) *Moderated neutrons*				
Aluminium†	$^{27}Al(n,\gamma)^{28}Al$	2.24 m	aluminosilicates, sedimentary rocks, aluminium ores and dressing products	40, 41
Copper	$^{65}Cu(n,\gamma)^{66}Cu$ $^{63}Cu(n,\gamma)^{64}Cu$	5.1 m 12.8 h	copper ores, copper-bearing drill core and other copper-bearing products	40 42–44
	$Cu(n,\gamma)$	prompt γ	porphyry copper deposits	45
Dysprosium	$^{164}Dy(n,\gamma)^{165}Dy$	139 m	phosphorite ore	46
Gold	$^{197}Au(n,\gamma)$	prompt γ	gold ore	47
Indium	$^{115}In(n,\gamma)^{116m}In$	53.7 m	multi-metallic ores and concentrates, sphalerite	48, 49
Iron	$^{56}Fe(n,\gamma)$	prompt γ	iron ores, concentrates and filings	37, 50, 51
Manganese	$^{55}Mn(n,\gamma)^{56}Mn$	155 m	manganese ores (pyrolusite), sedimentary rocks and soils, wolframite, hibnerite-bearing rock, multi-metallic ores, ferromanganese	52–58

Table 2.1—*continued*

Element determined	*Activation reaction*	*Half-life*	*Type of ore analysed*	*Ref.*
Nickel	Ni(n,γ)	prompt γ	low-grade nickel ore (>0.1%), nickel laterites	19, 45
Rhenium	$^{187}Re(n,\gamma)^{188}Re$	16.8 h	molybdenite concentrates	20, 59
Silver	$^{107}Ag(n,\gamma)^{108}Ag$	145 s	silver ore	
	$^{109}Ag(n,\gamma)^{110}Ag$	24.4 s	phosphorite ore	46
Tungsten	$^{186}W(n,\gamma)^{187}W$	23.9 h	wolframite, hibnerite-bearing rock, tungsten concentrates, mixtures of scheelite + fluorite, hibnerite, ferberite, Mo-containing scheelite ores and concentrates	58, 60, 61
Vanadium	$^{51}V(n,\gamma)^{52}V$	3.75 m	titanium–magnetite ores, limonite	48
(*c*) *Combination of fast and moderated neutrons*				
Aluminium and	$^{27}Al(n,\gamma)^{28}Al$	2.24 m	bauxite, nepheline, clay, kaolin,	48, 56, 62–66
silicon	$^{28}Si(n,p)^{28}Al$	2.24 m	manganesian rocks	

* Minimum of moderator used; activation essentially with epithermal neutrons
† See also section (*c*) of this table

prompt γ-counting from nickel ores[19], or rhenium assay in molybdenite concentrates[20], Ge(Li) detectors have been preferred.

Sample sizes are usually rather large, at least 10, sometimes a few hundred grams or even kilogram amounts. This requires careful standardisation, e.g. study of thermal neutron attenuation (especially in Mn, Cu and W ores) and γ-attenuation, if reliable results are to be obtained. It has been shown that a reproducibility of $\pm 0.5\%$ relative can be obtained for the Mn assay in pyrolusite; the results are of the same quality as titration methods (pyrophosphate or bismuthate method), with the advantage of not requiring the dissolution of the sample[52].

For details of the determination of other elements, the original literature should be consulted, or a recent review article[16]. We finally note that a number of commercial systems are offered by American, Finnish, Russian and Hungarian companies or organisations[21-24].

2.1.4 Applications using a neutron generator

The voluminous literature on 14 MeV neutron activation analysis (more than 800 references) has been abstracted in a recent annotated bibliography[67].

Table 2.2 Neutron activation analysis of geological samples with the aid of a neutron generator

*Material analysed**	*Elements determined*†
Apatite, phosphorite, phosphate rock	P, Al, Si, F, Fe, Sr
Bauxite	Al, Si, Fe, O
Coal, coal ash	O, Al, Si, C, Fe
Copper ore, concentrates and tailings	Cu, Si, Al, O, Fe, Mn, P
Copper–nickel ores, Cu–Ni–Te mixtures	Se
Copper–zinc flotation products	Cu, Zn, Fe, Si
Fluorite, fluorspar	F, Si, Ba
Gold-bearing ores and solutions	Au
Iron ores, minerals and slags	Fe, Si, Al, O, Mg
Manganese ore	Mn
Mica	F
Nepheline	Al, Si, Fe
Polymetallic ores	Pb, Cu, Zn
Rare earth element ores, gadolinite, monazite	Ce, Pr, Y
Rare earth element concentrates	Y, Ce, Pr, Nd, Sm, Tb, Er
Taconite and concentrates	Si, Fe, Al, O
Tar sand	Al, Si
Zirconium concentrates	Hf
Rocks (igneous, sedimentary, metamorphic), rock-forming minerals, soils, meteorites and lunar rocks and soils	O, Si, Al, Fe, Mg

* Including ore concentrates and other materials of industrial interest
† References to the original literature can be found in Refs. 16 and 67

Table 2.3 Nuclear reactions most frequently employed in 14 MeV neutron activation analysis of ores, minerals and rocks

Element	*Nuclear reaction*	*Threshold energy*/MeV	*Isotopic cross-section* $\times 10^{27}$/cm^2	*Half-life produced*	*Main γ-ray transition energies*/MeV
Oxygen	$^{16}O(n,p)^{16}N$	10.3	40	7.12 s	6.130; 7.112
Fluorine	$^{19}F(n,p)^{19}O$	4.2	15	29.1 s	0.197; 1.356
	$^{19}F(n,2n)^{18}F$	11.0	60	109.7 m	0.511 (β^+)
	$^{19}F(n,\alpha)^{16}N$	1.6	57	7.12 s	6.130; 7.112
Magnesium	$^{24}Mg(n,p)^{24}Na$	4.9	200	15.0 h	1.368 2.754
Aluminium	$^{27}Al(n,p)^{27}Mg$	1.9	80	9.46 m	0.844 1.015
	$^{27}Al(n,\alpha)^{24}Na$	3.3	120	15.0 h	1.368 2.754
Silicon	$^{28}Si(n,p)^{28}Al$	4.0	250	2.24 m	1.779
Phosphorus	$^{31}P(n,\alpha)^{28}Al$	2.0	140	2.24 m	1.779
	$^{31}P(n,2n)^{30}P$	12.8	10	2.5 m	0.511 (β^+) 2.24
Iron	$^{56}Fe(n,p)^{56}Mn$	3.0	115	155 m	0.847 1.811 2.113
Copper	$^{63}Cu(n,2n)^{62}Cu$	10.7	500	9.9 m	0.511 (β^+)
Zinc	$^{64}Zn(n,2n)^{63}Zn$	12.0	100	28.1 m	0.511 (β^+)

The state of the art of this technique is well described in a recent monograph[68]. About 120 papers deal with the analysis of geological samples, and it would be impractical to discuss even a part of them here; for more details and references, see Ref. 16.

Table 2.2 gives a survey of the elements that have been determined so far in ores and minerals using 14 MeV neutron activation; most applications deal with the determination of silicon, oxygen, aluminium, fluorine, iron, copper and phosphorus. In the bibliography referred to above, one finds for igneous, sedimentary and metamorphic rocks, rock-forming minerals, sediments and soils, 53 papers on the determination of silicon, 37 on oxygen, 34 on aluminium, 16 on iron, 15 on magnesium, and a few on the determination of fluorine, potassium, silver, sodium, zinc and gold.

The most frequently employed nuclear reactions in 14 MeV neutron activation analysis of ores, minerals and rocks are shown in Table 2.3.

2.1.4.1 *Analysis of common terrestrial and lunar rocks*

14 MeV neutron activation analysis is probably the most suitable analytical technique available for the determination of oxygen. The measurement of ^{16}N can easily be performed in the presence of many other radionuclides because of its hard γ-rays. Interfering reactions occur practically only in the presence of fluorine, $^{19}F(n,\alpha)^{16}N$, boron, $^{11}B(n,p)^{11}Be$ ($T_{\frac{1}{2}} = 13.6$ s, high energy β- and γ-radiation), and perhaps uranium (fission products having a wide range of β- and γ-energies).

For the determination of silicon the method is also extremely useful, and excellent results, comparable with gravimetry, can be obtained within a few minutes, by counting the 1.779 MeV γ-rays of 2.24 min ^{28}Al. Nuclear interference arises from the presence of phosphorus due to the $^{31}P(n,\alpha)^{28}Al$ reaction; this will only be important in the analysis of phosphate rocks, etc. Interference from $^{27}Al(n,\gamma)^{28}Al$ is usually negligible when irradiating with unmoderated 14 MeV neutrons. A common γ-spectrometric problem is the presence of ^{56}Mn activity produced by the $^{56}Fe(n,p)^{56}Mn$ reaction, because its 1.811 MeV γ-ray cannot be resolved from that at 1.779 MeV when using a NaI(Tl) detector. A correction is, however, possible by simultaneously measuring another ^{56}Mn γ-ray, e.g. at 0.847 or 2.113 MeV; the 2.56 h ^{56}Mn contribution can also be established by a second counting, after decay of ^{28}Al (after *ca.* 20 min). A correction can also be calculated if the iron content is known from an independent analysis. The two peaks at 1.779 and 1.811 MeV are resolved when a Ge(Li) detector is used.

Aluminium and iron can also be determined, via ^{27}Mg and ^{56}Mn, respectively. The 0.844 and 0.847 MeV photopeaks are, of course, not resolved and present a γ-spectrometric interference; the radionuclides have, however, very different half-lives, allowing the problem to be solved. Alternatively, the ^{27}Mg activity can be measured using the (less intense) 1.015 MeV photopeak. For magnesium, the method is less suited (usually low Mg content and interference from aluminium!) except for ultra-basic rocks.

14 MeV neutron activation analysis of rocks and meteorites has been reviewed by Morgan and Ehmann[69] and is also discussed by Gijbels[16]. In most

of the published papers the concern has been mainly analytical. The techniques are now at a point where the precision and accuracy are satisfactory, especially for oxygen and silicon in rocks, and these methods should be applicable to actual problems.

The direct determination of oxygen may indeed be useful in studying various kinds of metamorphism, for instance, or simply as a check of the classical rock analysis. The strange situation exists where there are probably as many good oxygen analyses available for lunar rocks and meteorites as for terrestrial rocks. Ehmann and Morgan[70] found a positive O–Si correlation in Apollo 12 and Apollo 11 crystalline rocks. In both cases the slope is, however, significantly different from that given by Eugster[71] for terrestrial igneous rocks. The data suggest the existence of a highly reducing system at the time of crystallisation of the lunar basalts.

2.1.4.2 Analysis of ores and minerals

The basic ideas underlying the 14 MeV neutron activation analysis of ores and minerals can be summarised as follows.

(a) *Aluminium, silicon and iron*—The determination of these elements in various ores and minerals, such as bauxites, nephelines and iron ores, poses basically the same problem as for common rocks. Again the reactions $^{27}Al(n,p)^{27}Mg$, $^{28}Si(n,p)^{28}Al$ and $^{56}Fe(n,p)^{56}Mn$ are usually employed. The γ-spectrometric interferences ^{28}Al–^{56}Mn and ^{27}Mg–^{56}Mn can be solved as discussed above. Ivanov *et al.*[62] have analysed bauxites and nephelines and concluded that activation with a sufficiently intense isotopic neutron source is to be recommended because of its simplicity if only Al and Si are to be determined. When the iron content is also wanted, these authors used a neutron generator.

(b) *Phosphorus*—To-on[72] has determined phosphorus in phosphate rocks (4–13% P) via the reaction $^{31}P(n,\alpha)^{28}Al$. The interference from $^{27}Al(n,\gamma)^{28}Al$ appeared to be negligible, whereas the nuclear interference from silicon, $^{28}Si(n,p)^{28}Al$, was avoided by treating the sample first with HF ($SiF_4\uparrow$) and $HClO_4$ ($HF\uparrow$).

Several papers have appeared in the Russian literature dealing with the determination of phosphorus in apatite, phosphorite and phosphoric raw materials, all utilising mobile neutron activation equipment[9-12]. There is a tendency to use large sample sizes, i.e. up to 50 kg. The reaction $^{31}P(n,2n)^{30}P$ was used[10] and the β^+-annihilation radiation counted with a γ–γ coincidence array. Correction for pair production by hard γ-emitters (in particular from ^{28}Al: 1.779 MeV) was required. Since apatite contains fluorine, interference from the reaction $^{19}F(n,2n)^{18}F$ can be expected; the positron emitter ^{18}F has, however, a much longer half-life (109.7 min against 2.5 min for ^{30}P) so that a correction can easily be applied. Interference from other β^+-emitters is possible, e.g. $^{54}Fe(n,2n)^{53}Fe$ ($T_{\frac{1}{2}} = 8.5$ min). Kasatchin *et al.*[10] reported an accuracy of $\pm$3–5% relative for P_2O_5 contents of 15% and a sensitivity of *ca.* 0.5% P_2O_5 when using a sealed neutron tube, with a yield of 1–3×10^8 neutron s^{-1}.

(c) *Fluorine*—The use of a 14 MeV neutron generator permits the production of both ^{19}O and ^{18}F for the determination of fluorine, and has been applied to fluorspar ores, apatite and mica. Knight *et al.*[73], Bartošek *et al.*[74] and Weber and Guillaume[75] used 29.1 s ^{19}O, Jeffery and Bakes[76], Nargolwalla *et al.*[77], Strain *et al.*[78] and Botvina *et al.*[79] used 109.7 m ^{18}F. Dibbs and McMahon[80] measured both isotopes.

A disadvantage of ^{19}O is its low-energy γ-radiation (mainly 0.197 MeV) which gives rise to γ-absorption in large samples. The high background from Compton scattering worsens the counting statistics, and the presence of a back-scatter peak can give rise to positive errors. The use of 109.7 m ^{18}F invariably requires longer analysis times and, since this radionuclide is a β^+-emitter, it lacks specificity.

The reaction $^{19}F(n,\alpha)^{16}N$ cannot readily be used with a 14 MeV neutron generator, since the high-energy neutrons also produce ^{16}N from the reaction $^{16}O(n,p)^{16}N$; this is not the case with unmoderated isotopic source neutrons[81].

(d) *Copper, zinc*—In all cases 9.9 min ^{62}Cu is counted, formed by the sensitive $^{63}Cu(n,2n)^{62}Cu$ reaction; ^{62}Cu is a positron emitter, allowing one to count the annihilation γ-rays with a coincidence array[10, 82, 83]. This results in a more favourable signal-to-background ratio and is especially useful for low copper contents ($<1\%$). Soviet workers have employed a portable neutron generator (sealed neutron tube, $1\text{–}3 \times 10^8$ neutron s^{-1}) for the determination of copper in ores under field conditions[9-11], e.g. at the Kounradskij deposits in Kazakhstan, and reported a sensitivity of 0.01–0.05% Cu.

Gorski *et al.*[82] have determined copper in lower Silesian deposits (0.01–12% Cu) with a coefficient of variation of *ca.* 2%. Interfering radionuclides formed from K, Zn, Ag, Mo and Sb gave an apparent copper content of *ca.* 0.05%.

Pradzynski[84] examined fast neutron and photonuclear activation with a betatron for the analysis of copper in ores, tailings and concentrates, via the reactions $^{63}Cu(n,2n)^{62}Cu$ and $^{63}Cu(\gamma,n)^{62}Cu$, respectively. Satisfactory results were obtained with both methods, although photon activation was preferred because of the somewhat better precision and because of the possibility of changing the photon energy, thus excluding some interfering reactions.

When zinc is present, interference from the reaction $^{64}Zn(n,2n)^{63}Zn$ occurs with 14 MeV neutrons, since ^{63}Zn is also a positron emitter. Correction is, however, possible because of its longer half-life (28.1 m). In fact, both elements can be determined in this way. Navalikhin *et al.*[85] determined copper and zinc in polymetallic ores down to 0.08% by 0.511–0.511 MeV coincidence counting after two cooling periods, i.e. *ca.* 8 min and 35 min.

2.1.4.3 Analysis of meteorites

In more academic research, the first serious applications of the 14 MeV neutron activation methods were to the analysis of meteorites, particularly chondrites, for silicon and oxygen[69]. Meteorite samples are scarce and nondestructive analysis has obvious advantages; the activity induced by this technique is small. The determination of oxygen in chondrites is of interest because of the systematic variations of oxidation state going from the highly oxidised carbonaceous chondrites, through the more reduced 'ordinary' chondrites, to the very reduced enstatite chondrites. Wing[86] and Ehmann *et al.*[87-89] accumulated considerable data on the oxygen and silicon abundances in stony meteorites and showed that chondrite groups were distinguishable on the basis of these two elements. The Si/O ratio was found to vary from 1.4 to 4.2. More recent work on recently fallen chondrites indicates that earlier oxygen abundances may be in error because of terrestrial oxidation.

2.1.5 Applications using reactor neutrons

Reactor neutron activation via (n,γ) reactions is particularly suitable for trace analysis or when dealing with small samples, such as lunar specimens or meteorites. Trace element analysis may be of interest for studying origin, genesis and paragenesis of minerals and other geochemical problems. Data for trace elements such as uranium, gold, rhenium, selenium, tantalum, lanthanides, etc. may, however, also be of technological interest. There exists an impressive literature on reactor neutron activation analysis of geological samples, some 400 papers according to a recent bibliography[6].

2.1.5.1 Radiochemical neutron activation analysis

Earlier work was mainly by radiochemical separation of the elements of interest. Procedures for many elements can be found in the literature, and reference is made to bibliographic surveys such as those from De Soete *et al.*[90] and Lutz *et al.*[6], and systematic treatments such as those from Bowen and Gibbons[91], the *ORNL Master Manual*, or the series *Radiochemistry of the Elements*. Even today, many elements can only be determined after radiochemical separation, if they are present in too low concentrations or if their spectra are masked by dominant activities from other species. Radiochemical neutron activation analysis (RNAA) can be used at very low concentrations without pre-concentration. It should be emphasised that blank problems are virtually non-existent, contrary to other analytical techniques. Turkevich *et al.*[92] even decided to use neutron activation for the determination of ^{204}Pb by the fast neutron-induced $^{204}Pb(n,2n)^{203}Pb$ ($T_{\frac{1}{2}} = 52$ h) reaction; the very low ^{204}Pb content of Apollo 11 lunar fines (2.2–7.1 p.p.b.) could be quite well determined by this technique, which had the advantage of being less likely to be invalidated by contamination, as is possible with isotope dilution mass spectrometry.

RNAA can lead to quite accurate results if the analyst is aware of the underlying principles and limitations of the technique. To give an idea of the sensitivity which can be obtained, we cite the determination of iridium in USGS standard silicate rocks. The average values found by Greenland *et al.*[93] and Gijbels *et al.*[94] are as follows:

Peridotite	PCC-1	4.6 p.p.b.
Dunite	DTS-1	1.0 p.p.b.
Basalt	BCR-1	0.004 p.p.b.
Andesite	AGV-1	0.008 p.p.b.
Granodiorite	GSP-1	0.015 p.p.b.
Granite	G-2	0.017 p.p.b.

2.1.5.2 Instrumental neutron activation analysis (INAA)

There has been a definite tendency in recent years to use purely instrumental techniques, especially γ-ray spectrometry of the irradiated sample, without chemical separations, or with simple group separations only. When dealing with precious materials such as meteorites or lunar samples, it is often important to have procedures giving the maximum possible information in a very small sample. There has been a tendency to combine for that purpose various instrumental and radiochemical techniques, as will be discussed later.

(a) *Scintillation γ-ray spectrometry*—Analytical applications of direct scintillation γ-ray spectrometry are very limited owing to the inferior resolution of NaI(Tl) detectors. In some cases, however, the spectra are sufficiently simple to use this technique, e.g. for determining Sc and Cr in pyrope, Ta in rutile and columbite, Na, Sc and Co in rocks, Hf in zircon, Ag and Sb in galena, V in magnetite, Sc and Ta in greisen, etc. Thoresen[95] determined Al, Br, Mn, La, Cu, K, Fe, Sc and Na in soils and rocks from different parts of Norway by combining short and long irradiation and cooling times. The complex NaI(Tl) pulse-height spectra were resolved by the method of least squares. Brunfelt and Steinnes[96] analysed nine USGS standard rocks for Cr, Co, Mn, Sm, Sc and Na by thermal neutron activation, followed by γ-ray scintillation spectrometry. Cr, Al, Mn, V and Cu were determined in chromium ores by Kramer *et al.*[97], whereas Wyttenbach and Dulakas[98] described the determination of Al, Mg, Ca, Na, Mn and V in rocks.

(b) *High-resolution γ-ray spectrometry*—Since the advent of high-resolution Ge(Li) detectors, more complex spectra can be tackled. Lamb *et al.*[99] determined Mn, In, Ga, Zn, As, Sb, Cu, Na, Fe and Au in a sulphide ore. Cobb[100] determined the abundances of Mn, Sc, Th and seven lanthanides (La, Ce, Sm, Eu, Dy, Yb and Lu) in granites, basalts, limestone and one garnet sample.

For full details of the activation procedures, the detection systems, treatment of the data, γ-ray energies of interest and interferences in the instrumental neutron activation analysis of silicate rocks, reference should be made to the comprehensive paper of Gordon *et al.*[101]. Abundances were determined in six USGS standard rocks, a sub-alkaline basalt and a rhyolite for 23 elements: Na, K, Rb, Cs, Ba, La, Ce, Sm, Eu, Tb, Tm, Yb, Lu, Th, Zr, Hf, Ta, Mn, Co, Fe, Sc, Cr and Sb. Some relevant data are summarised in Table 2.4.

The method can also be extended to short-lived radionuclides. A few minutes after irradiation, the following activities can be observed in silicates: ^{28}Al, ^{56}Mn, ^{27}Mg, ^{24}Na, ^{165}Dy and ^{52}V, and sometimes ^{139m}Ba, ^{60m}Co and

Table 2.4 Properties of analytically useful radionuclides observable with a 'conventional' Ge(Li) detector in reactor-irradiated silicate rocks

Element	*Product radio-nuclide and half-life*†	*Best time after irradiation for counting*	*Best photopeaks used in determination* /keV§ (absolute intens. %)	*Interference or remark**
Na	15.0 h ^{24}Na	1 day	1368.65 (100)	—
			1731.9 (DE)	—
K	12.4 h ^{42}K	1 day	1524.7 (18)	—
Rb	18.66 d ^{86}Rb	2–4 weeks	1076.6 (8.8)	^{59}Fe 1291.58 Compton edge
Cs	2.05 y ^{134}Cs	few months	604.7 (98)	*^{124}Sb 602.7*
			795.8 (88)	—
Ba	12 d ^{131}Ba	2–4 weeks	216.0 (19)	*^{160}Tb 215.65*
			373.2 (13)	^{233}Pa 375.4
			496.3 (45)	—
La	40.23 h ^{140}La	4–5 days	328.7 (21)	^{140}La 487.0 Compton edge
			487.0 (46)	^{181}Hf 482.0
			815.8 (24)	—
			1596.2 (97)	—
Ce	32.53 d ^{141}Ce	2–4 weeks	145.5 (49)	*^{59}Fe 142.5*;
				^{175}Yb 144.7;
				^{233}Pa 145.4
Nd	10.99 d ^{147}Nd	2–4 weeks	91.0 (27)	^{131}Ba 92.3
				^{182}Ta 100.1 + 84.7
				^{170}Tm 84.3
				^{169}Yb 84.2 + 93.6
				^{175}Hf 89.6
				^{153}Sm 89.5 + 97.5 + 103.2
				^{160}Tb 87.2 + 93.9
				^{233}Pa 86.8 + 103.8
				+ *U* Kα$_1$ *(98.5)*
				+ *U* Kα$_2$ *(94.7)*
				^{153}Gd 97.5 + 103.2
			531.0 (13)	—
Sm	46.5 h ^{153}Sm	4–5 days	103.23 (27)	^{153}Gd 103.23
				^{233}Pa U X-rays
Eu	13 y ^{152}Eu	few months	121.83 (29.4)	*^{131}Ba 123.73*
				^{154}Eu 122.91
			244.54 (7.6)	^{154}Eu 247.73
				^{131}Ba 246.8
			779.08 (14.5)	—
			1408.11 (22.9)	—
	8.6 y ^{154}Eu	few months	723.29 (21)	*^{95}Zr 724.24*
			1274.53 (37)	^{160}Tb 1271.87
Gd	241.5 d ^{153}Gd	few months	97.43 (51.5) }	^{233}Pa U X-rays
			103.18 (39) }	^{182}Ta 100.10
Tb	72.3 d ^{160}Tb	few months	298.58 (26.3)	*^{233}Pa 300.2*
			962.08 (12.6) }	*^{152}Eu 964.21*
			966.10 (27.6) }	
Dy	2.35 h ^{165}Dy	few hours	94.7 (3.7)	^{233}Pa U X-rays
				^{233}Th Pa Kα$_1$

Table 2.4—*continued*

Element	*Product radionuclide and half-life*†	*Best time after irradiation for counting*	*Best photopeaks used in determination*/keV§ (absolute intens. %)	*Interference or remark**
Tm	129 d ^{170}Tm	few months	84.26 (3.4)	^{160}Tb 86.79 *^{182}Ta 84.68* ^{233}Pa 86.59
Yb	31 d ^{169}Yb	1 month	63.5 (57.3)	*^{182}Ta 65.72*
			177.0 (20.2)	^{182}Ta 179.39
			197.8 (34.9)	*^{160}Tb 197* ^{182}Ta 198
	4.19 d ^{175}Yb	1–2 weeks	282.6 (3.7)	—
			396.1 (6.0)	*^{160}Tb 392.51* *^{233}Pa 399.6*
Lu	6.71 d ^{177}Lu	2 weeks	208.36 (7.5)	^{131}Ba 216.00 ^{160}Tb 215.65 ^{239}Np 209.73
Th	27.0 d ^{233}Pa	2–4 weeks	311.9 (34.0)	*^{169}Yb 307.5*
			415.9 (1.5)	^{152}Eu 411.06
Zr	65.5 d ^{95}Zr	few months	724.24 (43.1)	^{154}Eu 723.29
			756.87 (54.9)	^{154}Eu 756.78
	35.1 d ^{95}Nb		765.80 (100)	^{160}Tb 765.19
Hf	42.4 d ^{181}Hf	1 month	133.02 (40) 136.25 (6) 136.86 (1.7) }	^{131}Ba 133.54 ^{169}Yb 130.7
Ta	115 d ^{182}Ta	few months	67.75 (41.7)	*^{169}Yb 63.5* ^{153}Gd 70
			100.10 (14.2)	*^{233}Pa 94–98* ^{153}Gd 97.43 103.18
			1121.22 (35.0)	—
Sb	2.72 d ^{122}Sb	1 week	564.0 (66.3)	^{134}Cs 563.2 569.33
	60.20 d ^{124}Sb	few months	1691.0 (50)	—
Mn	2.582 h ^{56}Mn	few hours	846.78 (99.0)	^{152m}Eu 841.39 (^{27}Mg 844.0)
			1810.96 (29.7)	—
Co	5.272 y ^{60}Co	few months	1173.226 (99.9)	^{160}Tb 1178.12
			1332.483 (100.0)	—
Fe	44.6 d ^{59}Fe	few months	1099.27 (56)	—
			1291.58 (44)	^{152}Eu 1292.62 ^{182}Ta 1289.13
Sc	83.8 d ^{46}Sc	few months	889.25 (100.0)	^{46}Sc 1120.50 Compton edge
			1120.50 (100.0)	*^{182}Ta 1121.22* ^{152}Eu 1112.20
Cr	27.71 d ^{51}Cr	1 month	320.08 (9)	^{233}Pa 311.9 ^{147}Nd 319.4

* Prominent peaks which interfere with the analytical peak listed in column 4 are italicised in column 5. In other cases the interference is usually negligible or can be eliminated by suitable cooling time, e.g. 9.6 min ^{27}Mg (844.0) interference on 2.582 h ^{56}Mn (846.78 keV)

† Holden, N. E. and Walker, F. W. (1972). *Chart of the Nuclides* (Schenectady, N.Y.: General Electric Company)

§ Pagden, I. M. H., Pearson, G. J. and Bewers, J. H. (1971). *J. Radioanal. Chem.*, **8**, 127, 373; **9**, 101

Table 2.5 Properties of analytically useful radionuclides observable with low-energy photon Ge(Li) or Ge detectors in reactor-neutron-irradiated silicate rocks[119]

Element	*Product radionuclide and half-life*‡	*Irrad. and decay time**	*Best photopeak used for analysis/* keV (absolute intensity %)§	*Interference or remark*†
La	40.23 h ^{140}La	(a)	328.7 (21)	^{194}Ir‖
Ce	32.53 d ^{141}Ce	(b)	145.4 (49)	sepd. from ^{59}Fe 142.5 count after decay of ^{175}Yb 144.7
Nd	10.99 d ^{147}Nd	(b)	91.0 (27)	—
Sm	46.5 h ^{153}Sm	(a)	69.68 (4.6)	*Au $K\alpha_1$ 68.8*
			103.23 (27)	^{239}Np (Pu $K\alpha_1$) 103.6
Eu	13 y ^{152}Eu	(b)	121.83 (29.4)	^{147}Nd 120.5
			244.54 (7.6)	—
Gd	241.5 d ^{153}Gd	(b)	97.43 (51.5)	*^{233}Pa (U $K\alpha_1$) 98.5* ^{153}Sm 97.4
			103.18 (39.0)	*^{233}Pa 103.8* ^{153}Sm 103.2
Tb	72.3 d ^{160}Tb	(b)	86.8 (12.0)	*^{233}Pa 86.6*
			298.6 (26.3)	*^{233}Pa 299.9*
Dy	2.35 h ^{165}Dy	(c)	94.7 (3.7)	—
			47.5 (Ho $K\alpha_1$)	^{152m_1}Eu (Sm $K\beta_1'$)
			46.7 (Ho $K\alpha_2$)	45.4 + 46.6
Ho	26.8 h ^{166}Ho	(a)	80.57 (<0.1)	Au $K\beta_2'$ 80.2 + ?
Tm	129 d ^{170}Tm	(b)	84.26 (3.4)	*^{182}Ta 84.7* ^{169}Yb 84.2
Yb	31 d ^{169}Yb	(b)	63.5 (57.3)	^{175}Hf (Lu $K\beta_2'$) 62.9
			130.7 (11.2)	^{140}La 131.0
			177.0 (20.2)	—
			50.7 (Tm $K\alpha_1$)	^{170}Tm (Yb $K\alpha_2$) 51.3
			49.7 (Tm $K\alpha_2$)	^{152}Eu (Gd $K\beta_2'$) 50.0
Lu	6.71 d ^{177}Lu	(b)	208.36 (7.5)	^{169}Yb 207.4 ^{239}Np 209.8
Ta	115 d ^{182}Ta	(b)	67.75 (41.7)	^{182}Ta (W $K\beta_1'$) 67.2 *Au $K\alpha_2$ 67.0* *^{181}Hf (Ta $K\beta_2'$) 67.0*
			100.10 (14.2)	^{233}Pa (U $K\alpha_1$) 98.5
			152.44 (7.2)	—
			222.11 (8.0)	—
			59.3 (W $K\alpha_1$)	^{169}Yb (Tm $K\beta_2'$) 59.0 (small correction)
U	2.350 d ^{239}Np	(a)	106.13 (21.9)	^{155}Eu 105.4**
Th	27.0 d ^{233}Pa	(b)	311.9 (34.0)	—
			98.4 (U $K\alpha_1$)	*^{153}Gd 97.5* ^{239}Np (Pu $K\alpha_2$) 99.5 ^{182}Ta 100.1
			94.7 (U $K\alpha_2$)	*^{169}Yb 93.6* ^{160}Tb 93.9
Cr	27.71 d ^{51}Cr	(b)	320.08 (9)	^{147}Nd 319.4
Fe	44.6 d ^{59}Fe	(b)	142.45 (0.8)	—
			192.23 (2.5)	—
Ba	83.3 min ^{139}Ba	(c)	165.8 (21.7)	—
	11.7 d ^{131}Ba	(b)	123.73 (26)	*^{154}Eu 123.3*
			216.0 (19)	*^{160}Tb 215.6*

Table 2.5—*continued*

Element	*Product radionuclide and half-life*‡	*Irrad. and decay time**	*Best photopeak used for analysis*/ keV (absolute intensity %)§	*Interference or remark*†
Hf	42.4 d ^{181}Hf	(b)	133.02 (40.0)	—
			136.25 (6.0) } 136.86 (1.7) }	^{177}Lu 136.7

* Irradiation and decay time: (a) 6 h irradiation, 7 d cooling time, 30 min counting time; (b) 6 h irradiation, 15–35 d cooling time, 2–8 h counting time; (c) 5 min irradiation, 1 h cooling time, 30 min counting time

† Prominent peaks which interfere with the analytical peak listed in column 4 are italicised in column 5. In other cases the interference is usually negligible or can be eliminated by suitable cooling time

‡ Holden, N. E. and Walker, F. W. (1972). *Chart of the Nuclides* (Schenectady, N.Y.: General Electric Company)

§ Pagden, I. M. H., Pearson, G. J. and Bewers, J. M. (1971). *J. Radioanal. Chem.*, **8**, 127, 373; **9**, 101

|| This occurs in special samples only, e.g. in irradiated diamonds

** Formed by reactions $^{154}Eu(n,\gamma)(n,\gamma)^{155}Eu$ and/or $^{154}Sm(n,\gamma)(\beta^-)^{155}Eu$

^{51}Ti. The ^{27}Mg 844 keV photopeak contains such a large contribution from the 2.58 h ^{56}Mn 847 keV peak that is necessary to use the 1014 keV peak instead. In practice, however, Mg determinations are restricted to ultramafic rocks and stony meteorites, which contain more than 10% of Mg [102]. γ-Ray spectrometry of short-lived species poses several problems, such as a decreasing fractional dead-time during the measurement, and also changes in the average dead-time per pulse. In addition, pulse pile-up problems may arise. These topics are discussed in detail in Chapter 1, Section 1.1 of this volume.

Similar methods have been developed and/or applied to terrestrial, meteoritic and lunar samples by Frána *et al.*[103], Goles *et al.*[104, 105], Jérome *et al.*[106, 107], Kharkar and Turekian[108], Kuncíř *et al.*[109], Melsom[110], Řanda *et al.*[111], Schock[112], Smales *et al.*[113], Vobecký *et al.*[114] and many others. Some papers have been more specifically directed towards information on a limited number of elements only, e.g. Fe, La, Ce, Eu, Dy and Lu in Mid-Atlantic Ridge sediments[115], Sc in clays[116], Na in ultrabasic rocks[117], and Ta, Co, Hf, Sc, Eu, Fe, Cr and Zr in minerals and rocks[118], to mention just a few.

The precision and accuracy of INAA can be greatly enhanced by using a planar Ge(Li) or Ge detector, which exhibits an improved resolution compared with 'conventional' co-axial detectors. These so-called 'low-energy photon detectors' (LEPD) are specially useful for measuring low-energy photons, i.e. γ- and X-rays: this is of special interest for determining the rare earth elements, U and Th, as can be seen from Table 2.5[119]. By combining short- and long-lived radionuclides, and counting with both a co-axial and a planar Ge(Li) detector, the following elements can be determined in basaltic rocks: Na, Mg, Al, K, Ca, Sc, Ti, V, Cr, Mn, Fe, Co, Rb, Sb, Cs, Ba, La, Ce, Nd, Sm, Eu, Gd, Tb, Dy, Ho, Tm, Yb, Lu, Hf, Ta, Th and U. In more acid rocks an additional number of elements can be determined, such as Zn, Sr and Zr, but in ultrabasic rocks the situation is less favourable. In addition to the above elements, one can also determine silicon via the reaction $^{29}Si(n,p)^{29}Al$ if the neutron flux is not too well thermalised. Phosphorus can be determined in silicate rocks by β-counting, using a 210 mg cm^{-2} Al absorber[120, 121].

INAA is obviously also applicable to ores and concentrates. De Lange *et al.*[122] determined Au and U in Witwatersrand ores, Turkstra *et al.*[123] determined Pd, Ir, Pt and Au in ores, matte and lead assay beads, whereas Schroeder *et al.*[124] determined Ag in minerals and ores, etc. Generally, the analytical possibilities of INAA with a Ge(Li) detector are determined by the composition of the sample, i.e. by the major elements (e.g. Fe), but also by the minor or trace elements, such as Mn, Na, La, Sc, Co, Cs, Eu, Ta, . . . which may give rise to dominant peaks and corresponding Compton continua in the γ-ray spectrum of the activated sample. A systematic survey for a number of ores and minerals was conducted by Vobecký *et al.*[125]. They found that a number of elements (not including those giving rise to short-lived radioisotopes) can easily be determined by INAA of minerals, such as Na, K, Sr, Mn, Ba, La, Sm, Eu, Dy and Au in diamond, Ag and Sb in galena, Zn, Ag and Sb in sphalerite, Fe, Co, Zn, Se and Ag in chalcopyrite, Co, Zn and Se in pyrite, Sc, Cr, Se, Cs, Hf, Ta, W and Re in molybdenite, La, Ce, Nd, Sm, Eu, Tb, Yb and Lu in fluorite, Sc, Co, Cs, Ce, Eu, Tb, Yb and Hf in limestone, Sc, Fe, Ce, Eu and Yb in magnesite, Sc, La, Ce, Nd, Sm, Eu, Tb, Yb, Lu, Th and U in monazite, Cr, Sr, La, Ce, Nd, Sm, Eu, Tb, Yb, Lu and Th in apatite, Sc, Cr, Fe and Co in garnet (pyrope), Sc, Fe, Ce, Tb, Yb, Ta and Th in tourmaline, Sc, La, Ce, Nd, Sm, Eu, Tb, Yb, Lu and Th in orthite, Sc, Fe, Zn, Rb, Sb, Ca, Ta and Th in cinvaldite, Sc, Cr, Fe, Co in serpentine, Sc, Ta and Th in wolframite and cassiterite, Sc, Cr, Fe, Co, Cs, Ce, Eu, Hf, Ta and Th in coal and brown coal. A number of elements could also be instrumentally determined in separated rock-forming minerals: Na, Mg, Al, Ca, Sc, Ti, V, Cr, Mn, Fe, Co, Ba, La, Ce, Nd, Sm, Eu, Tb, Dy, Ho, Tm, Yb, Lu, Hf, Th in feldspars[126,127], Na, Mg, Al, K, Ca, Sc, Ti, V, Cr, Mn, Fe, Co, Zr, (Ba), La, (Ce), Sm, Eu, (Tb), Dy, Ho, (Tm), Yb, Lu, Hf, (Ta), W, (Au), in (clino-) pyroxene[127], etc.

(c) *Coincidence counting*—Other counting techniques can, of course, be used instead of Ge(Li) γ-ray spectrometry, especially if the analysis is aimed at a particular element. Iridium can, for instance, be determined in various materials on a purely instrumental basis by using triple γ coincidence counting of ^{192}Ir. For slags from copper–lead production, the sensitivity is 20 p.p.b. ($2 \times 10^{-6}\%$), for silicate rocks, *ca.* 1 p.p.b., and for plagioclase feldspars even less[93, 94, 128]. Applications of coincidence counting to geological samples have been reviewed by Ehmann *et al.*[129], with special emphasis on meteoritic studies.

2.1.5.3 Group separations

The applicability of NAA can be substantially improved by introducing some rather simple group separations, followed by Ge(Li) γ-ray spectrometry. Several approaches for geological samples appear in the literature, some of which will be discussed here.

In the group separations system of Morrison *et al.*[130], the sample is dissolved in H_2SO_4 and HF, the volatiles are trapped in concentrated NaOH and the residue dissolved in 8 mol l^{-1} HCl. Na and Ta are absorbed on hydrated antimony pentoxide (HAP) and the effluent of the HAP column is passed through a Dowex 1 × 8 column. The effluent and wash (8 mol l^{-1} HCl) are extracted with tributyl phosphate (TBP): Sc, Zr, Hf and Pa appear in the organic phase, whereas the aqueous phase contains the rare earths, Cr, Ca, Ba, Sr, K, Rb and Cs. The anion-exchange column is eluted with 0.5 mol l^{-1} HCl. The eluate contains Fe, Co, Cu, Ni, Ga, W, Mo and Np, while Zn, Cd, Sb, Re, In, Hg and Au are counted on the resin. This method was applied to lunar rocks, soils and fines[131-133].

Another group separation system is described by Laul *et al.*[134, 135]. The elements Au, Ga, Ge, Mo, Sb, Te and Tl are extracted from 6 mol l^{-1} HCl with diethyl ether, the organic phase purified from unwanted activities (Fe primarily) and the desired elements counted in HCl as a group after precipitation of Tl. From the aqueous phase the elements As, Hg, Re, Sb and Se are distilled at 200 °C as the bromides, and Os as OsO_4. After removal of the excess bromine in the distillate which would interfere with the chemical yield determinations by re-activation [^{82}Br(!)], γ-ray group spectra are accumulated from the elements in HCl solution, or after precipitation as the sulphides. In the pot liquid all elements which form insoluble sulphides in acid medium, such as Bi, are precipitated. In the supernatant liquid, Co, Sc and Zn are precipitated (basic sulphides) and further purified. This treatment leaves the alkali metals in solution (Cs).

Allen *et al.*[136] have described a procedure for the NAA of 39 elements in geological samples. The elements are separated into 12 chemical groups (four of which are single elements), the order of separation being dictated by the short half-lives of some of the radionuclides of interest: group 1 (Ag, Cl, Br), group 3 (Mn), group 2 (Au, Ba, Pt, Ta), group 4 (In), group 5 (Cu, As, Se, Sb, Pd), group 6 (Fe), group 7 (lanthanides), group 8 (Cr, Hf, Sc), group 9 (Ni), group 10 (Co, Ga, Zn), group 11 (Ca, Sr) and group 12 (K, Rb, Cs). The twelfth group is separated and ready for radioassay within 7 h after the end of irradiation, when the work is done by three persons. Chemical yields are typically between 50 and 90% and are determined by reactivation, except for Fe and Ni.

Several authors have described the determination of the lanthanides, after separation as a group. The sample is first dissolved by an alkaline fusion, and the rare earths separated by hydroxide and oxalate or fluoride precipitations. Chemical yields are determined by reactivation[137-141]. Brunfelt and Steinnes irradiate the samples with epithermal neutrons instead of reactor neutrons[142].

2.1.5.4 Selective activation by means of resonance neutrons

A quantitative treatment of the advantage of applying epithermal neutrons in activation analysis of geological materials has been given by Steinnes[143]. The resonance effect (i.e. resonance peak in the $\sigma(E)$ curve) is utilised to some extent after removing thermal neutrons (below about 0.4 eV) by means of a thin cadmium shield surrounding the sample. It is obvious that the determination of an element is easier to perform if the element of interest has a nuclide with a prominent resonance activation peak, and the interfering nuclides have not. One can define an advantage factor[144]:

$$F = (\sigma_0/I)_i/(\sigma_0/I)_d$$

where d refers to the nuclide to be determined and i to the interfering nuclides; σ_0 is the thermal neutron activation cross-section and I the resonance activation integral. Relevant cross-section data and F values are tabulated by Steinnes[143]. A recent compilation of σ_0/I values is given by Van der Linden *et al.*[145-147].

The applications of epithermal neutron activation analysis (ENAA) to geological material may be divided into three or four groups.

(a) *Improvement of precision and sensitivity in INAA*—The advantages are obvious for determining Rb, Sr, Sb, Cs, Ba, Tb, Hf, Ta, Th and U in silicate rocks; in some cases it is even possible to determine Au at the p.p.b. level. The dominant interfering ^{46}Sc, ^{59}Fe, ^{60}Co and ^{51}Cr activities in basic rocks are much reduced with ENAA, allowing a number of trace elements to be determined with a better precision and sensitivity.

(b) *Reduction of high activity levels*—In geological samples, ^{56}Mn and ^{24}Na most often represent problems of this kind. As both these nuclides have low relative resonance activation rates, the high-activity problem can be considerably reduced by ENAA. Another example is ^{32}P formed in phosphate rocks.

(c) *Reduction of fission interference*—Steinnes has shown that the interfering reaction $^{235}U(n,f)^{99}Mo$ can be minimised by ENAA, thus improving the accuracy of the Mo determination in rocks[143].

Some sources of error, such as resonance absorption of neutrons within the sample, are discussed by Steinnes[143].

2.1.5.5 Neutron capture γ-ray analysis

In theory, this technique offers a greater sensitivity than activation analysis, but in practice this sensitivity cannot be achieved due to the difficulties of operating γ-ray detectors in regions of high neutron flux. However, capture γ-ray analysis can complement INAA for the determination of a few elements which cannot be determined by the latter method because their isotope products are too short-lived, stable or pure β-emitters. With the neutron fluxes available in reactor beam tubes, the technique is suitable for determining boron and selected rare earths at trace levels, and of hydrogen in the percentage range.

Table 2.6 shows some results obtained by Whitley *et al.*[148] for boron in peralkaline lavas from the Main Ethiopian Rift.

On irradiation with thermal neutrons, ^{10}B is converted into Li by an (n,α) reaction, for which the isotopic cross-section is 4000 b. The ^{7}Li is excited and decays with the emission of a 477 keV γ-ray. Potential sources of interference are sodium (472 keV), chlorine (518 keV) and background radiation (511 keV).

With a Ge(Li) detector, the peak of interest is well separated from the 511 keV peak, but its lower sensitivity necessitates larger samples and longer

Table 2.6 Determination of boron in peralkaline lavas[148]

Sample No.	B/p.p.m.		
	Optical spectrogr.	*Spectrophotometry*	*Neutron capture*
Syenite 1	100	93	91 ± 6
Y 322	5–25	—	22 ± 9
Y 408	—	16	17 ± 7
Y 606	5–25	8	16 ± 6
Y 852	5–25	22	28 ± 8

counting times. The sodium interference requires a correction, or spectrum stripping.

The analytical use of neutron capture γ-rays, with special emphasis on boron and the rare earth elements, has been discussed by Henckelmann and Born[149, 150]. Results are given for the La, Nd, Sm, Eu, Gd, Dy and Th content of monazite samples (0.06–16%). Using a counting time of 1000 min, the authors could determine boron and all the rare earths, except Ce, in the Canadian standard syenite rock S1. The results are (in p.p.m.): B (15), La (230), Pr (170), Nd (330), Sm (235), Eu (14), Gd (70), Tb (35), Dy (115), Ho (120), Er (60), Tm (10), Yb (75), and Lu (6). There is a satisfactory agreement with the optical spectrographic data of Sine *et al.*[151].

2.1.5.6 Delayed neutron counting

Delayed neutron counting is a specific method for determining uranium and thorium. If a nuclide having $Z > 90$ (e.g. isotopes of thorium, uranium, plutonium and other transuranic elements) undergoes fission, some of the fission-product nuclides decay by β-emission, with half-lives in the range 1–60 s to highly excited states of daughter nuclei which, in their turn, decay by emission of a neutron. These neutrons are emitted almost instantaneously after the precursor nuclide has emitted a β-particle, and consequently they exhibit an apparent half-life in their emission which coincides practically with that of the precursor nuclide. These neutrons are known as delayed neutrons, in contrast to those neutrons which are emitted promptly during the process of fission. In order to induce fission, samples containing uranium and thorium are exposed to reactor neutrons. ^{235}U gives fission with slow neutrons, ^{232}Th and ^{238}U with fast neutrons. If two determinations are made for each sample, one in a mixed (fast and slow) neutron flux, one with the slow neutrons screened out by a cadmium shield, both uranium and thorium may be determined.

Delayed neutron emission from irradiated samples can be measured in a paraffin-moderated BF_3 counter assembly of roughly 4π geometry. Assuming an irradiation time of 60 s at a thermal flux of 5×10^{12} neutron cm^{-2} s^{-1}, a delay time of 20 s and a counting time of 60 s, 1 μg of natural uranium gives 200–400 counts in such a set-up, for a counter background of 5–25 c.p.m. If the sample weight is *ca.* 20 g, uranium contents of 0.1 p.p.m. can be determined with a standard deviation of $\pm 5\%$. This shows that the method is both fast and sensitive. For thorium, the technique is less sensitive by more than an order of magnitude. An extensive study of delayed neutron counting was made by Amiel[152], including geological samples, by Dyer *et al.*[153] and by Gale[154].

2.1.5.7 Nuclear track counting

Some elements, such as uranium, thorium[155-159] and boron[156] can be determined by nuclear track counting. The sample is pressed against a dielectric detector in the form of a foil and irradiated. The boron analysis is based upon the $^{10}B(n,\alpha)^{7}Li$ reaction, where the recoiling α-particles penetrate into the foil (cellulose acetate butyrate 'Triafol BN'); for uranium and thorium, polycarbonate foils ('Lexan', 'Makrofol') or mica detectors are commonly used to catch the recoiling fission fragments. The radiation damage produced by fission fragments or α-particles can afterwards be observed by etching the foil with a suitable reagent. The nuclear tracks have a characteristic form and can be counted under a microscope. Manual optical microscopy is still most widely used as a method for data acquisition, as appears from a 1972 survey on track registration[160]. The counting can, however, also be performed with automatic scanning devices, for instance, or facilitated by projecting the picture on to a screen, or by counting the tracks on a photomicrograph, etc. It is obvious that standardisation can be carried out by simultaneously irradiating a sample with a known U, Th or B content.

A surprisingly simple and inexpensive technique for counting tracks in very thin foils (thickness less than the range of the fission fragments or α-particles) has been developed by Cross and Tommasino[161] and is known as 'spark counting'; see also Johnson *et al.*[162] and Agard[163] for instance.

The determination of uranium by the fission track method is superior to standard neutron activation analysis in all cases where it is desirable to obtain information on the spatial distribution of the uranium content. In that case, the sample should of course not be powdered. The method allows the determination of uranium contents down to the 10^{-3} p.p.b. range over areas less than 10^{-6} cm^2, and is therefore suitable for determining very small quantities of uranium, e.g. in mineralogical samples.

The nuclear track technique is used in such diversified areas as geological studies, cosmic ray studies, dating studies, studies of meteorites and lunar specimens, studies of the solar system origin, solar flare studies, geochemistry of the heavy elements and the search for superheavy elements, to mention just a few in the context of this chapter[160].

Fission-track dating of impact glasses, obsidians, tektites, apatites, zircons, etc. may well complement other dating methods; it is based on observing the fossil tracks from spontaneous fission of ^{238}U in the specimen itself, after etching with a suitable reagent[164]. If a similar sample is irradiated with thermal neutrons, after annealing of the fossil tracks by heating, 'fresh' tracks from neutron-induced fission of ^{235}U can be observed after etching. The 'fission-track age' t can be calculated from the equation:

$$t = \frac{P_s}{P_i} n \frac{\sigma_F I}{\lambda_{SF}}$$

where

P_s and P_i = number of spontaneous and induced tracks per unit area, respectively;
I = ratio of the natural isotopic abundances of ^{235}U to ^{238}U (= 0.0072/0.9927);
σ_F = cross-section for the reaction $^{235}U(n,f)$ (= 577×10^{-24} cm^2);
λ_{SF} = partial decay constant of ^{238}U for spontaneous fission (= 8.42×10^{-17} y^{-1});
n = neutron dose (neutron cm^{-2}).

This method of age determination assumes that the latent fossil tracks are conserved over the geological time. Evidence for a complex thermal history of the sample can be found by comparing the parameters of spontaneous and induced (fresh !) tracks (track length and/or pit diameter). If the sample has been reheated in geological times, both the mean track diameter and the track density P_s will be lowered, and t must be corrected[165]. The study of the thermal history of geological samples is quite a unique aspect of the fission track method.

2.1.5.8 Combination of various techniques

In connection with meteoritic and especially lunar specimen studies, a tendency has developed to combine various instrumental techniques (14 MeV neutron activation, photon activation, short and long thermal and/or epithermal neutron activation, followed by direct Ge(Li) counting or other instrumental measurements, such as delayed neutron counting), and/or radiochemical techniques.

The analytical scheme proposed by Brunfelt and Steinnes[166, 167] is based on seven different irradiations (A–G), involving five independent portions of the sample. Four of these runs are based on purely instrumental work, the remaining three irradiations are followed by radiochemical separations. Most radioactivity measurements are carried out by Ge(Li) γ-ray spectrometry (see Table 2.7).

Schmitt *et al.*[168] used the following irradiation scheme: 14 MeV neutron activation (O, Si), thermal neutron activation, followed by Ge(Li) spectrometry (Al, Na, Mn, K, Cr, Fe, Co, Ba, Sc, Cu) and 23 MeV photon activation (Ca, Ti, Ni). The method has subsequently been extended (RNAA) and applied to lunar samples[169, 170].

Goles *et al.*[104, 105] made use of essentially the same scheme for lunar rocks, but determined in addition La, Ce, Sm, Eu, Tb, Ho, Yb, Lu, Zr, Hf and Ta by INAA. Uranium was determined by delayed neutron counting. The latter technique was also employed by Řanda *et al.*[111]; these authors determined Ti, Fe, Sr, Y and Zr in lunar rocks by non-dispersive X-ray fluorescence.

Table 2.7 Survey of experiments included in the scheme of Brunfelt and Steinnes[166]

Experiment	*Irradiation*	*Irrad. time*	*Delay before start expt.*	*Type*	*Detector used*	*Sample weight*	*Elements determined*
A	thermal	30 s	no delay	INAA	Ge(Li)	10 mg	Al, Ti, V, Ca
B	thermal	5 m	2 h	INAA	Ge(Li)	(= A)	Mn, Na, (Dy)
C	epithermal	2 d	3–5 d	INAA	Ge(Li)	100 mg	Fe, Br, Sr, Ba, Tb, Hf, Ta, Th; (Ca, Sc, Co Rb, Sb, Cs, La, Sm, Eu, Au, U)
D	thermal	1 d	5 d	INAA	Ge(Li)	(= C)	Sc, Cr, La, Ce, Eu, Yb, Lu, (Fe, Co)
E	thermal	15 m	20 m	RNAA	NaI + Ge(Li)	50 mg	Cl, In, Dy, Sm, Ho, Er
F	thermal	20 h	30 h	RNAA	NaI + Ge(Li)	50 mg	As, Sb, Ta, Pa (Th), Np (U), K, Co, Cu, Zn, Ga, W
G	thermal	7 d	7 d	RNAA	NaI + Ge(Li)	200 mg	Se, Rb, Ag, Sb, Cs, Ir, Au

Kharkar and Turekian[108] combined INAA (Na, Mn, Dy, Sm, Lu, La, Yb, Ce, Eu, Tb, Sc, Hf, Cr, Co, Fe, Ti, Ta) with fission-track counting and RNAA, again for lunar rocks.

Smales *et al.*[113] determined Na, Sc, Cr, Mn, Sm, Eu, Dy, Yb, Al, La, Tb, Ho, Lu and Hf by INAA, Co, Cu, Zn, Rb, Ag, Cs and Ba by RNAA, and F, Sr, Fe, Ca and Ti by photon activation. The lunar samples were also assayed by mass spectrometry–isotope dilution (K, Rb, Sr, Cs, Ba, Nd, Sm, Eu), emission spectrography (Ni, V, Cu, Zr, Y) and X-ray fluorescence (Si, Al, Fe, Mg, Ca, Ti).

Information on cosmic-ray-produced ^{21}Na, ^{22}Na, ^{26}Al, ^{53}Mn, ^{59}Ni, ^{60}Co, (from Na, Mg, Al, Si, Cr, Fe, Co, Ni and Mn 'targets') was obtained by 14 MeV, thermal and fast reactor neutron irradiation[171].

Wänke *et al.*[172] made use of the following scheme: 14 MeV neutron activation (O, Si); 14 MeV neutron activation analysis of Fe, Al and Mg with a large volume Ge(Li) detector; reactor neutron activation (6 h at 7×10^{11} neutron cm^{-2} s^{-1}) for the determination of Na, K, Sc, Ti, Cr, Mn, Fe, Co, La, Ce, Nd, Sm, Tb, Yb, Lu, Hf and Ta by Ge(Li) spectrometry, and a similar irradiation followed by radiochemical separations for determining Ni (^{58}Co), Cu, Ga, Ge, Rb, Sr, Pd, In, Cs, Ba, Pr, Gd, Dy, Ho, Er, W, Ir, Au, Th (^{233}Pa) and U (^{135}Xe).

Filby *et al.*[173, 174] determined Al, Mn, Na, V, Ti and Ca by INAA via short-lived radioisotopes, and Sb, Co, Zn, Sc, Cr, Eu, Th, U, Ba, Ta and Fe via long-lived species. The high ^{24}Na activity, which in most irradiated silicate rocks and meteorites masks several (n,γ)-activation products of interest, was eliminated by adsorption on hydrated antimony pentoxide; this allowed ^{42}K, ^{72}Ga, ^{69m}Zn, ^{140}La, ^{64}Cu, ^{87m}Sr, ^{139}Ba and ^{152m}Eu to be measured shortly after a 2 h activation at a thermal flux of 8×10^{12} neutron cm^{-2} s^{-1}.

2.2 FORENSIC NEUTRON ACTIVATION ANALYSIS*

2.2.1 Introduction

The introduction of neutron activation analysis (NAA) into a law court was first attempted in France by Griffon and Barbaud[175] in 1951. This analytical technique was brought to remarkable perfection later on, enabling the qualitative and quantitative determination of most elements in a wide variety of materials. The various fields of criminalistics that can be covered by NAA are: identification problems, toxicology, gunshot residues and falsifications. The essential problem is always the interpretation of the results of the chemical analysis and its relevance to a particular legal issue.

2.2.2 Identification problems

There exists a constant need to identify materials as diverse as hair, textile fibre, paint, glass, alcohol, drug, soil, pesticides, food, etc. These various items are either left behind in the vicinity of a crime or carried away from it. They often form important evidence, as they may provide a link between the suspect and the scene of the crime.

Comparison of objects, with a view to proving the identity of origin, first requires a careful visual inspection and confirmation of the similarity in major constituents. A final proof resides in their minor constituents. It is at this stage of the expertise that neutron activation analysis proves to be a most useful tool, but only once all difficulties inherent to sample preparation are surmounted. First of all, some considerations are necessary about the use of trace analysis for identification in connection with each particular kind of material.

2.2.2.1 Hair

The identification of hair through its oligo-elements is probably one of the most widely investigated—but at the same time controversial—fields of forensic activation analysis. The trace elements in hair originate either from the nutritional medium, provided by the blood stream, or from external absorption, due to various forms of contamination. Absorption of metals due to shampooing, hair preparations, scalp medications (e.g. —Se— containing) are only a few of the many possible influences[176]. All this is, by its nature, unpredictable. Table 2.8 illustrates the concentrations of eight elements in

* See also Chapter 5.

Table 2.8 **The variation of the concentration of eight elements in the hair of M.H. as a function of time (21 samples collected chronologically)**[177]

Date	Zn /p.p.m.	Cu /p.p.m.	Au /p.p.m.	Mn /p.p.m.	Hg /p.p.m.	Sb /p.p.m.	As /p.p.m.	Se /p.p.m.
4–5–68	152	30.4	0.029	0.261	0.113	0.175	0.106	N.D.
19–10–68	153	20.5	0.044	0.156	0.101	0.097	0.135	N.D.
7–12–68	150	24.0	0.032	0.168	0.167	0.140	0.125	1.26
25–1–69	160	19.9	0.013	0.104	0.084	0.049	0.086	N.D.
15–3–69	190	24.3	0.038	0.532	0.058	0.093	0.423	N.D.
26–4–69	161	19.7	0.022	0.144	0.116	0.111	0.199	0.89
7–6–69	167	19.2	0.020	0.221	0.070	0.087	0.126	N.D.
18–10–69	156	22.4	0.019	0.185	1.36	0.073	0.177	N.D.
29–11–69	160	23.4	0.026	0.194	0.086	0.075	0.123	N.D.
14–3–70	157	23.0	0.025	0.279	0.051	0.238	0.160	4.98
18–4–70	168	24.1	0.013	0.121	0.078	0.149	0.114	8.61
30–4–70	158	22.4	0.045	0.218	0.161	0.232	0.119	N.D.
7–5–70	168	22.9	0.038	0.131	0.058	0.135	0.084	N.D.
16–5–70	185	19.0	0.013	0.150	0.095	0.118	0.50	9.47
6–6–70	149	19.2	0.013	0.087	0.037	0.072	0.078	10.7
31–7–70	146	23.2	0.013	0.099	0.072	0.056	0.106	22.3
26–9–70	127	16.8	0.020	0.346	0.049	0.157	0.214	17.0
7–11–70	130	24.1	0.015	0.213	0.058	0.087	0.187	12.4
5–12–70	153	19.7	0.022	0.232	0.048	0.058	0.141	7.53
16–1–71	143	19.7	0.018	0.105	0.031	0.068	0.159	0.75
20–3–71	138	21.3	0.020	0.186	0.090	0.061	0.147	0.73
Concentr. range	127–190	16.8–30.4	0.013–0.045	0.087–0.532	0.031–1.36	0.049–0.238	0.078–0.50	0.00–22.3
Median	156	22.4	0.020	0.185	0.078	0.093	0.135	0.75
Mean	156	21.9	0.024	0.197	0.142	0.111	0.167	4.60

female hair over a period of three years[177]. Selenium appears to be present only occasionally.

Much publicity has been given to this method of hair individualisation by Guinn[178] and others. Extensive work in this field was performed by Perkons and Jervis[179]. They consider the trace pattern of hair as a most useful tool in investigating identity. Coleman *et al.*[180, 181] made very important basic investigations in the same field. They studied the variation over the head of a person as well as over the population of England and Wales. Their results were statistically interpreted by the aid of Parker's statistical method[182, 183]. This allowed them to make an explicit decision in identifying the origin of the hair and also to express the probability that a random member of the population shows the same characteristics.

Numerous other research workers have tried to evaluate the method of trace identification by the aid of NAA[184, 185]. One of the main problems encountered in this analysis of hair is in sample preparation. Contamination through improper handling can cause substantial errors. It was thought that surface contamination can be removed by careful washing. However, as hair behaves like an ion exchanger, washing changes the trace pattern[186, 187]. This is illustrated, for example, by the work of Bate and Dyer[188, 189], who compared various hair-washing procedures. Coleman *et al.* also report on hair washing, recommending a cleansing technique through refluxing in diethyl ether[180]. Van den Berg, De Bruin and Houtman[190, 191] studied the sorption and desorption behaviour of trace elements in hair. From all these experiments, it is in fact debateable whether washing hair is of any real help at all, and indeed, a series of cleaning experiments performed by Cornelis and Speecke[192] led them to discard all treatment. For the same reaons, namely ion exchange, hair that has been blood-stained or remained in water or other fluids should not be retained for identification purposes through the estimation of its trace contents.

Another assessment made by various research workers is that hair does not possess a uniform trace composition all over the head. Even in a single hair the trace element pattern varies markedly along its length[193, 194]. This variation is of the same order of magnitude as the variation from hair to hair across a single head. However, these variations can be estimated by analysing a large number of hairs from the same head. If they turn out to be 'large', i.e. as large as the variability in a given population, individualisation becomes impossible.

An unpredictable factor is the evolution of the trace elements as a function of time. Some authors put time limits (e.g. 1 week, 1 year)[179, 180] beyond which no hair comparison should reasonably be made. Others consider this variation to be an insuperable factor[177, 195]. They express rather pessimistic views about the actual usefulness of hair identification by trace determination.

The general opinion is that in each particular case very careful consideration must be given to the evaluation of the significance of trace comparison of hair. Those engaged in this study are inevitably confronted with the difficulties inherent to the chemical complexity of hair, which is complicated still further by possible surface contamination. In the light of present knowledge, *irrefutable* evidence for hair individualisation derived from its trace composition remains wishful thinking.

2.2.2.2 *Textile fibres*

A preliminary, simple and non-destructive classification of fibres is indispensable. If resemblance in type and colour is established, the analysis of the traces becomes meaningful. These traces come either from the raw material or may have been introduced during processing. Over a period of time, they too are probably altered by repeated washing or cleaning.

Research into the minor constituents of fibres looks rather promising[196, 197]. Tables 2.9 and 2.10 list the concentrations of certain trace elements in sweaters

Table 2.9 Mean concentrations of elements found in sweaters before wear*[196]

Sweater	*Sodium*	*Chlorine*	*Manganese*	*Iodine*	*Bromine*	*Size* in./	*Batch numbered as received*
1	306	1116	62	42	49	42	1
2	366	1209	71	41	52	42	
3	334	1524	62	43	49	42	
4	319	1137	62	38	56	42	
5	425	1015	21	32	76	38	2
6	460	1090	23	34	34	38	
7	388	1043	22	31	39	38	
8	402	1010	22	34	63	38	
9	394	1146	20	37	73	38	
10	465	1079	22	35	37	38	
11	392	1163	0.9	4.7	359	38	3
12	451	1037	0.8	3.9	356	38	
13	446	1655	0.9	2.4	15	42	4
14	477	1514	0.5	1.4	12	42	
15	572	1638	0.7	1.3	14	42	
16	457	1636	0.6	1.3	12	42	
17	402	1622	0.7	1.3	11	42	
18	294	1505	0.9	19	28	42	
19	604	1093	85	37	32	38	5
20	632	1142	47	33	39	38	
21	93	1046	1.73	30	67	38	
22	173	1146	1.36	19	36	38	

* Figures given are concentrations in p.p.m.

before and after wear. However, problems similar to those encountered with hair have come to haunt the scientist trying to assess beyond doubt the individuality of a particular fibre.

Table 2.10 Mean concentration (p.p.m.) of trace elements found in sweaters after wear[196]

Sweater	*Sodium*	*Chlorine*	*Manganese*	*Iodine*	*Copper*	*Bromine*	*Method of cleaning*†	*Frequency of cleaning*†	*Frequency of wear*†
1	3755	560	(3.3)‡	35.4	116	20.6	SP	W	D
2	2355	955	3.6	40.2	60.5	28.0	D	M	SD
3	1810	1021	11.3	44.5	N.D.	30.2	D	O	O
4	300	1065	1.8	43.3	40.7	27.6	D	M	OPW
5	1384	921	(0.6)	29.2	N.D.	44.2	SP	W	OPW
6	1730	3325	18.4	37.4	N.D.	28.0	DC	O	SD
7	2095	3368	30.4	35.5	N.D.	35.4	DC	O	D
8	1398	1363	3.6	40.6	49.7	27.2	SP	O	SD
9	5076	1076	N.D.§	35.2	N.D.	18.7	D	W	D
10	886	996	20.8	31.3	(64.2)	17.4	DC	O	OPW
11	6450	702	2.4	N.D.	157	243	SP	M	D
12	3219	783	4.1	N.D.	(86.1)	298	SP	M	SD
13	5602	1126	(1.8)	4.9	157	N.D.	SP	O	SD
14	2785	1136	1.5	N.D.	90	N.D.	SP	M	SD
15	145	1266	2.9	N.D.	119	4.4	D	W	SD
16	467	1541	1.1	(1.10)	86	9.1	N	N	O
17	383	899	(0.4)	N.D.	(69.1)	4.1	SP	M	SD
18	3580	980	N.D.	N.D.	68.1	N.D.	SP	O	SD
19	1028	1097	6.0	50.0	85	24.0	D	O	O
20	1981	3354	66.4	50.1	(222)	39.6	DC	M	D
21	380	1067	1.8	34.9	98	42.0	DC	M	SD
22	1883	1372	3.2	38.5	93	32.7	SP	O	O

† SP = soap powder; D = detergent; DC = dry cleaned; N = never; W = weekly; M = monthly; O = occasionally; D = daily; SD = several days/week; OPW = once per week
‡ The figures in parenthesis are for these elements not detected in all of the readings
§ N.D. indicates these elements not detected

Table 2.11 Concentration of trace and minor constituents in a few paint samples selected at random[198]

	Concentration in samples/p.p.m.							
Element	1	2	3	4	5	6	7	8
Na	250 ± 15	175 ± 10	675 ± 35	110 ± 10	310 ± 15	1100 ± 55	1020 ± 50	570 ± 25
K	155 ± 10	20 ± 5	160 ± 20	30 ± 5	85 ± 12	54 ± 17	130 ± 18	14 200 ± 500
Sc	2.1 ± 0.2	<0.04	0.29 ± 0.09	<0.2	<0.2	<0.2	<0.2	<0.8
Cr	350 ± 20	2050 ± 100	<12	<7	<8	17 ± 3	<8	38 600 ± 2000
Fe	7600 ± 800	2000 ± 200	<1250	<910	<800	1150 ± 300	<900	<3200
Co	13.4 ± 1.0	<0.6	500 ± 25	610 ± 30	325 ± 15	220 ± 10	1400 ± 70	2100 ± 100
Cu	<66	<44	110 ± 30	<27	3250 ± 250	<120	1380 ± 100	<140
Zn	92 000 ± 5000	98 ± 9	75 ± 20	26 ± 5	<100	175 ± 20	110 ± 28	57 000 ± 2500
Ga	<0.7	13.2 ± 0.5	<1	<0.3	<0.8	4.0 ± 0.3	13.5 ± 1	3.5 ± 0.3
As	<4	<0.3	<0.5	0.17 ± 0.05	<0.8	<0.6	<6	<0.6
Br	<5	<5	<7	<2	<8	<8	17 ± 3	<8
Sb	5.3 ± 0.2	1.4 ± 0.2	6.0 ± 0.3	0.28 ± 0.07	4.8 ± 0.5	11.5 ± 0.3	28 ± 1	<1.6
Ba	114 000 ± 5000	<100	1850 ± 200	<250	<300	<350	15 800 ± 800	<2000
La	1.0 ± 0.1	<0.3	<0.6	<0.25	<0.3	<0.6	<0.5	<0.5

Element	*Detection limits*/p.p.m.							
Ni	60	4	25	20	20	20	20	20
Se	25	1.5	2.0	10	15	10	15	30
Mo	63	17.5	40	10	90	40	65	65
Ag	20	1.6	9.0	7	6	6	5	30
Cd	90	10	45	25	30	35	40	170
Cs	15	1	6.6	5	4	4.5	5	20
Ce	10	25	4	2.5	2.5	2.5	5	7.5
Sm	0.1	0.03	0.07	0.02	0.15	0.07	0.12	0.1
Eu	0.03	0.01	0.02	0.007	0.02	0.03	0.03	0.03
W	1.5	0.6	1.5	0.4	2.5	1.5	2.0	1.5
Au	0.2	0.07	0.17	0.05	0.3	0.17	0.25	0.20

2.2.2.3 Paint

Paint is a very common type of evidence, especially in car accidents. Most often it is recovered in the form of small chips or smears, the latter being extremely difficult to study at all. It seems advisable first to investigate the type of paint, its possible multiple layer structure and eventually the elements present. Some of these are major constituents, but others are just traces which do not affect the properties of the paint and hence are not kept under control. It is generally assumed that only relatively small variations occur between batches of the same paint from a given source. However, this does not exclude the possibility that many more paints show close trace-resemblance to the particular one under consideration.

Sampling is perhaps the most critical point of the expertise. The flakes have to be collected with extreme care so as to exclude contaminants. Furthermore, with multiple coat systems, great care must be taken not to mix primer and intermediate coats with the finish. This usually requires a painstaking manual separation of often strongly adhering layers. Other particularly annoying sources of contamination are dirt deposited on the paint, and substrate material scraped off together with the paint. Both these contaminants should, one way or another be eliminated. It is conceivable that exposure to rain, soot, car-wax, detergents, etc. introduces changes in the trace pattern of the paint. Possible restoration of the paint layer adds one more difficulty to the problem. Undoubtedly comparison of specks of paint is most advisably performed with samples very near to the original impact, if this happens to be exactly known. For all this, trace analysis research of the simple paint directly from the can does not contribute significantly to the paint identification problem.

Once all difficulties inherent to sample preparation are surmounted, NAA is without any doubt the best tool to assess the trace components of the paint and even some of its major constituents. As NAA is a multi-element, non-destructive technique, it can deal with minute specimens[198]. Table 2.11 lists the concentration of traces and minor constituents in a few samples selected at random. However, the relevance of paint identification more often than not remains rather limited. The same paint could have been placed on numerous similar objects (e.g. cars), so that only the number of possible culprits becomes restricted.

2.2.2.4 Glass

The comparison of glass fragments through NAA of trace components has been extensively studied by Coleman and Goode[199]. They determined at least 20 elements in glass and revealed some correlations between certain elements (e.g. Mg and Ca, Rb and Cs). Their results were statistically interpreted and led to the conclusion that this technique is adequate for the comparison of glass fragments.

NAA is so sensitive that it can easily cope with extremely small chips (this however, gives rise to another problem—the homogeneity of the glass. How small a sample can be taken before its trace element pattern is no longer representative for the whole slab?). Only a small number of independent manufacturers are involved in glass fabrication and consequently glass will show identical trace element patterns over many countries. This fact drastically limits the investigative value of the comparison. Exception can be made for old glass, which became very rare and indeed possesses a characteristic trace pattern, thus excluding nearly all coincidence[200].

2.2.2.5 *Alcohol*

NAA has been applied to indicate the distillation of illicit spirits due to alarmingly high Pb and Hg contents, as well as to determine the common origin of samples. A considerable amount of research in this field has been done by Pro *et al.*[201], who characterised 'moonshine whiskys' through their elemental composition. The traces originate from the raw material as well as from poor quality processing equipment.

Another type of counterfeit has been investigated by Lima *et al.*[202]. Cheap, locally made licit spirit was employed to refill bottles of imported whisky. These were subsequently sold at high prices as the genuine product. The spirits from both origins did not exhibit remarkably different trace patterns, neither did the cork rings. A solution was found in the analysis of the bottle caps. Those taken from the apprehended bottles proved to be of the same origin as those of local fabrication, but at the same time their elemental composition was totally different from that of the caps present on the imported bottles.

2.2.2.6 *Drugs and narcotics*

Jervis[203] and Pro *et al.*[201] have expressed the opinion that the trace analysis of drugs and narcotics could be used for differentiating between their sources. However, although a preliminary examination of the major constituents remains indispensable, absolute proof of origin is not possible solely through trace component analysis.

2.2.2.7 *Soil*

Soil is a very complex mixture of a rather heterogeneous nature, changing its composition over the surface as well as with depth. The problem of comparing soil is basically one of sampling. The known sample should be collected at almost the exact spot from which the unknown is thought to come. Bate[204] has adequately described all the problems which impede the use of minor elements as a method of identifying soils. Although NAA of soil yields accurate results for each particular sample, the significance of these results is questionable. Differences in trace element concentrations within a soil may be as great or greater than that between soils. According to Bate, the presence of an abnormally high or abnormally low concentration of a particular element in the soil under question is probably of more importance for identification than a knowledge of the concentrations of a whole range of elements that vary over normal ranges.

2.2.2.8 Miscellaneous materials

Identification of materials[178] as diverse as food, pesticides, oil, rubber tyres, dust, paper[205], ink, bullet lead and jacket material[206, 207], dental identification[208], telephone wire[209], Cu wires[210], etc. with the aid of NAA has been reported in the literature. Unfortunately, in many cases, one cannot help but feel that conclusions have been rather hastily drawn, so that the real value of NAA in these cases remains rather limited.

2.2.2.9 Conclusions

NAA of trace elements present in a particular material can yield final proof of identity. However, a systematic analysis of a large number of samples and a careful statistical interpretation of the analytical results[211] must be undertaken before firm conclusions can be drawn.

2.2.3 Toxicology

Whereas identification of materials requires a multi-element analysis, toxicological problems mainly deal with a single element. Neutron activation analysis is an excellent tool for the determination of minute quantities of Hg, As, Se, Sb, Cu, Zn, etc. see, for example, Ref. 212. Very often chemical separations are performed to speed up the analysis and yield more accurate results by eliminating interfering radionuclides. Unfortunately Pb and Tl are practically out of the reach of NAA because of their unsuitable nuclear characteristics.

The main interest of toxicology is in the detection of certain elements in human tissue, sera and excreta. The interpretation of the analytical results assumes a knowledge of the distribution of the normal contents over the population. Caution is needed, as is exemplified by the studies of Smith[213] who established six different causes for the presence of a particular element: essential element, normal environmental contaminant, abnormal environmental contaminant, exposure for medical reasons, for homicidal reasons

Table 2.12 Arsenic content in the entrails of a case of acute fatal poisoning

Tissue	As *content in fatal case** /p.p.m.	As *content in normal persons*[239] /p.p.b.
Liver	1290	10.9 ± 6.5
Stomach wall	610	—
Contents of small intestine	1930	—

*From R. Cornelis

and contamination during or between sampling. The most comprehensive studies of As and Hg contents in human tissues have been reported by Smith[214], Rodger and Smith[215], and Lenihan *et al.*[216]. An illustration of acute arsenic poisoning, as demonstrated by the analysis of a victim's intestines, is given in Table 2.12.

Determination of residues of toxic material in food can be dealt with exactly in the same way[217]. It is worthwhile indicating that NAA possesses considerable advantages over all other existing analytical techniques for the detection of the elements mentioned previously.

2.2.4 Gunshot residues

Firearm discharge residues containing metal traces mainly originate from the primer and only to a lesser degree from the bullet or the cartridge case. Most primers contain amongst others Pb, Ba and Sb components, whereas old fashioned percussion primer mixtures contained mercury fulminate, stibnite, etc. NAA can be successfully employed to detect minute quantities of Sb, Ba and Hg. It has therefore proved to be a useful tool in indicating the hand which recently fired a handgun or in determining bullet holes and estimating the firing distance.

A research team at Gulf General Atomic[178] has undertaken extensive studies to assess the Sb and Ba contents on the firing hand employing the non-firing hand as the normal blank. Other researchers have also tried to evaluate the method[218, 219] and develop it[220-222]. Cowan *et al.*[223] have applied the method developed by General Atomic to more than a hundred cases (suicidal and homicidal) where handguns were involved. Their conclusions have been confirmed by another study[224] on gun-firing detection based on Sb, Ba, Hg and Pb deposits on the hand, with a view to evaluating the credibility of the test. Table 2.13 gives the amounts of Sb, Ba, Pb and Hg collected from both hands in eight cases of confirmed suicide. In each case the amounts conformed with expectations. Conclusions could be summarised as follows: only in a clear-cut case do high amounts of residues confirm gun firing, in the other cases they provide proof of the presence of gunshot residues. The final interpretation depends entirely on the circumstances of the case. The absence of residues by no means implies that there has been no firing, as the residues might obviously have disappeared with time or been wiped or washed off. Care should be taken to discriminate between gunshot residues and high occupational blanks.

A bullet hole can be distinguished from an ordinary rip or tear by the detection of Sb and Ba[225]. Again high amounts are relevant, although the quantities deposited vary with the type of gun, ammunition, shape of bullet and other factors. The distance also is important as the material travelling along with the bullet spreads out and drops. Its radial distribution around the hole can be shown by autoradiography[226]. Fairly accurate estimates could be made for firing distances shorter than one metre[227-231].

Pillay and co-workers[232] suggest the tagging of primers with rare-earth traces to improve the specificity of detection.

Table 2.13 Amounts of Sb, Ba, Pb and Hg collected from both hands in eight cases of confirmed suicide[277]

Case	*Gun and ammunition*	*Hand*	*Sb*			*Ba*			*Pb*			Hg		
			µg	*µg per hand*	*% per hand*	µg	*µg per hand*	*% per hand*	µg	*µg per hand*	*% per hand*	µg	*µg per hand*	*% per hand*
1972/99	Shotgun bore 12, small-	R		0.82	33.6		8.80	100.0		50.0	92.6		0.069	71.5
P.P., 23 y	shot charge Legia	L		1.77	66.4		0.00	0.0		4.0	7.4		0.027	28.5
	Star High Speed													
1972/196	Shotgun bore 12	R		0.31	72.1		5.15	50.1		3.55	80.0		N.D.	
D.A., 63 y	Dina Star F.N.	L		0.12	27.9		5.12	49.9		0.89	20.0		N.D.	
1972/207	Pistol 0.32	R		3.97	93.4		9.52	88.4		60.6	97.9		0.057	80.4
W.A., 75 y	RWS ammunition	L		0.28	6.6		1.25	11.6		1.3	2.1		0.014	19.6
1972/271	Pistol 0.25	R—upper	0.21	0.56	50.5	2.62	5.70	59.6	0.97	2.04	64.6	0.15	0.45	75.4
V.B., 18 y	ammunition F.N.	palm	0.35			3.07			1.07			0.30		
	Hg fulminate	L—upper	0.14	0.55	49.5	1.92	3.87	40.4	0.47	1.12	35.4	0.064	0.15	24.6
		palm	0.41			1.95			0.65			0.086		
1973/40	Pistol 0.38	R—upper	0.09	0.14	82.4	1.00	2.96	77.9	1.62	2.88	67.4	0.12	0.21	95.4
D.G., 40 y	ammunition British Army	palm	0.05			1.96			1.26			0.09		
	43, Hg fulminate	L—upper	0.007	0.03	17.6	0.34	0.84	22.1	0.21	1.39	32.6	0.00	0.01	4.6
		palm	0.023			0.50			1.18			0.01		
1973/44	G. P. 0.38	R—upper	1.30	1.72	62.8	2.30	3.80	79.2	3.31	4.56	86.5	0.08	0.09	64.3
B.G., 23 y	F.N. ammunition	palm	0.42			1.50			1.25			0.01		
		L—upper	0.70	1.02	37.2	0.50	1.00	20.8	0.38	0.71	13.5	0.05	0.05	35.7
		palm	0.32			0.50			0.33			0.00		
1973/83	Revolver 0.38	R—upper	0.98	1.61	9.7	1.17	2.39	25.0	3.66	5.37	5.4	0.24	0.40	12.3
D.M., 38 y	Arminius special	palm	0.63			1.22			1.71			0.16		
	ammunition: 12×	L—upper	6.53	15.05	90.3	3.46	7.15	74.9	83.5	99.8	94.6	1.71	2.84	87.7
	Gevelot, 1× Norma sp.	palm	8.52			3.69			16.3			1.13		
1973/91	Pistol F.N. 0.32	R—upper	0.45	0.53	82.0	2.78	3.81	43.2	2.79	4.03	48.4	0.64	0.89	52.7
J.F., 61 y	Browning German	palm	0.082			1.03			1.24			0.25		
	ammunition DWM	L—upper	0.050	0.12	18.0	2.85	5.00	56.8	2.34	4.29	51.6	0.36	0.80	47.3
	Hg fulminate	palm	0.066			2.15			1.95			0.44		

2.2.5 Falsifications

Art and antique collectors and dealers are much concerned about the authenticity of ancient coins, paintings, pottery and other objects they either possess or want to acquire. Because ancient coins are made with less pure alloys than nowadays, their trace patterns differ markedly. Even ancient falsifications consisting of a copper core with a silver outer layer have been brought to light[233]. The same principle of minor constituent analysis applies to dated paintings[234, 235] and pottery[236]. NAA analysis has the big advantage that it is non-destructive (coins) or requires only minute samples. Authentication of paintings through neutron activation autoradiography combined with γ-ray analysis is another technique that yields useful information on the characteristics hidden beneath the surface[237].

2.2.6 Conclusion

The ability of NAA to detect traces in various kinds of materials is now a well-established fact. However, caution is necessary in the interpretation of the chemical analyses relative to the facts of a particular case.

The main point of interest of forensic NAA is the identification of items through the estimation of their minor constituents. Many difficulties arise because most traces are only present fortuitously and the results are very liable to external contamination. An elaborate background knowledge of the distribution of the elements in the material is essential, as well as a statistical interpretation of the data. Toxicological problems related to the elements (As, Hg, Sb, Cu, Se, Zn, etc.) are less equivocal but again a detailed knowledge of the normal amounts in the materials involved is essential. NAA also provides a valuable forensic tool for the detection of gunshot residues. As this depends on the presence of additional amounts of elements coupled with the firing act, its application seems less controversial. The same holds true for the possible detection of falsification of works of art or the demonstration of their authenticity.

In all cases, the application of forensic activation analysis as a probative method requires important statistical background data combined with a realistic consideration of all of the possibilities[238]. Only with these often complex secondary requirements in mind is it possible to use this extremely sensitive analytical method to its fullest potential.

2.3 TRACE ELEMENTS IN BIOLOGICAL MATERIALS

Biomedical science continues to show growing interest in concentrations of minor and trace constituents in biological fluids and tissues. In medicine, ever-increasing efforts are being made to find correlations between disease and these elements, to detect synergic or antagonistic behaviour between various metals. This was explicitly put forward in a panel meeting of experts organised by the International Atomic Energy Agency in Teheran (Iran,

1968), where reports were presented on relations between trace elements and, amongst others, protein–calorie malnutrition[240], cardiovascular diseases[241], pregnancy[242], cancer[243], etc.

Although in the past few years atomic absorption spectroscopy has taken a growing part of the work in this field, activation analysis continues to be the method of last resort as to sensitivity, accuracy and applicability. This overall situation is clearly illustrated in the assembled papers of the IAEA Symposium on Nuclear Activation Techniques in the Life Sciences, held in Bled (Yugoslavia) in 1972.

It is thus quite astonishing that in 1971 a survey paper was published on methods of analysis and normal concentrations in blood and urine for essential and abnormal elements, completely overlooking activation analysis[244]. The authors, well known in this field of research, advance atomic absorption as a suitable method, although it lacks sensitivity for at least five or six essential elements in the concentration ranges now thought to be normal.

From the literature of the past four years on activation analysis of biological materials, no sensational innovation or important achievement can be singled out. The general conclusion, however, is twofold: greater, yet not sufficient, care is devoted to contamination problems and, apart from the well-known 'normal values' (Fe, Cu, Zn), trace concentrations tend to decrease by orders of magnitude in the various matrices.

2.3.1 Elements and materials

The trace elements considered by more or less common consensus to be essential, F, V, Cr, Mn, Fe, Co, Ni, Cu, Zn, Se, Mo, I, together with other abnormal (toxic) or non-essential elements, such as As, Cd, Sb, Hg, Pb or Rb, Br, Ag, Cs, Au, have been determined in the widest variety of matrices[245].

In botanical studies, all possible materials from grass and fodder to fruits and cigarette tobacco have been studied. Even an example of geological prospection can be mentioned, whereby Au was determined in grass as an indicator for gold-bearing regions[246].

In the animal world many materials have been analysed, very often laboratory animals, or parts of them, used in clinical experiments. Sometimes extremely small samples, such as one newt egg cell, have been analysed for ng amounts of Na[247]. Fish represents a frequent subject, but mostly for ecological purposes. With humans, every part, fluid, tissue or bone has been investigated; most often, blood has been studied with respect to red cells and plasma or serum, followed by urine, these being the most readily available biological fluids with promising medical characteristics. In addition, however, studies on liver and kidney from autopsies, hair, skin and cerebrospinal fluid, etc. have also been made. In some cases the extreme possibilities of NAA with regard to multi-element results and great sensitivity have been almost fully realised. Swanson *et al.*[248] determined 17 elements in the four constituents of normal and cataractous lens tissue. In an effort to develop a diagnostic procedure for early detection of cystic fibrosis, Fite *et al.*[249] analysed fingernail clippings of newborn children for Cu. The human body as a whole has also been used as a sample to be analysed; this is discussed later on.

It should also be noted that although two of the elements given above, Pb and F, do not lend themselves very well to activation analysis, considerable effort has been made to include them in the scope of this method. Hislop *et al.* used high-energy γ-photons to determine Pb in biological samples[250], as did Chattapadhyay *et al.*[251] for environmental samples. Both authors could not use purely instrumental methods at the concentrations encountered and had to carry out extensive chemical separations.

Activation analysis of F in bone by cyclotron-produced fast neutrons and steam distillation of $H_2Si^{17}F_6$ was proposed by van der Mark[252], while Ohno[253] determined the same element in urine after a lengthy pre-separation of the fluorine and activation with 18 MeV photons.

These four proposed methods use highly complex nuclear machines, which are even less available than nuclear reactors. Also, at the desired sensitivity for blood, urine and bone, no purely instrumental analysis mode is adequate. When thought is given to the many papers dealing with atomic absorption methods for Pb, where small samples are handled very often without isolation of Pb and with excellent sensitivity[254-258], or to the very sensitive ion-selective electrode methods for the determination of fluorine in blood serum and urine[259-261], considerable doubts arise concerning the practical, apart from theoretical and scientific, usefulness of the efforts of activation analysts in these particular instances.

2.3.2 Sampling and storing of material

Apart from the sensitivities obtainable, the most publicised advantage of activation analysis is its freedom from contamination. This applies only after irradiation has taken place, i.e. no reagent blanks must be taken into account. This means also that if not all possible precautions are taken during sampling and storing of the materials to be irradiated, activation analysis is of no greater merit than any other method with respect to true values.

It is remarkable that so few papers published give the impression that their authors were aware of the almost certain introduction of unwanted additions to their samples. The literature (not only on activation analysis) abounds with expressions such as 'metal-free containers', 'proved blank materials', 'chemically clean glass', 'metal-free HNO_3 for pre-separation', 'all glassware was rendered metal-free'. It is obvious that in a normally clean laboratory the above expressions are no threat to accurate sampling and storage if the trace contents in the biological fluids and tissues are of a certain level, e.g. 10^{-6}*. In cases such as Cu and Zn in blood or liver, the amount of metal picked up can be insignificant compared to the analysis error and may cause negligible bias on the biological variations. When, however, concentrations of the order of 10^{-9} are determined, as in the case of Mn, Cr and Co in blood, the whole situation is changed and extreme care and working conditions are essential. Versieck *et al.*[262, 263] very extensively studied the contamination introduced when taking blood samples by venepuncture, or liver biopsies with surgical blades or needles. In order to obtain quantitative results, the surgical tools

* $10^{-6} = \mu g\,g^{-1}$ and $10^{-9} = ng\,g^{-1}$.

Table 2.14 Contamination introduced expressed as concentration per wet weight

	Concentration (10^{-6})						*Concentration* (10^{-9})†				
	*Blood**		*Liver*				*Blood**		*Liver*		
	Needle	*'Normal'*	*Blades*	*Needle*	*'Normal'*		*Needle*	*'Normal'*	*Blades*	*Needle*	*'Normal'*
Fe	0.03	1.1	0.3–5	1–20	190	Cr	10	20–0.7	3–20	480–1100	9
Cu	0.0005	1.1	0.020	0.5–2	8	Mn	0.1–0.02	0.58	3	20–650	1500
Zn	0.0005	0.9	0.001–0.010	2–17	58	Co	0.2	0.4	0.1–1.5	15–240	<26
						Ni	10	38	2–64	1200–12 000	<65

* Blood samples after rejecting first 20 ml; 'normal' value is given for serum

† Most of the 'normal' values (10^{-9}) are at the present stage estimates rather than true values; see text

were made highly radioactive and the clinical sampling procedure simulated. The introduced activity was then expressed as metal concentration of normal samples. The most important results are summarised in Table 2.14.

It was immediately concluded that *ca.* 60 ml of blood must rinse the needle before a valid sample is taken and that Menghini needles, and probably all similar material, are useless for taking uncontaminated liver samples. Complementary work by the same authors on polythene containers, quartz tubing, cleaning procedures, etc. resulted in the creation of a 'dust-free' room for handling the samples and a working procedure[264] which at great expense and effort led to blood sampling with *ca.* one unexplained 'high' result (for Mn) in twenty.

That biological fluids and tissues are apt to take up metals was proved by Lux *et al.*[265, 266] when studying metallosis from osteosynthesis metals. Coleman *et al.*[267] detected raised Co and Cr levels in blood, urine and hair in patients with metallic hip replacements: cobalt in blood went from 0.6 to 7–20 ng ml^{-1} and chromium from 1.4 to 6–15 ng ml^{-1}. A possible check on contamination by steel tools would be the detection of ^{182}Ta because of its high sensitivity and of ^{46}Sc, probably produced by $^{46}Ti(n,p)^{46}Sc$. Both Ti and Ta, which are minor alloying elements in these special steels, are practically undetectable by other methods and disappear when adequate precautions are taken[262, 263].

Whereas liquid samples are rather easy to collect, soft tissues must be dissected and hard samples like bone must be sawn or crushed. Gibbons *et al.* opened up a novel path by proposing the use of laser beams of adapted energy to cut samples without excessive heating, melting or charring[268].

Analysing very pure quartz for irradiation containers, Razeghi *et al.*[269] found for Mn metal contents up to 460 μg g^{-1} and for Sb, Sc, Co and Au metal contents of 120, 77, 95 and 15 ng g^{-1}, respectively. This precludes all analyses in the containers after irradiation and endangers the quality of the final solution if the activated biological material is digested in the presence of the quartz; this was very clearly observed for chromium[358].

Carefully selected and cleaned storage vessels can still have a double action on (liquid) samples. Traces can be leached out of the walls and raise the concentrations; this interference can be virtually eliminated by selecting the right material. The second source of error, adsorption on the walls, was studied by Struempler[270] for Ag, Pb, Cd, Zn and Ni on glass, polypropylene and polyethylene. He found the last material to be the best, except for lead where glass would be a possibility. This problem, however, is drastically reduced when the liquid samples are converted, e.g. by lyophylisation, to dry material; this is, of course, mandatory for longer irradiations at higher neutron fluxes in a reactor.

It is therefore essential to remember that in the analysis of rather small samples with extremely low contents, as with most biological materials, continuous care and rigid discipline are necessary together with repeated checks on the quality of all equipment used.

These considerations raise questions about pre-irradiation separation or concentration as described by Rottschafer *et al.*[271]. Extreme care must obviously be taken and all reagents continuously checked to be of exceptional purity for the element sought. Methods must be adapted so as to use a minimum quantity of reagent for a maximum amount of sample. Situations, however, exist where such a procedure must be adopted. Levstek *et al.*[272] and Darmsgaard *et al.*[273] determined vanadium in this way using the ^{52}V 3.75 m half-life isotope in Bowen's kale powder. They obtained values of 414 $\pm$ 13 ng g^{-1} (six determinations) with a 10 ng blank and 322 $\pm$ 10 ng g^{-1} (nine determinations) with a 5 ng blank, mostly from nitric acid.

2.3.3 Standards and standardised reference materials

The standards used by various investigators can be divided into two classes:
(i) Standards in the usual sense, when the activity of known amounts of the element is compared with the activity of the same element in the unknown, both being irradiated at the same time.
(ii) Monitors which act as flux meters, when the activity of the element in the unknown sample is compared with a specific activity of the same element being activated at another moment.

The simplest method of using flux monitors is acceptable when irradiation conditions, particularly the (reactor) neutron spectrum, can be assumed to be constant. This method was used by Versieck *et al.*[264] and Dubinskaya *et al.*[274], who employed Cu wire in the determination of Mn, Cu and Zn in blood. However, as Dubinskaya[274] pointed out, the irradiation period must be virtually constant over the period of work, because the saturation factors for ^{56}Mn and ^{64}Cu–^{69m}Zn are sufficiently different to cause important positive or negative errors for Mn when, during the analysis, the samples are irradiated for a shorter or longer time than during the establishment of specific activities.

For irradiation in high-flux reactors, more sophisticated flux monitors must be used, as shown by Van der Linden *et al.*[275], but this is not always considered necessary[276]. The ideal standard in activation analysis would be identical to the unknown, as in this way sources of error such as neutron shadowing, flux depression, differences in total or specified radioactivity [very important in modern Ge(Li) computerised γ-spectrometry], or differences in behaviour when standard and sample must be treated chemically are eliminated.

As no such material is available for biological samples, Cornelis *et al.*[277] investigated the possibility of preparing such a standard by doping a mass of serum with known amounts of elements to be determined. They succeeded in obtaining a standard for 10 elements producing long-lived isotopes with adequate accuracy and excellent homogeneity at the 20 mg level. Anderson[278] prepared a Hg reference material with doped gelatin, and noted that gelatin is free of large metal concentrations, but contains trace quantities of elements expected to be found in plant and animal tissues.

In checking systematic errors of a new method of analysis, reference materials are of the greatest help. They should resemble as closely as possible the samples to be analysed regarding the matrix, and their trace content should be as near as possible to a real sample.

Before they are issued as Standard Reference Materials (SRMs), their unknown trace concentrations should be determined by as many methods and analysts as possible. In the biological field the most publicised SRMs are:

Bowen's kale powder
NBS 1571 orchard leaves
NBS 1577 bovine liver
Univ. of Kentucky 1R1 tobacco
IAEA dried animal blood
IAEA wheat (Hg treated or not)

Probably the best defined reference material is Bowen's kale powder, but its various trace contents have been changing with time from an early mean value[279] to a best value[280] and a latest version[281]. In this respect, discrepancies for Al or Si are always present[282-284] as soil particles will easily be distributed inhomogeneously amongst the kale leaves. For NBS 1571 orchard leaves, a more serious situation exists. The certified value[285] for As of 14 ± 2 $\mu g\ g^{-1}$ has been challenged by Darmsgaard *et al.*[286], who found a value of $9.9 \pm 0.15\ \mu g\ g^{-1}$ after thorough investigation of an established analysis procedure[287]. This same value of $10\ \mu g\ g^{-1}$ was also found by Morrison *et al.*[288]. One of the reference materials listed is practically useless as it contains from 10 to 100 times the normal content of a real sample.

Nevertheless, and fortunately, many authors have included these materials in their investigations, allowing other analysts to judge the quality or improvement of a proposed procedure.

2.3.4 Irradiation

Practically all the reported investigations have used nuclear reactors for activation of the samples. Neutron fluxes have been selected according to availability and desired sensitivity, the irradiation periods varying accordingly.

Apart from the irradiations reported in the section on *in vivo* analysis, other facilities have been proposed, e.g. photonuclear activation of N in grain by Brune *et al.*[289], cyclotron-produced prompt atomic and nuclear reactions on microquantities of biological samples by Demortier *et al.*[290]. Magnesium has been determined in plant material by 14 MeV neutrons[291].

2.3.5 Post-irradiation treatment

Biological samples present special problems after irradiation because practically the whole of the total radioactivity can be attributed to the matrix (^{24}Na, ^{42}K, ^{32}P). Hence, two ways are possible of obtaining radioisotopes of interest in conditions suitable for measurement.

(i) If these isotopes have half-lives 10 or 20 times those of ^{24}Na or ^{42}K, a decay period of 10 days presents the easiest solution to the problem and has, therefore, been adopted by most investigators. Longer waiting times are necessary for very long half-lives, which are also present at lower concentrations (Rb, Fe, Zn and later Co, Cs, Ag in blood).

(ii) If, however, the isotopes of interest have half-lives shorter than the matrix activity or are present in small quantities with approximately equal half-lives, separations must be carried out according to generally accepted methods such as ion exchange or solvent extraction (Mn, Cu, Zn in blood).

Two further considerations also enter into the argument: (a) the generally available high-resolution Ge(Li) detectors make practically all separations (individual or group) obsolete; and (b) many efforts have been made to simplify the separation of the matrix activity with the help of so-called inorganic filters after the successful application of Na-clearing HAP by Girardi *et al.*[292], subsequently investigated further by Ralston *et al.*[293] and by Török *et al.*[294]. Other filters have been proposed, such as zirconium phosphate[295] and Al_2O_3 [296] for P, and potassium phosphotungstate for K [297], which are all more or less efficient. Brominated ion-exchange resins may also be used to free solutions from interfering ^{82}Br activity[298].

It is rather surprising that so much energy has been devoted to complicated, tedious and delicate group or sequential elemental separation schemes, such as an ion-exchange procedure by Edgington *et al.*[299], Plantin[300] and Schumacher *et al.*[301], all culminating in a combination of distillation, precipitation, extraction, ion-exchange and thin layer chromatography as proposed by Tölg[302].

Finally, mention should be made of Samsahl's[303] automation of his ion-exchange–partition chromatography[304] process.

An individual separation of an element is justified only if the highest precision is required for its analysis. Another problem situation lies in the determination of radio-elements whose γ-ray peaks are obscured by other (matrix) activities. This is the case in the detection of ng g^{-1} levels of chromium in serum, where chemical intervention such as distillation or extraction frees the ^{51}Cr (320.0 keV, 27.8 d) from the important bremsstrahlung arising from ^{32}P (14 d).

It should also be pointed out that too many authors attempt to prove the accuracy of their method by recovery experiments. To a (known or not) quantity of the element in the matrix under investigation, amounts of the trace are added, sometimes 100 times more than actually present. A yield of near 100% is accepted as justification for a rather hazardous extrapolation.

Finally, it should be noted that very often biological samples suitable for purely instrumental analysis require dissolution to ensure reproducible counting geometry or complete recovery from the irradiation containers. The most widely used digestion mixtures are combinations of nitric, sulphuric and perchloric acids. The oxygen combustion flask method is also very useful for attacking small samples to produce a solution that is easy to handle. An efficient and fast destruction at low temperature was obtained by Thomas by vapour-phase oxidation with concentrated nitric acid[305].

2.3.6 Measurement and presentation of results

The nuclear measuring equipment for biological samples does not differ from that for other activation analysis methods and no further attention will be given to this subject, or to the generalised use of computers for data retrieval.

The presentation of results lacks uniformity, i.e. they are mostly presented as $\mu g\ g^{-1}$ wet or dry weight. The least useful way is to give the concentrations as percentage ash, since this necessitates a dry ashing step which is very seldom necessary (in activation analysis). Presenting the results as $\mu g\ g^{-1}$ protein or $\mu g\ g^{-1}$ N could be the most exact method but it also represents an additional analysis. The easiest weight to obtain is that of wet or fresh samples, but this may lead to a dispersion of the results as the water content of fluids and tissues fluctuates more than its dry content[306]. Thus, it seems advisable, particularly for tissues, to give all results as $\mu g\ g^{-1}$ dry weight since this quantity is quite easy to obtain and usually it is necessary to have dry samples for irradiation[307].

2.3.7 *In vivo* analysis

A very special application of activation analysis is the *in vivo* determination of major and minor constituents of the human body. For this purpose nuclear reactors, with thermal[308], epithermal and fast neutrons[309], isotopic neutron sources[310, 311], 14 MeV neutron generators[312, 313] and cyclotrons[314-316], have been used to determine elements such as N, Na, P, Cl, Ca, Cu and I.

Whereas irradiation and measurement geometry, neutron spectrum changes and standardisation have been discussed in detail as possible sources of errors in classical activation analysis, in this particular application these problems are still major subjects for study. Measurements of radioactivity are mostly made by the use of reliable whole-body counters from health physics studies [317], and standardisation also often uses the man-like phantoms from the same set-ups[312, 319]. A very active group at the University of Birmingham (U.K.) has extensively investigated these difficulties. Ettinger[315] using cyclotron thick-target neutrons found that the human body is such an efficient moderator that at 10 cm depth a uniform neutron distribution is obtained. Chettle *et al.*[318] proposed a Ca determination by measuring the $(n,\alpha)^{37}Ar$ reaction product released in the air by breathing under controlled conditions.

The major problem, however, remains that of health physics, in that priority must be given to the doses received in order to obtain sufficient sensitivity; this is particularly so for fast neutrons, or γ-radiation in reactors[308].

To date, few if any results of determinations have been presented, and the method is still in the development stage with medical applications remaining as later possibilities[319, 320].

2.3.8 Results and discussion

Since more than 200 papers on activation analysis in the life sciences have been published since 1969, a review must necessarily be brief and rather incomplete. Only some salient points will be presented here.

First of all, work has continued on the determination of iodine as this is probably the most frequent application of the method in medicine. The range of determination, however, has been extended to matrices other than blood and thyroid and very small samples have been treated. By aqueous Čerenkov counting, Hoch *et al.*[321] determined 2 ng of I in 10 mg samples of liver mitochondria and Cornelis *et al.*[322] found 35 ng I per g in 20 mg samples of human skin.

Most papers, however, have dealt with Cu and Zn determinations in blood (serum and red cells) and liver. Probably because of the high concentrations encountered and the relative ease of activation, no difficulties arise when comparing results between authors. Also the agreement with other methods is very satisfactory. From a medical point of view, the well-known correlation of high liver Cu levels and Wilson's disease has been established many times, but often on very small biopsies[323-325] of the order of 1 mg. It should be pointed out that perhaps not enough attention has been given to the representative value of such small samples for inhomogeneous organs such as liver or kidney[326].

A less satisfactory situation arises in the determination of Mn in biological fluids. Although in 1966 Cotzias *et al.*[327] published an important activation analysis study on the determination of Mn in serum and plasma and gave 0.587 ± 0.183 ng g^{-1} as a normal value for 14 persons, various papers continued to be published giving much higher results. The low (0.5–1.2) values were first suggested by Fernandez *et al.*[328] by a catalytic method in 1963 and were confirmed by Heydorn *et al.*[329] as 0.54 ± 0.05 ng g^{-1} for 11 normal persons. Speecke[330] reported a value of 0.56 ± 0.10 ng g^{-1} for 50 persons with a range of 0.35–1.05 and no significant difference between male and female, and presented a normal frequency distribution of the Mn concentration in serum of the form usually expected for all essential trace elements[331]. At this stage, Hagenfeldt *et al.*[332] published a paper in 1973 on the influence of the menstrual period on the Cu, Zn and Mn concentration of plasma with seven females and found a cyclic variation for plasma Mn from 1.8 to 3.2 ng g^{-1}. These results are at variance with the findings of Versieck *et al.*[264, 333], who found a mean concentration of 0.55 ± 0.14 ng g^{-1} for 25 females with a range of 0.38–0.70 and one value of 1.04.

The present knowledge of the chromium content in blood or soft tissue is similar to the manganese situation some 5–10 years ago. Chromium is accepted as an essential element[244], necessary to the glucose metabolism[240] and possibly preventing arteriosclerosis[241]. However, when reviewing the published data being aware of contamination problems and accepting results of work in progress[334], great suspicion must arise about the foundation of these medical conclusions on the importance and the role of an element such as Cr. In Table 2.15 the literature data for Cr in blood, red cells, plasma or serum are presented, together with the method of analysis and the year of publication.

From this survey it is obvious that no definite values for normal Cr contents are available. There is no agreement between methods nor within the same method. Contamination can explain the high results, and even the category of about 10 ng g^{-1} could be erroneous in the light of the results given in Table 2.14.

Much more work has to be done, as probably only neutron activation analysis with high flux reactors will enable the desired sensitivity and accuracy to be achieved. Indeed, the results quoted in Ref. 334 are a mean of values obtained as upper limits after irradiation of 200 mg of lyophilised serum

Table 2.15 Chromium concentration in ng g^{-1} wet weight

Blood	*Cells*	*Plasma*	*Serum*	*Method**	*Year*	*Ref.*	*Remarks*
47				color	1950	337	
		22 (17)		spec.	1956	338	
		180 (25)		spec.	1956	339	
24 (6)§	24 (6)	30 (8)		color	1959	340	
23 (40)	22 (40)	24 (39)		spec.	1959	341	
23 (61)	19 (61)	26 (61)		spec.	1960	342	
			10 (21)	X-ray	1961	343	
	110		345	color	1962	344	(m: 520; f: 170)
1.2 (8)				aa	1964	345	†
700				mss	1967	346	
			2.5	aa	1967	347	0.7–3.2
	26 (129)	29 (129)		at a	1967	348	
			18	spec	1968	349	1–80
			1	aa	1968	350	tentative value
			28	(spec?)	1969	351	50% <detection‡
	35 (1)		38 (1)	aa	1969	352	
			36 (2)	aa	1969	353	
			30	at a	1969	335	best lit. value
			40 (20)	glc	1970	354	
		38		—	1970	240	lit. review
			20	at a	1970	355	
			13 (36)	spec	1971	356	
		28		at a	1971	244	
			30	spec	1971	357	50% <10‡
			3	aa	1971	358	0.4–5
			13 (21)	glc	1972	359	2.7–24
			9.3	aa	1972	360	
			10	aa	1972	361	
		12	7	glc mss	1972	362	
28		45	32	aa	1972	363	
			2.2 (37)	color	1972	364	80% <2‡
		5.1 (7)		at a	1972	365	
15				at a	1972	366	
1.4 (20)				aa	1973	267	
		4.7 (10)		at a	1973	367	females
			0.7 (15)	aa	1973	334	see text

* Abbreviations: color = spectrophotometry; spec = emission spectrography; X-ray = X-ray fluorescence; aa = activation analysis; mss = mass spectrometry; at a = atomic absorption; glc = gas–liquid chromatography

† The value of 1.2 ng g^{-1} for whole blood has been recalculated for a given value of 6.2 ng g^{-1} for dry blood; if, however, the published correction for the $^{54}Fe(n,\alpha)^{51}Cr$ threshold reaction is taken into account (which the author explicitly does not do), a negative value of 0.6 ng g^{-1} Cr would be found; obviously the nuclear interference has been over-estimated (presence of chromium in 'pure' ferric oxide) and the 1.2 ng g^{-1} value is an upper limit

‡ Of the samples analysed, the indicated percentages were below the given value

§ Values in parentheses indicate the number of individuals

samples for 10 d at a flux of 10^{13} neutron cm^{-2} s^{-1}. Work in progress[334], with pre-ashed samples of *ca.* 1 g serum and a neutron flux 10 times higher, gives an indication of even lower contents. This is in sharp contrast to a review by Christian[335] on traces and atomic absorption, and a paper by Hansen *et al.*[336] on gas–liquid chromatography, who both accept concentrations of 20–50 ng g^{-1} of Cr as normal physiological levels in blood.

It should also be noted that at present a similar situation exists for chromium in liver and kidney, etc., and for elements such as cobalt, nickel, molybdenum, etc. in the same biological materials. As it is an aim of biomedical sciences to correlate the trace element concentrations in biological fluids or tissues with diagnostic or therapeutic considerations, it is an absolute necessity to have *real* 'normal values'. This has been shown dramatically by Versieck *et al.*[368]. Thus, after having established 'normal concentrations' of Mn in serum as reported above, an analysis was undertaken of sera from patients with liver diseases or malfunctions. As a result, a well-defined interrelationship was found between Mn and SGOT* in hepatitis; for example, during the initial stages of acute hepatitis the Mn level increases to five or six times the normal value, reaching 2–4 ng g^{-1} and returns to normal during recovery, a situation hitherto unobserved. This observation was made possible with the help of (neutron) activation analysis using extremely clean experimental conditions.

2.4 SYSTEMATIC ERRORS IN 14 MeV NEUTRON ACTIVATION ANALYSIS

Since the first publication on the subject, in 1956, 14 MeV neutron activation analysis has rapidly taken its own place amongst new analytical methods. Although a major part of the literature is concerned with the determination of oxygen, it has also been applied in other areas†. It is not possible, however, to review its numerous applications here. An exhaustive survey covering the literature up to the end of 1971 may be found in a recent annotated bibliography[369], while a detailed treatment of instrumentation is given elsewhere[370].

As 14 MeV NAA, in distinction to reactor activation, is most frequently used for major or minor element determinations, extreme care must be taken to avoid systematic errors. The present section deals with possible sources of systematic error. Indeed, when the standard is not exactly the same in composition or dimensions as the sample, errors may be introduced owing to different neutron, γ-ray and β-ray attenuation. Moreover, when a flux monitor is used to correct for variations of the neutron output between successive irradiations, it can be influenced to a different extent by the sample and the standard.

Obviously, the influence of these effects can be overcome by using standard reference materials of the same bulk composition as the sample and previously analysed by other techniques. An example is the determination of silicon in steel[383], whereby chemically analysed steels are used as standards. This requires, however, that techniques with a higher accuracy and precision than

* SGOT = serum glutamic oxalacetic transaminase.

† See also Section 2.1.4.

14 MeV NAA are available for the determination of the particular element in the matrix of interest. Sometimes this is not possible. Mostly, a standard prepared from stoichiometric or easily analysed compounds of the element is then used. Silicon dioxide has been used, for instance, as a standard for the determination of oxygen and silicon in rocks[371] and a vanadium carbide–oxalic acid mixture for the determination of the vanadium to oxygen ratio in vanadium oxides[372]. The inevitable differences in density and composition between the sample and the standard can then cause significant systematic errors, so that a quantitative knowledge of the correction factors becomes important.

The above can be illustrated by the determination of oxygen in non-ferrous metals[373]. Matrices which differ by almost a factor of ten in density (aluminium–tungsten) can be analysed using one carefully prepared oxygen standard[374] consisting of a Fe_2O_3–graphite pellet pressed into a cylindrical steel box of the same outer dimensions as the sample. The use of an artificial standard is inevitable since other analytical techniques for oxygen in metals offer a precision and accuracy which is, at best, only comparable with 14 MeV NAA. By applying suitable corrections for attenuation and geometry effects, an independent determination of oxygen is provided relative to an 'absolute' standardisation.

Several irradiation and counting systems have been applied in practice[370]. One method consists of irradiating the sample while the neutron output is monitored. This is most frequently realised by means of: (a) a direct neutron counter, e.g. a low geometry boron trifluoride long counter[375], or (b) an oxygen-containing flux monitor irradiated simultaneously with the sample but in a different geometry. The flux monitor can, for example, be irradiated in the neutron shadow of the sample[374,375]. Subsequently a standard is irradiated and counted under the same conditions as the sample while the same monitoring system is used. Another method consists of irradiating the standard and the sample simultaneously by means of a spinning mechanism[377,379] so that both are exposed to the same external flux, no flux-monitoring being required.

A general equation taking all the above mentioned effects into account may be written as:

$$W_x = \frac{W_s}{K} \cdot \frac{A_x}{A_s} \cdot \left(\frac{C_s}{C_x}\right)_n \cdot \left(\frac{C_s}{C_x}\right)_\gamma \cdot \left(\frac{C_s}{C_x}\right)_\beta \cdot \frac{S_x}{S_s} \cdot \frac{D_s}{D_x} \cdot \frac{\Phi_s}{\Phi_x} \tag{2.1}$$

where:

W_x = weight of the analyte element in the sample;

W_s = weight of the analyte element in the standard;

K = activity ratio of the standard to the flux monitor (= 1 when no flux monitor is used);

A_x = activity of the sample;

A_s = activity of the flux monitor or activity of the standard when no flux monitor is used;

$\left(\frac{C_s}{C_x}\right)_n, \left(\frac{C_s}{C_x}\right)_\gamma, \left(\frac{C_s}{C_x}\right)_\beta$ = correction factors for different neutron, γ-ray and β-ray attenuation in the sample and standard, respectively;

$\frac{S_x}{S_s}$ = correction factor for the different influence of the sample and standard on the induced activity in the flux monitor;

$\frac{D_s}{D_x}$ = ratio of the counting efficiency of the standard to that of the sample;

$\frac{\Phi_s}{\Phi_x}$ = ratio of average neutron fluxes in the void space occupied by the standard and by the sample.

Although most experimental work quoted in the literature considers systematic errors in the determination of oxygen through the $^{16}O(n,p)^{16}N$ reaction, the results can, in most cases, be applied with slight modifications to other systems.

2.4.1 Neutron attenuation

Several solutions have been proposed to account for 14 MeV neutron attenuation.

When a neutron collides with an atom of the sample, the interaction can result in a nuclear reaction whereby the neutron is absorbed, in inelastic scattering so that its energy is decreased to a value below the effective threshold of the $^{16}O(n,p)^{16}N$ reaction (~11 MeV), or in elastic scattering which results in only a slight change in the direction of motion. From these considerations, as pointed out by Adams *et al.*[370] and Nikolaenko *et al.*[376], it follows that elastic scattering should not be included in the cross-section used to correct for neutron attenuation. Hence, the *non-elastic scattering cross-section* (σ_x), which is the sum of the inelastic scattering and the reaction cross-section, seems the appropriate quantity to apply. Nargolwalla *et al.*[377] and Adams *et al.*[370] have recommended the use of a semi-empirical *removal cross-section* (σ_R) which does not differ substantially from the non-elastic scattering cross-section, but takes into account small changes in the direction of the neutron during an elastic collision. In a recent paper[378], experimental evidence has been presented for the use of non-elastic scattering cross-sections. An oxygen-containing flux monitor was irradiated in the neutron shadow of different metal discs, thus allowing the determination of the appropriate cross-sections. In Table 2.16 the experimental results are compared with the literature data which have been used to correct for neutron attenuation. Once the appropriate cross-section has been chosen, the ratio of the ^{16}N activity in the standard and the sample should be related to it.

The attenuation of 14 MeV neutrons may be described by an exponential law of the type:

$$\frac{\phi}{\phi_0} = \exp(-\Sigma d_n) \tag{2.2}$$

where

ϕ = average neutron flux (neutron cm^{-2} s^{-1}) within a sample;

ϕ_0 = average neutron flux over the same volume, but without a sample present;

Σ = macroscopic removal cross-section (cm^{-1});

d_n = effective attenuation sample thickness (cm).

Table 2.16 Experimental and literature values of cross-sections used for the description of 14 MeV neutron attenuation[378]

Metal	σ*	σ_R† (14.5 MeV)	σ_x‡ (14.0–14.5 MeV)	σ_x§	σ_T‖ (14 MeV)
Al	0.93 ± 0.07	1.03	1.00		1.75
Fe	1.31 ± 0.05	1.39	1.36	1.58	2.55
Ni	1.51 ± 0.05	1.52	1.40		2.7
Cu	1.52 ± 0.05	1.55	1.47	1.42	2.9
Zr	1.61 ± 0.12	1.86	1.72		4.0
Mo	1.48 ± 0.08	1.91	—	1.59	4.1
Ag	1.69 ± 0.14	2.05	1.83		4.15
Cd	1.49 ± 0.15	2.08	1.90		4.2
Sn	1.74 ± 0.18	2.12	1.90		4.5
Sb	1.94 ± 0.18	2.15	1.96		4.6
Ta	2.26 ± 0.07	2.63	2.0		5.3
W	2.17 ± 0.08	2.65	2.41	2.59	5.3
Bi	2.65 ± 0.28	2.82	2.55		5.5

* Experimental results from Ref. 378

† From Nargolwalla *et al.*[379] interpolated from Avery *et al.*[384]

‡ Mean value of experimental data, from Allen *et al.*[385]

§ Data mentioned by Nikolaenko and Shtan[376]

‖ Experimental data from Hughes and Schwartz[386]

Appropriate d_n values, which depend on the irradiation geometry and on the dimensions of the sample, can be determined in several ways:

(i) Nargolwalla *et al.*[377, 379] determined the ratio of the specific ^{16}N activities in a number of oxygen compounds with different Σ values to that in a standard sample and plotted log (C_x/C_s) *versus* $\Delta\Sigma$. The value of d_n is then given by the slope of the straight line fitted through the experimental points. The samples should, however, be carefully selected so that their linear γ-ray attenuation coefficients (cm^{-1}) are the same. These authors also showed that the relationship log (C_x/C_s) *versus* $\Delta\Sigma$ established for oxygen compounds via the $^{16}O(n,p)^{16}N$ reaction may also be used for reactions with a much lower threshold such as $^{31}P(n,\alpha)^{28}Al$ (E_T = 1.95 MeV) and $Si(n,p)^{28}Al$ (E_T = 2.04 MeV).

(ii) Vandecasteele *et al.*[375] approximated the definition of d_n in an experimental way by considering a cylindrical sample (ϕ = 26 mm, thickness = 9 mm) composed of nine discs, each 1 mm thick, so that:

$$\exp(-\Sigma d_n) = \frac{\sum_{i=1}^{9} A_i}{\sum_{i=1}^{9} A_{0i}} \tag{2.3}$$

where

A_i = induced activity in the ith layer of the sample;

A_{0i} = induced activity in the same layer but without neutron attenuation.

The A_i values were determined from a simultaneous irradiation of the nine discs, while the A_{0i} values were determined from a separate irradiation of the nine discs placed in their normal geometry, the neutron output being monitored with a BF_3 counter. Normalisation to the same activity in the first layer allowed the calculation of d_n from equation (2.3). The results obtained for three different nuclear reactions on three different matrices (Al, Cu and Fe) confirmed that d_n was independent of the matrix. The correction $(C_s/C_x)_n$ amounted[375] to 0.99 and 1.03, respectively, for cylindrical aluminium and tungsten samples of 9 mm thickness and 26 mm diameter and for the irradiation geometry described in Ref. 375.

For a simple target to sample geometry, it is also possible to calculate the correction by numerical integration[375, 376] or by means of mathematically established escape probabilities[380, 381]. These calculations can, however, only be considered as approximate. In practice, it is impossible to take into account effects such as anisotropy in the neutron production and energy, neutron removal by the target support and the cooling water, the exact shape of the deuteron irradiated target area, etc. even for the simplest experimental configurations.

2.4.2 γ-Attenuation

The same situation as to the choice of the appropriate absorption coefficient exists for γ-ray attenuation as for neutron attenuation. Indeed, depending on the width of the energy window used for counting the 6.1 MeV ^{16}N γ-rays, a different fraction of the secondary γ-rays resulting from Compton interaction in the sample is counted. This results in absorption coefficients that are apparently lower than the total absorption coefficients.

However, it has been shown[378] that this effect is not very important, despite the observations of Nargolwalla *et al.*[379] who noted a discrepancy of a factor of two. This view is further supported by theoretical calculations of the appropriate absorption coefficients[376, 378].

A relation similar to equation (2.2) can be used to correct for γ-attenuation. Experimental methods for the determination of d_γ are essentially the same as for d_n. For a detailed discussion reference should be made to the original literature[375, 379]. It should also be noted that, for high-density samples, the γ-attenuation correction is more important than the neutron attenuation correction. It varies[375] from 0.98 to 1.26, respectively, for aluminium and tungsten samples 9 mm thick and 26 mm in diameter.

2.4.3 β-Attenuation

Beside 6.1 MeV γ-rays (68%), ^{16}N also emits 10.4 MeV β-rays (26%). Owing to varying degrees of self-absorption of β-rays in samples of different density, additional errors are possible. In contrast to γ-attenuation, the effect is most important for low-density samples and standards. Indeed, when the density is high, only relatively few β-rays leave the sample, so that the errors introduced by the differences in density are less important. By irradiating several oxides of varying density, packed in cylindrical polyethylene containers with walls 1 mm thick, it has been demonstrated by absorption experiments that without any shielding of the detectors (apart from the 1.5 mm aluminium pneumatic transfer tube and the *ca.* 0.2 g cm^{-2} aluminium detector can), the β-contribution to the total count rate in the 4.5–6.5 MeV energy range decreases from 16 to 8% when the density increases from 1.4 to 4.0 g cm^{-3}. The simplest way to overcome this source of error is to shield the detector from the β-rays, with a 1 cm aluminium or a 2 mm copper absorber[378]. This solution has also been adopted by Huysmans *et al.*[371], who showed that failure to insert a suitable absorber can give rise to an error of 1.3% when analysing 1.3 g rock samples with a silicon dioxide standard of the same dimensions but weighing 1.1 g. For samples with the same apparent density as the standard, the effect of the absorber appeared to be negligible.

Leal Horta *et al.*[380] corrected for β-attenuation using calculated escape probabilities for 10.4 MeV β-rays. In their approach, the β-absorption coefficient is calculated from the empirical relation:

$$\frac{\mu_\beta}{\rho} = \frac{17}{E_{max}^{1.14}} \tag{2.4}$$

where

μ_β/ρ = the mass absorption coefficient ($cm^2\ g^{-1}$);

E_{max} = the maximum energy of the continuous β-spectrum.

This relation is, however, only based[387] on data for nuclides with $E_{max} < 4$ MeV, reported as mass-absorption coefficients in aluminium. When nuclides other than ^{16}N, e.g. ^{28}Al ($E_{max} = 2.8$ MeV), are measured, no special shielding is usually required since most of the β-particles are stopped in the sample, in the 'rabbit', in the detector can, or in the transfer tube.

2.4.4 Flux monitoring

When, for the purpose of flux monitoring, a simultaneously irradiated oxygen-containing material is used, the standard and the sample can influence the activity induced in the monitor to a different extent as a consequence of neutron attenuation and scattering. The effect is largest when the flux monitor is irradiated in the neutron shadow of the sample. In this case, the correction can be determined experimentally[374, 375] or calculated theoretically[376, 378], and it can be shown[378] that it is given by

$$\frac{S_x}{S_s} = \frac{\exp(-\Sigma_x f t_x)}{\exp(-\Sigma_s f t_s)} \tag{2.5}$$

where

Σ_x, Σ_s = non-elastic scattering cross-sections for the sample and the standard;

t_x, t_s = sample and standard thickness;

f = correction for the longer path in the sample and the standard traversed by the obliquely impinging neutrons.

Even if not placed in the shadow of the sample, a BF_3 long counter used as a flux monitor can also be differently influenced[375] by the sample and the standard, owing to the inelastic scattering of the neutrons. The sample and the standard indeed act as nearly isotropic secondary neutron sources. The effect is most important for high-Z, high-density materials and may amount to a few per cent under the conditions described in Ref. 375.

2.4.5 Geometry errors in irradiation and counting

When possible, sample and standard are irradiated and counted under the same geometry, so that no correction is required. Sometimes, for practical reasons, they are irradiated under a slightly different geometry. This is, for example, the case when metal samples are analysed for oxygen. An oxygen standard[374] is then conveniently prepared by pressing an appropriate oxygen-containing mixture into a steel box of the same outer dimensions as the sample. Samples are preferably irradiated as such, to overcome the possible oxygen blank for the box and to increase the simplicity and reliability of the method. The important non-linear lateral and axial flux gradients[388] around the target can then cause important systematic errors.

The correction factor for a change in the irradiation geometry, Φ_s/Φ_x, can be obtained by integrating[374,375] the experimental or the theoretical flux distribution around the target over the sample and standard volume. As expected, the correction factor decreases when the target diameter increases. For the experimental set-up described in Ref. 375, it decreases from 1.093 to 1.034 when the target diameter changes from 10 to 30 mm. This is mainly due to the less important radial flux gradients with the larger target.

For theoretical flux calculations, the same remarks can be made as for neutron attenuation calculations. An extensive comparison of theoretical flux distribution calculations with experimental results is given in Ref. 388.

Obviously, the same geometry effect exists at the detector. Since the active surface of the NaI(Tl) crystals (mostly 7.5 cm diameter) is usually larger than that of the tritium target, the effect is less important than that of irradiation geometry. Appropriate corrections can be deduced[378] from calculated or experimental γ-ray detection efficiencies for NaI(Tl) crystals[389]. The fact that the samples are not homogeneously activated owing to the existing flux gradients around the target must, however, be taken into account in these calculations.

2.4.6 Conclusion

14 MeV NAA results for oxygen in metals have been compared[374,377] with those from other techniques. The oxygen content in synthetic[378] metal samples, made by sandwiching mylar foils between metal discs, and in stoichiometric materials with a precisely known oxygen content[380] has been determined, applying suitable corrections, and compared with the expected value.

In addition, standard reference materials have been analysed not only for oxygen, but also for elements such as silicon[371,379], magnesium[379], silver, etc. From such comparisons and from the analysis of the errors involved in the experimental determination of the correction factors, it appears that the observed results are probably accurate to within *ca.* 1%.

2.5 CHARGED PARTICLE ACTIVATION ANALYSIS

2.5.1 Introduction

Charged particle activation analysis (CPAA) can be considered as a useful complement to thermal and fast neutron activation. Until now, light particles (^{1}H, ^{2}H, ^{3}H, ^{3}He, ^{4}He) of energy up to *ca.* 20 MeV (40 MeV for α-particles) have been used exclusively for activation analysis. These particles are usually obtained in a cyclotron or a Van de Graaf accelerator. Although the principles of charged particle activation analysis have been understood for some time, until recently applications in the p.p.m. and sub-p.p.m. concentration region have been rather rare. Nevertheless, it is in this concentration range that CPAA offers, as a consequence of the high particle fluxes available (typically from 10^{12} to 10^{14} particles per second) and the reasonably high cross-sections of the reactions used (a few hundred mb), the most interesting possibilities. It is indeed clear that, owing to the necessity for highly sophisticated particle accelerators, CPAA should be thought of as a technique for special problems which are difficult to solve by other analytical methods, rather than as a general purpose method.

Recently, some effort has been devoted to prompt radiation analysis[390], whereby scattered particles, particles or γ-rays emitted during the nuclear reactions, or ion-induced X-rays are counted during the irradiation. Also, in addition to bulk characterisation, activation analysis with charged particles of low energy offers interesting possibilities for surface analysis[391].

The present review will not cover these applications but will treat the applications of CPAA to 'classical' in-depth trace analysis, emphasising the determination of light elements (from lithium up to fluorine). These elements are indeed insensitive to thermal neutron activation and the sensitivity of CPAA for these elements is usually higher than by other nuclear methods such as 14 MeV neutron or photon activation analysis. This survey is based on the literature between 1968 and 1973, although occasionally some earlier references will be included.

2.5.2 Standardisation methods

Standardisation methods in CPAA differ from and are usually more complex than those applied in NAA because of the limited range R of these particles (for example, $R_{Si} = 0.126$ g cm^{-2} for 34 MeV α-particles; 0.0381 g cm^{-2} for 15 MeV helium-3 ions; 0.166 g cm^{-2} for 10 MeV protons; 0.101 g cm^{-2} for 10 MeV deuterons) and because of the energy degradation of the incident particles, which results in a continuous variation of the reaction cross-section σ.

Practical relative standardisation procedures can be divided into two main groups[392,393]: those making use of a real standard of the matrix to be analysed containing a known amount of the element under investigation, and those using a standard of any material containing a known amount of the element to be determined.

To the first group belongs the technique in which one or several samples and a standard with the same bulk composition as the sample are irradiated simultaneously and under the same conditions, e.g. by means of an appropriate spinning mechanism. This procedure results in a decrease in induced activity, whereas the reproducibility of the results depends on the beam stability. An obvious solution to these drawbacks consists of irradiating the sample and the standard separately while monitoring the flux by means of an external or internal flux monitor. For trace analysis, these methods seem to be of limited value since reliable standard reference materials are usually not available.

Whilst in NAA the nature of the matrix can be an important source of errors as a consequence of different neutron attenuation in the sample and the standard, in CPAA attenuation of the particle flux can mostly be neglected owing to the lower cross-sections and the small range. However, the nature of the matrix is of considerable importance as it determines the energy loss per unit path length. Thus, when the standard is of a different bulk composition from the sample, corrections are required to allow for differences in stopping power. This can be achieved in two ways:

(i) When the sample thickness exceeds the range of the bombarding particles, the disintegration rate per second at saturation is given by:

$$D = In \int_0^R \sigma_t \, dt = In \int_{E_0}^0 \sigma_E \frac{dt}{dE}(E) \, dE \qquad (2.6)$$

where

I = beam intensity expressed as the number of particles incident on the sample per second (s^{-1}). This quantity can be related to the beam current in μA (1 μA = 6.24×10^{12} particles per second for a unit electrical charge);

n = number of target atoms per mg (mg^{-1});

$\sigma_t = \sigma_E$ = reaction cross-section (cm^2) at depth t (mg cm^{-2}) or energy E (MeV);

R = range (mg cm^{-2}) of the incident particles of energy E_0.

The 'integral cross-section' $\int_0^R \sigma_t \, dt$ depends on the nuclear reaction, on the particle energy E_0 and on the target material. The *average cross-section* is defined[394,395] as:

$$\bar{\sigma} = \frac{\int_0^R \sigma_t \, dt}{\int_0^R dt} = \frac{\int_0^R \sigma_t \, dt}{R} \quad (2.7)$$

As shown by Ricci *et al.*[394,395], equation (2.6) can be transformed into

$$D = I n \bar{\sigma} R \quad (2.8)$$

where $\bar{\sigma}$ is approximately independent of the target composition, leaving R as the only matrix-dependent factor. A detailed experimental verification of this assumption for a variety of proton and deuteron reactions on B, C, N and O has recently been presented by Kuin *et al.*[396]. The experimentally observed $\bar{\sigma}$ values for the same reaction and particle energy for matrices containing light and heavy elements (e.g. Al_2O_3 and Ta_2O_5 for reactions on oxygen) agreed within the limits of the experimental uncertainty due to fluctuations in the counting equipment, beam current measurement, inaccuracy of the range values and of the light element contents. A separate irradiation of a sample and of a 'thick' standard (e.g. a SiO_2 standard[397-401] for the determination of oxygen in silicon) allows the calculation of the unknown concentration in the sample. Indeed, it follows from equation (2.8) that:

$$n_x = n_s \frac{D_x}{D_s} \cdot \frac{I_s}{I_x} \cdot \frac{R_s}{R_x} \quad (2.9)$$

where the subscript x refers to the sample and s to the standard. The ratio I_s/I_x can be obtained from the activity induced in an appropriate flux monitor[402,403] (e.g. a thin foil) placed in front of the sample and standard. Kuin *et al.*[403] used, for example, graphite pellets, made by mixing pure aluminium nitride and boron carbide powder with 'Specpure' graphite, as standards in the determination of boron and nitrogen in silicon carbide. A tantalum foil (*ca.* 15 mg cm^{-2}) placed in front of the sample was used as a flux monitor. Other workers have preferred to integrate the ion current on the sample and the standard contained in an insulated sample holder[397-400]. It must be borne in mind that errors in the current measurement may arise from secondary electron emission, when no special precautions are taken, e.g. using a modified Faraday cup as a target holder[396].

As shown by Chen *et al.*[404], this 'average cross-section method' can sometimes further be simplified by using an 'internal standard' as flux monitor. ^{18}F formed through the $^{17}O(d,n)^{18}F$ reaction can, for example, be used as an internal standard in the determination of carbon in alumina through the reaction $^{12}C(d,n)^{13}N$. An alumina sample and a lead carbonate standard are irradiated separately, and the unknown carbon concentration can then be calculated from:

$$\left(\frac{n_C}{n_O}\right)_{Al_2O_3} = \frac{(D_C/D_O)_{Al_2O_3}(n_C/n_O)_{PbCO_3}}{(D_C/D_O)_{PbCO_3}} \quad (2.10)$$

where the subscript C indicates carbon (or ^{13}N) and O oxygen (or ^{18}F).

(ii) A second way of accounting for variations in stopping power requires the determination of the activation curve, i.e. the relative activity as a function of the depth of penetration in the matrix. An activation curve is generally determined in one of the following ways. Increasing thicknesses of the investigated matrix are irradiated while sandwiched between a few thin foils containing the element studied, e.g. mylar or mica foils for the determination of an activation curve for a reaction involving oxygen. Determination of the activity ratio of these foils and plotting *versus* the matrix thickness yields the activation curve. This technique is rather time-consuming as a large number of irradiations are required, but it is entirely free from any assumptions regarding charged particle ranges.

An experimentally established activation curve for a given reaction in a given matrix can also be transformed to another matrix[405,406] by means of the appropriate range–energy data. In the case of reactions on oxygen, the experimental results can be obtained by irradiating a stack of thin (*ca.* 10 μm) oxygen-containing foils such as mica or mylar with a total thickness exceeding the range. From published charged particle ranges[407] in the pure elements, a range–energy table for mica or mylar can be constructed by applying the empirical equation:

$$\frac{1}{R} = \sum_i \frac{f_i}{R_i} \quad (2.11)$$

which has been shown to agree satisfactorily with experiment[408]. In this equation, f_i is the weight fraction of the element i in the chemical compound, and R_i is the range in the pure element i (mg cm^{-2}). Each point of the plot of the relative activity *versus* the depth in mylar or mica is then converted to the corresponding energy by means of this table, and to a depth in any other matrix using the range–energy data for this matrix. This method has the advantage that it requires only one irradiation per nuclear reaction and that it allows the construction of detailed activation curves. It is based, however, on published range–energy data and on the validity of equation (2.11). The good agreement obtained between activation curves for example for the reactions $O \xrightarrow{\alpha} {}^{18}F$ [405,406,409] and $O \xrightarrow{^3He} {}^{18}F$ [409], when foils of different composition are used, confirms the validity of this method. Figures 2.1 and 2.2 give examples of the activation curves in silicon for the $O \xrightarrow{\alpha} {}^{18}F$ and the $O \xrightarrow{^3He} {}^{18}F$ reactions as determined by Vandecasteele *et al.*[409] with the aid of both mica and mylar foils. When no thin foils containing the element under consideration are available, use can be made of thin targets of the considered element deposited on metal foils, e.g. sulphur[410], and Tl_2SO_4 or Bi_2O_3 [411] vapour deposited on aluminium.

In practical activation analysis, a few thin foils of an appropriate material

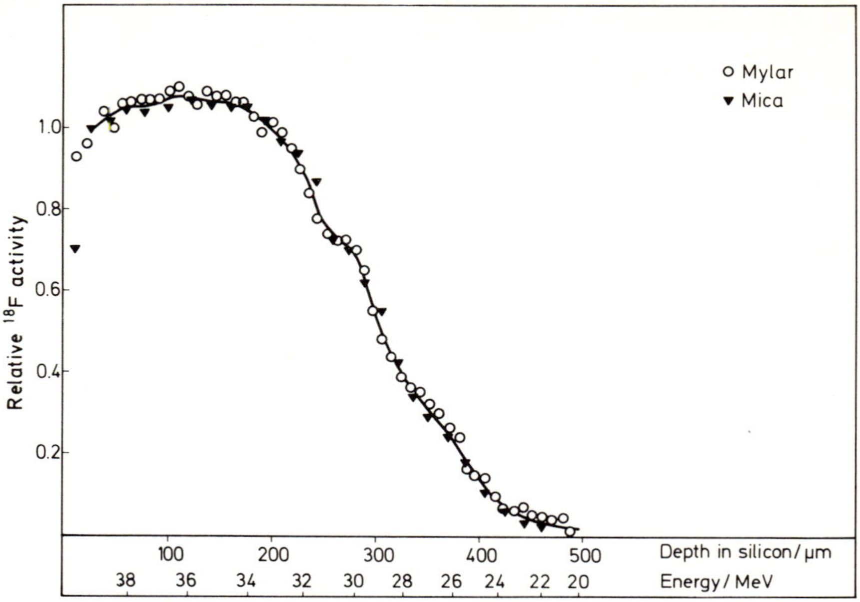

Figure 2.1 Activation curve in silicon for the O $\xrightarrow{\alpha}$ ^{18}F reaction[409]

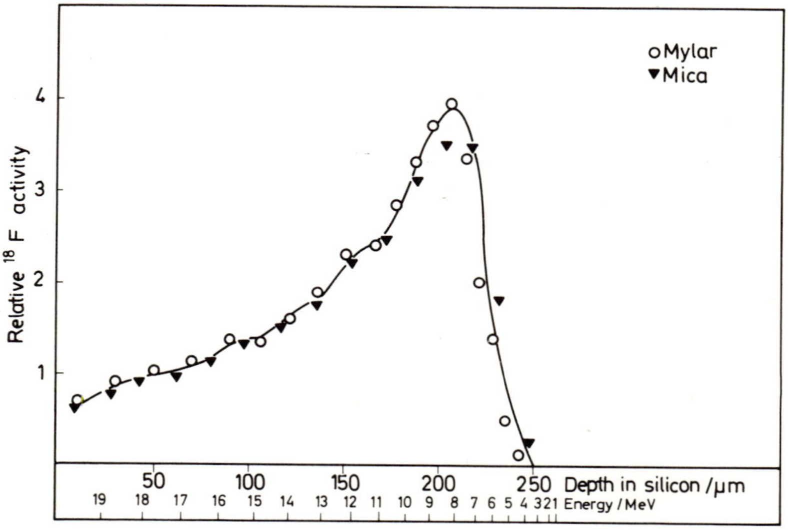

Figure 2.2 Activation curve in silicon for the O $\xrightarrow{^{3}He}$ ^{18}F reaction[409]

containing the element to be studied may be placed in front of the sample[409,412-414]. For the foil used as a standard, the disintegration rate can be written as:

$$D_s = I n_s \sigma_s t_s \qquad (2.12)$$

and for the sample

$$D_x = I\, n_x \int_0^R \sigma_t \, dt \qquad (2.13)$$

where σ_s is the cross-section at the average energy E_s of the particles in the thin standard foil and t_s is the thickness of this foil (mg cm^{-2}). Hence,

$$\frac{D_x}{D_s} = \frac{n_x}{n_s} \cdot \frac{\int_0^R \sigma_t \, dt}{\sigma_s \, t_s} \qquad (2.14)$$

A knowledge of the activation curve allows the calculation of the equivalent thickness:

$$e = \frac{\int_0^R \sigma_t \, dt}{\sigma_s} \qquad (2.15)$$

In this method the first foils have to be discarded owing to recoil losses[392] and a surface layer has to be removed from the sample after irradiation to eliminate atoms that recoil from the standard into the sample[392,409].

As an alternative, a thick standard can be irradiated separately with a particle beam of the same energy, but a lower intensity, while the flux is monitored by means of an appropriate flux monitor[410,411] or the current on the sample and standard is measured. In this case, it follows from equations (2.12) and (2.13) that:

$$\frac{D_x}{D_s} = \frac{I_x}{I_s} \cdot \frac{n_x}{n_s} \cdot \frac{\int_0^{R_x} \sigma_t \, dt}{\int_0^{R_s} \sigma_t \, dt} \qquad (2.16)$$

The ratio of the two integrals in this expression can be determined from the activation curve in the sample and the standard.

When techniques based upon a knowledge of the activation curve are used, the influence of the energy distribution of the particle beam has to be considered. Discrepancies between activation curves determined with particle beams with a different energy spread have been observed[416], resulting in different e or $\bar{\sigma}$ values. It is thus not justified to use equivalent thicknesses or average cross-sections deduced from published activation curves, without taking the influence of the energy resolution of the particle beam into account. The distribution in energy arises not only from the accelerator itself, but also from statistical fluctuations in the energy loss (straggling) accompanying the degradation of the particle energy when the beam traverses the samples. A detailed study by Tran *et al.*[408] has shown that the contribution of straggling to the total energy spread is dependent on the energy loss, but approximately independent of the nature of the target. Activation curves determined experimentally in one matrix can thus be used for another matrix (after range transformation) on condition that the energy spread of the incident beam remains unchanged.

2.5.3 Interferences

In CPAA, nuclear interferences from different sources are more important than in any other activation analysis technique, and can only be avoided or minimised by a judicious choice of the activating particle and its energy.

The first group of interfering reactions consists of 'parasitic' reactions which produce, from the matrix itself, the same radionuclide as the one used for analysis. Such spallation reactions sometimes seriously interfere with the determination of light elements in matrices formed of low-*Z* elements (e.g. Si, Al). Giroux *et al.*[417] studied the $Si \xrightarrow{\alpha} {}^{18}F$ reactions and showed their importance for determining low levels of oxygen in silicon through the $O \xrightarrow{\alpha} {}^{18}F$ reactions. When silicon is irradiated with 54 MeV α-particles, the resultant ${}^{18}F$ activity decreases very rapidly from the surface to a depth of *ca.* 500 µm (corresponding to an energy of *ca.* 36 MeV) with no apparent correspondence with the activation curve for $O \xrightarrow{\alpha} {}^{18}F$. Analogous profiles were found for the production of ${}^{24}Na$ and ${}^{29}Al$. An interference-free trace-oxygen determination can thus only be obtained by using α-particles of energy not exceeding 36 MeV. The same authors indicated further than when the α-beam is contaminated with deuterons, the apparent oxygen content increases to an even greater extent. This contamination effect seems, however, to be related to the use of a synchro-cyclotron and does not occur with modern machines. Similar interfering reactions are possible when one determines carbon or oxygen in silicon through $C \xrightarrow{{}^{3}He} {}^{11}C$ and $O \xrightarrow{{}^{3}He} {}^{18}F$. Nozaki *et al.*[399,400] measured the activation curve for $Si \xrightarrow{{}^{3}He} {}^{11}C$ and $Si \xrightarrow{{}^{3}He} {}^{18}F$. Both reactions have a negligible yield below *ca.* 20 MeV. Reactions of the type ${}^{27}Al(\alpha,x){}^{18}F$ have been shown[405,406] to interfere with the determination of oxygen in aluminium at energies in excess of 33 MeV. A simplified interpretation of the emission of light fragments similar to α-disintegration has been suggested[399,400], which assumes that fission of the compound nucleus takes place via tunnelling through the coulomb barrier. Such an approximate theory gives an indication of the choice of the incident energy which must be associated with a very low penetrability (e.g. 10^{-10} if impurity contents as low as 1 p.p.b. are to be determined). In a practical analysis, however, the absence of any interferences due to the reactions mentioned above should be ascertained by irradiating the sample at different energies[409,418,419], since the relative importance of the interfering reactions can be expected to rise sharply as the energy increases. Whenever possible, different activating particles should be used[409,414]. Giroux *et al.*[420] have recently reviewed possible interpretations of the above reactions. They have pointed out that although similar interfering reactions, producing ${}^{18}F$ from the matrix, are energetically possible in the case of high-Z elements such as germanium, the coulomb barrier for the emerging particles renders these reactions negligible below about 50 MeV for α- and 42 MeV for helium-3 particles.

Secondly, the impurities associated with the matrix can also interfere with the determination. Fortunately, selection of the incident energy in order to minimise such interferences is often possible. Data relative to possible interferences in analyses for light elements have been published[421,422]. The same sort of interferences can also occur when heavy elements are determined. The $^{204}Pb(p,pn)^{203}Pb$ reaction can, for example, interfere with the determination of Tl through $^{203}Tl(p,n)^{203}Pb$[402,411]; $^{108}Cd(p,\alpha)^{105}Ag$ with $^{105}Pd(p,n)^{105}Ag$[402]; and $^{76}Se(p,\alpha n)^{72}As$ with $^{72}Ge(p,n)^{72}As$[402].

In addition to interferences resulting from the production from two different elements of the isotope which serves as the analytical indicator, interference to the measurement of a particular isotope may also arise from the production of another radionuclide having similar β-energy, γ-energy or half-life depending on the characteristic used for discrimination against unwanted activity. This sort of interference is very likely to occur when the determination is based on annihilation radiation measurements, as is often the case in analyses of light elements. Indeed, a large part of the radionuclides produced by charged-particle bombardment decay via β^+-emission. In this case, either decay-curve analysis or chemical separation can be applied.

2.5.4 Applications

Although charged particle activation has been successfully applied to the determination of major and minor elements, e.g. in sedimentary rocks[423] and in ancient coins[424], in recent years it has almost exclusively been used for trace element determinations.

2.5.4.1 Determination of light elements ($3 \leqslant Z \leqslant 9$)

A large part of the literature on CPAA deals with the determination of light elements since most cannot readily be determined in ultra-trace concentrations ($<1\ \mu g\ g^{-1}$) by other analytical methods, including thermal neutron activation analysis. In many high-purity materials, however, light element (C, N, O) concentrations can be several orders of magnitude higher than those of other impurities and, although the influence of light elements on the physical properties of metals and semiconductors is often not clearly understood, it is believed that they partly determine the quality of these materials[425,426].

(a) *Using long-lived isotopes* ($T_{\frac{1}{2}} \geqslant 10$ min)—The measurement of isotopes with a reasonably long half-life has the advantage of allowing a careful and easily controlled surface removal after irradiation. This feature can be particularly important in trace analyses for oxygen[427], carbon[428-430] and nitrogen, and is one of the advantages of nuclear methods over other analytical techniques for these elements. This explains, together with the difficulties of fast radiochemical separations, why reactions such as $^{14}N(d,n)^{15}O$[431] and $^{12}C(\alpha,n)^{15}O$ leading to 2.03 min ^{15}O, and $^{16}O(d,n)^{17}F$[432] leading to 66 s ^{17}F, have only rarely been applied in trace analysis, although they should, *a priori*, offer good selectively and sensitivity. With a few exceptions, e.g. the determination of lithium[433] via the measurement of 53.1 d ^{7}Be produced by deuteron or proton bombardment, most practical applications are based on the measurement of 10.0 min ^{13}N, 20.3 min ^{11}C or 109.8 min ^{18}F, which decay exclusively via β^+-emission.

For analyses based on the use of ^{13}N and ^{11}C, fast radiochemical separations have been developed. Nozaki *et al.*[434] have investigated in detail the behaviour of ^{11}C, produced from carbon in high-purity silicon, during the dissolution and the fusion of the matrix. They applied[399,400] a method whereby the sample is dissolved in a HF–HNO_3–KIO_4 mixture. The CO_2 gas is absorbed in a NaOH solution containing $KMnO_4$ and a Na_2CO_3 carrier. After acidification, it is carried into a $Ba(OH)_2$ solution where it is precipitated as $BaCO_3$. Kuin[403] separated ^{11}C from important ^{13}N and ^{30}P activities produced in a silicon carbide matrix, using a special combustion apparatus. The sample was burned in an oxygen stream in a furnace with lead borate as an accelerator flux. Afterwards, the emerging gas was successively passed through a column filled with active manganese dioxide, one filled with activated copper oxide and finally through a carbon dioxide absorption flask filled with soda asbestos and Anhydrone. Separation of ^{11}C from ^{13}N and ^{30}P was achieved to an extent better than 1 in 10^7. A simultaneous separation of ^{13}N and ^{11}C by oxidising fusion, whereby $^{11}CO_2$ is trapped on Ascarite and $^{13}N_2$ on a molecular sieve cooled in liquid nitrogen, has been described[435]. This technique should, in principle, be useful for the simultaneous determination of two light elements, e.g. carbon and nitrogen, boron and carbon, etc. For ^{18}F, more conventional separation methods, such as solvent extraction[436-438], distillation[397-401,418,419,439], precipitation as PbFCl[397-400,439] or LaF_3 [401,420] and selective fixing on compounds such as hydrated antimony pentoxide[440], alumina[441] and hafnium dioxide[464] have been applied.

(i) *Boron and nitrogen*. In CPAA for boron and nitrogen, the mutual interference of these two elements must be carefully considered. The most interesting possibilities for their determination are proton and deuteron activation.

When protons are used [^{14}N(p,α)^{11}C or ^{11}B(p,n)^{11}C], the ^{11}C activity produced from boron[421,422,431,442] is about two times that produced from nitrogen in the 10–15 MeV energy range, increasing to *ca.* 10 times at 5 MeV. Nitrogen can thus only be determined accurately by this method when the boron content of the sample is known or can be ignored. This was, for example, the case when semiconductor silicon was analysed for nitrogen by Nozaki *et al.*[399,400], who corrected the apparent nitrogen content for boron. The boron concentration was deduced from electrical measurements. Nitrogen concentrations as low as 1 p.p.b. may be determined[443] after chemical separation of ^{11}C. In addition, when semiconductor graphite[442] prepared from nitrogen-containing compounds is analysed for nitrogen, the nitrogen content can safely be considered to be much higher than the boron content. Another solution[403,444] is to carry out two irradiations with protons of different energies for the determination of both boron and nitrogen. Kuin[403] determined nitrogen and boron in silicon carbide after irradiation with 5 and 14 MeV protons. It is clear that, to be applicable, the method requires that the concentrations of the two elements should be of the same order of magnitude. By irradiating with 14.5 MeV protons, Engelmann *et al.*[445] have determined 'apparent' boron contents of 0.5 p.p.b. in polycrystalline silicon and *ca.* 3 p.p.b. in monocrystalline samples. These values in fact give an upper limit for both boron and nitrogen.

According to other workers[431], deuteron activation [$^{10}B(d,n)^{11}C$ and $^{11}B(d,2n)^{11}C$] is a more useful method for boron than proton activation. Indeed, the ratio of the ^{11}C activity from boron to that produced from the same amount of nitrogen is *ca.* 30 for 10 MeV deuterons, increasing to infinity at 5.9 MeV [threshold of the $^{14}N(d,\alpha n)^{11}C$ reaction], so that the method can easily be made specific for boron without greatly reducing the energy. Deuteron activation also allows the determination of nitrogen[431] through the $^{14}N(d,n)^{15}O$ reaction.

(ii) *Carbon.* The most important possible interferences[421,422] in the determination of carbon via $^{12}C(d,n)^{13}N$ are nitrogen and oxygen. The $^{12}C(^{3}He,\alpha)^{11}C$ reaction is interfered with by beryllium, boron, nitrogen and oxygen. A thorough discussion of nuclear methods for the determination of carbon (and nitrogen) in silicon and germanium can be found in Ref. 446.

Martin and Haas[428] used 2.3 MeV deuterons [$^{12}C(d,n)^{13}N$] produced in a Van de Graaf machine for carbon analysis in silicon. The effective range of these particles is only 37 μm. This makes the technique more sensitive to surface contamination than techniques using higher energy particles. The importance of errors due to carbon injected into the specimen from the accelerator has indeed been clearly demonstrated[429]. Post-irradiation removal of a surface layer of at least 3.5 μm proved effective in overcoming this problem, so that concentrations of 0.2 p.p.m. could be determined non-destructively. Other workers[447] obtained good agreement between results in the 1–0.1 p.p.m. concentration range for carbon in silicon using 10 MeV deuterons and 40 MeV photons. The same reaction has been used[404] to determine carbon in high-oxygen-content materials, making use of the induced ^{18}F activity as an internal standard.

The $^{12}C(^{3}He,\alpha)^{11}C$ reaction has also been applied[399,415,443,448] to the determination of carbon in silicon. For concentrations from 0.3 to 3 p.p.m., Gross *et al.*[415] reported results that agreed with values obtained through i.r. spectroscopy. The detection limit for a non-destructive determination can be made as low as *ca.* 10 p.p.b.[415,448], provided the carbon to oxygen ratio is not too unfavourable. Nozaki *et al.*[443] obtained the same sensitivity after chemical separation of ^{11}C. Applying the same method, Engelmann *et al.*[447] obtained results that are significantly higher than with deuterons or photons. The discrepancy was attributed to the presence of boron or beryllium in concentrations as high as 1 p.p.m.

(iii) *Oxygen.* Although other interesting possibilities[449] such as the $^{16}O(p,\alpha)^{13}N$ reaction exist, helium-3 and α-particles have been most frequently used for the determination of oxygen through the production of ^{18}F. Indeed, they offer a sensitivity of the order of a few p.p.b.[409,443] in favourable matrices and the 109.8 min half-life of ^{18}F allows even relatively lengthy chemical separations. The choice of the irradiation energy has already been discussed in connection with possible nuclear interferences from the matrix. In helium-3 activation, fluorine and sodium are the main sources of interference[421,422], while fluorine and to a lesser extent nitrogen and sodium can interfere in α-activation analysis. Since the practical activation threshold of the $O \xrightarrow{^{3}He} {}^{18}F$ reactions ($Q > 0$; coulomb barrier = 3.9 MeV) is very low compared with that of $O \xrightarrow{\alpha} {}^{18}F$ ($Q = -20.4$ MeV), helium-3 activation is an attractive method for the non-destructive determination of oxygen in metals of high atomic number, with a high coulomb barrier (*ca.* 15 MeV for $Z = 50$). Indeed, when the energy of the incident particles is limited to a value below this barrier, the production of interfering radionuclides from the matrix can be avoided or greatly reduced. Non-destructive determination of oxygen in molybdenum[450] and zirconium[450] is, for example, possible with 11 MeV helium-3 particles, while for the same determination α-activation[437,450] requires elaborate radiochemical separations. Although some destructive determinations have been carried out, non-destructive trace oxygen analysis is also possible in silicon (α [409,413,414,451] and helium-3 [398,401,409,415]) and in aluminium (helium-3 [418,419]), when γ–γ coincidence techniques are used together with decay-curve analysis. Other matrices such as gallium phosphide[397,398] and germanium[397,401,452] always require chemical separation even when low-energy helium-3 particles are used. When germanium[397,401] is irradiated with 10–15 MeV helium-3 particles, considerable activities are formed of ^{73}Se, ^{71}As and ^{69}Ge, so that several successive chemical separation steps become necessary to achieve the high decontamination factors required for the trace analysis of oxygen. In addition to the matrices already mentioned, oxygen has been determined in beryllium[453], sodium[454], chromium[455], iron[455], nickel[455], copper[456,457], niobium[450], hafnium[450], tungsten[450] and lead[438].*

When bulk analyses for oxygen are performed, it is of the utmost importance to remove surface and recoil contamination after irradiation. Although other methods, e.g. mechanical grinding[451], have been applied, most authors[397-401,409] remove a surface layer some 20 μm thick by chemical etching. However, this procedure can cause contamination of the surface by the removed material. Rook *et al.*[427] have thoroughly investigated the chemical etching problem for the determination of oxygen in silicon and aluminium. In the case of aluminium, the ^{18}F contamination appeared not to occur via physical adsorption, but rather by chemisorption enhancing the contamination by a factor of *ca.* 10^3. The problem posed by such phenomena can easily be overcome by applying several successive etchings in different solutions.

(b) *Using short-lived isotopes* ($T_{\frac{1}{2}} < 1$ *min*)—The impossibility of surface treatment after irradiation, which can be an important drawback in analysing high-purity matrices for such elements as carbon, nitrogen or oxygen using short-lived isotopes, is less important when dealing with lithium, boron or fluorine, since possible surface impurities can be removed before the irradiation and contamination during the irradiation is not likely to occur. In addition, the use of short-lived isotopes offers several important advantages: speed of analysis, lower cost per analysis, feasibility of higher beam intensities and possibility of repetitive irradiation and counting cycles.

Some relatively simple 'rabbit systems' useful for isotopes of $1\ s < T_{\frac{1}{2}} < 1$

* See note added in proof, p. 137.

min have been developed. Lee *et al.*[458] and Debrun *et al.*[459] used a system where a thin monitor and a sample, each mounted on a slide, are irradiated one behind the other. During irradiation these are held in position by an electromagnet which is de-activated automatically at the end of the irradiation. The slides fall by gravity to their respective counting positions in less than 1 s. When isotopes of still shorter half-life are used, transfer of the samples is impractical and the samples have to be counted in their irradiation position.

Schweikert *et al.*[452] used this 'quasi-prompt' approach to determine boron and lithium in silicon, germanium and glass using the $^{7}Li(d,p)^{8}Li$ and the $^{11}B(d,p)^{12}B$ reactions induced by 19 MeV deuterons. Both reactions yield very short-lived isotopes with half-lives of 850 ms and 20 ms, respectively, which emit high-energy β-particles (E_{max} = 13 and 13.4 MeV). For the selective counting of the β-particles, these authors developed a coincidence technique with two thin (3 mm) plastic scintillators placed one behind the other. The selectivity was further improved by placing appropriate absorbers (e.g. a 3 MeV absorber for the silicon analysis) in front of the detector to eliminate low-energy β-particles and by using low-energy threshold settings in both counting chains. Analysis of the resulting decay curves allowed the calculation of the net ^{8}Li and ^{12}B activities. Results obtained for lithium in silicon using thick H_3BO_3 pellets, containing LiF as a standard, compare well with those obtained by flame emission, while results for boron in silicon and glass using LiF pellets containing H_3BO_3 as a standard agree with certified values. Under the described experimental condition (i.e. 10 irradiations of 0.3 s with a 2 μA deuteron intensity), detection limits are estimated at *ca.* 0.1 p.p.m.

Lee *et al.*[458] determined fluorine through $^{19}F(^{3}He,2p)^{20}F$ ($T_{\frac{1}{2}}$ = 10.8 s) using the 1.63 MeV γ-ray of ^{20}F. In contrast to the use of reactions such as $^{19}F(^{3}He,\alpha)^{18}F$, this method is very specific for fluorine. At equal concentrations, oxygen interferes by only 0.2%.

2.5.4.2 *Determination of heavy elements* ($Z \geqslant 10$)

Although CPAA complements NAA for the determination of elements with $Z \geqslant 10$, only a restricted number of applications has been published, mostly for elements that are difficult to determine with thermal neutrons. Some typical applications are given in Table 2.17, restricted to papers where analytical results are mentioned and a detailed outline of the analytical method is given.

As shown by Debrun *et al.*[402], it is also possible to carry out multi-element determinations using proton activation coupled to Ge(Li) spectrometry. These authors non-destructively determined some 30 trace elements (detection limits from 0.01–2 p.p.m.) in cobalt, silver and aluminium after a 11 MeV proton irradiation. The technique can be applied to some matrices which are strongly activated during thermal neutron irradiation (Co, Mn, Ag, etc.); moreover, it allows the determination of elements that are insensitive to neutron activation (S, Ti, V, Ge, Nb, Sn, Tl, Pb)*.

* See note added in proof, p. 137.

Table 2.17 Determination of heavy elements by charged particle activation

Element determined	*Matrix*	*Reaction*	*Chemical separation*	*Conc. range + sensitivity (s)*	*Standard + flux monitor (F.M.)*	*Interferences*	*Ref.*
S	Fe Al	$S(p,x)^{34m}Cl$	AgCl precipitation; anion exchange; AgCl precipitation	30 p.p.m. 0.3 + 1.6 p.p.m. s: 0.1 p.p.m.	stainless steel standard (NBS) F.M.: $^{66}Zn(p,n)^{66}Ga$ in Zn foil	$^{35}Cl(p,pn)^{34m}Cl$	440
Ca	Si Al Mg	$^{40}Ca(\alpha,p)^{43}Sc$	non-destructive	<0.3 p.p.b. <7 p.p.b. 7 p.p.m.	CaF_2 standard	$^{41}K(\alpha,2n)^{43}Sc$ $^{41}K(\alpha,n)^{44}Sc$	460
Tl	glass	$^{203,205}Tl(p,xn)^{203}Pb$	dithizon extraction; $PbCrO_4$ precipitation	0.4–60 p.p.m. s: 1 p.p.b.	Tl_2SO_4 F.M.: $^{56}Fe(p,2n)^{55}Co$ $^{56}Fe(p,\alpha n)^{52}Mn$ in Fe foil	$^{204}Pb(p,pn)^{203}Pb$	411
Pb	Pd Ta	$^{206}Pb(p,n)^{206}Bi$		3 p.p.m. <0.1 p.p.m.			461
	glass	$^{206}Pb(d,2n)^{206}Bi$ $^{207}Pb(d,3n)^{206}Bi$	$Bi(OH)_3$ precipitation; Bi^{3+} extraction	2–34 p.p.m. s: 10 p.p.b.	Pb metal F.M.: $^{63}Cu(d,p)^{64}Cu$ in Cu foil	—	411
	Au		$Bi(OH)_3$ precipitation; reduction with $NaHSnO_2$	0.9 p.p.m.			
Bi	glass	$^{209}Bi(p,3n)^{207}Po$	deposition on Ni	0.16–0.36 p.p.m. s: 1 p.p.b.	Bi metal F.M.: see Tl		411
S	Al Fe	$^{34}S(p,n)^{34m}Cl$	AgCl precipitation; cation exchange; AgCl precipitation	<0.1 p.p.m. s: 20 p.p.b.		$^{35}Cl(p,pn)^{34m}Cl$	462
V	Al	$^{51}V(p,n)^{51}Cr$	non-destructive	s: 36 p.p.b.			462
Nb	Al	$^{93}Nb(p,n)^{93}Mo$	non-destructive	s: 3 p.p.b.		$^{94}Mo(p,pn)^{93}Mo$ $^{94}Mo(p,d)^{93}Mo$	462

Finally, mention should be made of the work of Debrun *et al.*[459], who presented a systematic survey of the possibilities for the determination of heavy elements by proton irradiation followed by Ge(Li) spectrometry, using short-lived isotopes ($T_{\frac{1}{2}} < 1$ min) and of McGinley et *al.*[463] who explored the possibilities of charged particle (p and d) activation coupled with X-ray counting (after irradiation) for the determination of elements with $34 \leqslant Z \leqslant 82$.

Note added in proof

The accuracy and precision of charged particle activation analysis for oxygen have been investigated by comparing results obtained by both α and ^{3}He activation. Table 2.18 summarises some analytical results, showing that a precision of *ca.* 10% can be realised at the 1 p.p.m. level and that the accuracy is of the same order of magnitude. For lead the coefficient of variation was higher, probably owing to a less homogeneous distribution of the oxygen. It was also shown that for copper and lead the interference of fluorine and sodium with the oxygen determination were negligible.

Table 2.18 Comparison of results* for oxygen obtained by both α and ^{3}He activation

Matrix	*α activation*	^{3}He *activation*
Si (czochralski)[409]	8.4 ± 0.6† (n = 10)	10.0 ± 0.4 (n = 2)
(floating zone)	0.051 ± 0.017 (n = 2)	0.062 ± 0.014 (n = 2)
Cu (O.F.H.C.)[465, 466]	1.17 ± 0.13 (n = 10)	1.08 ± 0.11 (n = 5)
Pb[467]	0.99 ± 0.21 (n = 8)	0.79 ± 0.36 (n = 8)

* Results in p.p.m.
† Standard deviation on one determination

Since the preparation of this review proton activation has been successfully applied to the analysis of Ta[468,469] and Nb[469]. Barrandon *et al.*[469] determined 38.7 ± 1.6 p.p.m. (n = 6) of Nb and 2.1 ± 0.4 (n = 6) p.p.m. of Mo in Ta and 71.5 ± 2.4 (n = 6) p.p.m. of Mo and 72 ± 6 p.p.m. of W in Nb. Upper limits (between 0.01 and 25 p.p.m.) were given for 34 other elements. A detailed study of the non-destructive analysis of Co[470] by activation analysis with 10 MeV protons showed that this technique yields detection limits less than 1 p.p.m. for 13 elements and less than 0.1 p.p.m. for 20 other elements. Fe, Cu, Ca, Ti, Cr, Ni, Zn, As, Se, Mo and Pt were detected in part of the samples in concentrations between 0.06 and 880 p.p.m.

References

1. Showalter, D. L. and Schmitt, R. A. (1972). *Advances in Activation Analysis*, Vol. 2, 185 (J. M. A. Lenihan, S. J. Thompson and V. P. Guinn, editors) (London, New York: Academic Press)

2. Heier, K. S. (1971). *Activation Analysis in Geochemistry and Cosmochemistry*, 11 (A. O. Brunfelt and E. Steinnes, editors) (Oslo: Universitetsforlaget)
3. (1970). *Science*, **167,** 449
4. Smales, A. A. (1971). *Activation Analysis in Geochemistry and Cosmochemistry*, 17, (A. O. Brunfelt and E. Steinnes, editors) (Oslo: Universitetsforlaget)
5. (1971). *Neutron Sources and Applications, American Nuclear Society National Topical Meeting, Augusta (Georgia) USAEC Rept. Conf., 710402*
6. Lutz, G. J., Boreni, R. J., Maddock, R. S. and Wing, J. (editors) (1972). 'Activation Analysis, a Bibliography through 1971', *NBS Tech. Note 467* (Washington, D.C.: NBS)
7. Berzin, A. K., Bespolov, D. F., Zaporozhets, V. M., Kantor, S. A., Leipunskaya, D. I., Sulin, V. V., Feldman, I. I. and Shimelevich, Y. S. (1966). *Atomic Energy Rev.*, **4,** 59
8. Belov, A. F., Egiazarov, B. G., Zyubko, V. A. and Karpukhin, O. A. (1969). *Proc. Sci. Techn. Conf., Budapest, Instrumentation in Activation Analysis*, 114 (Moscow: Comecon)
9. Berzin, A. K., Leipunskaya, D. I., Savosin, S. I., Shimelevich, Y. S. and Spiridonov, V. P. (1969). Ref. 8, p. 196
10. Kasatchin, V. A., Leipunskaya, D. I., Savosin, S. I. and Chulanov, Y. G. (1971). *Atomnaya Energiya*, **30,** 411
11. Savosin, S. I., Leipunskaya, D. I. and Spiridonov, V. P. (1968). *Tezisy Dokl. 2-ogo Vses. Soveshch., po Aktivatsionnomu Analizu*, 109
12. Simon, L. (1968). *Rudy (Czech.)*, **16,** 84
13. Senftle, F. E. and Hoyte, A. F. (1966). *Nucl. Instr. Meth.*, **42,** 93
14. Senftle, F. E., Hoyte, A. F. and Martinez, P., Jr. (1969). U.S. Pat. 3 463 922
15. Savosin, S. I., Berzin, A. K., Leipunskaya, D. I., Sulin, V. V. and Shimelevich, Y. S. (1968). *Yadernaya Geophys.*, **3,** 20
16. Gijbels, R. (1973). *Min. Sci. Eng.*, **5,** 304
17. Tolmie, R. W. and Thompson, C. J. (1969). *Nuclear Techniques and Mineral Resources*, 485 (Vienna: IAEA)
18. Janghorbani, M., Gillum, D. E. and Ehmann, W. D. (1973). *Analytical Methods Developed for Application to Lunar Samples Analyses*, ASTM Spec. Tech. Publ. 539, 128 (Philadelphia, Pa.: ASTM)
19. Senftle, F. E., Wiggins, P. F., Duffey, D. and Philbin, P. (1971). *Econ. Geol.*, **66,** 583
20. Linn, T. A. Jr., Byrne, J. M. and Sandquist, G. M. (1972). *Anal. Chim. Acta*, **61,** 303
21. Nuclear Chicago, Texas Nuclear Division: Model NA-80, Neutron analytical lab. analyzer; Model NA-79, 'Neutron Activation Analysis for Industrial Process Control' (Techn. Spec. Nos. 402 and 401)
22. Outokumpu Oy, Instrument Division, Tapiola, Finland
23. Vses. Obj. Isotop, Atomizdat, Moscow: 'Neutron activation analysis system RS-1'
24. Metrimpex (Hungarian Trading Company for Instruments), Budapest: 'Automatic Activation Analysis System for Al and Si', Type MTA-1527
25. Kuusi, J. (1969). *Proc. Symp. Gaithersburg, Modern Trends in Activation Analysis*, (J. De Voe, editor) NBS Spec. Publ. 312, 450 (Washington, D.C.:NBS)
26. Kuusi, J. (1970). *Nucl. Appl. Tech.*, **8,** 465
27. Lundán, A. (1973). *Proc. 3rd Symp. Recent Develop. Neutron Activation Anal., Cambridge, UK*
28. Bakes, J. M. and Jeffery, P. G. (1966). *Analyst.* **91,** 216
29. Bakes, J. M. (1973). *Proc. 3rd Symp. Recent Develop. Neutron Activation Anal., Cambridge, UK*
30. Jeffery, P. G. and Bakes, J. M. (1967). *Analyst.* **95,** 151
31. Prokopchik, V. I. and Subbotina, T. I. (1966). *Soviet Atom. Energy*, **21,** 1108
32. Prokopchik, V. I., Bushkov, A. P. and Bernadskii, K. (1969). *Soviet Atom. Energy*, **27,** 883
33. Lobanov, E. M., Chanyshev, A. I. and Chanysheva, T. I. (1962). *Izv. Akad. Nauk Uz. SSR, Ser. Fiz. Math. Nauk*, **9,** 68
34. Simon, L., Sok, V. and Filges, R. (1969). *Rudy (Czech.)*, **17,** 339
35. Pawaskar, P. B., Reddy, G. R. and Das Sankar, M. (1972). *Indian J. Technol.*, **10,** 80
36. Gorbachev, A. N., Karpunin, A. M. and Matukanis, L. F. (1971). *Z. Angew. Geol.*, **17,** 436

37. Christell, R. and Ljunggren, K. (1965). *Radiochemical Methods of Analysis*, Vol. 1, 263 (Vienna: IAEA)
38. Ashe, J. B., Berry, B. F. and Rhodes, J. R. (1969). *Proc. Symp. Gaithersburg*, *Modern Trends in Activation Analysis*, (J. De Voe, editor) NBS Spec. Publ. 312, 913 (Washington, D.C.: NBS)
39. Philbin, P. and Senftle, F. E. (1971). *Trans. Soc. Min. Eng. AIME*, **250**, 102
40. Glasson, V. V., Drynkin, V. I., Leipunskaya, D. I. and Putyatina, N. D. (1963). *Yad. Geophys.*, **3**, 599
41. Plaksin, I. N., Anchevski, E. V. and Belyakov, M. A. (1965). *Dokl. Akad. Nauk SSSR*, **163**, 1202
42. Evans, D. J. R. and Herage, T. I. (1971). *Activation Analysis in Geochemistry and Cosmochemistry*, (A. O. Brunfelt and E. Steinnes, editors) 129 (Oslo: Universitetsforlaget)
43. Gambaryan, R. G., Danilchenko, I. A. and Shtan', A. S. (1968). *Ind. Lab.*, **34**, 1170
44. Gambaryan, R. G. and Danilchenko, I. A. (1968). *Tr. Vses. Nauch.-Issled. Inst. Radiats. Tekh.*, 135
45. Nargolwalla, S. S. (1973). 'Nuclear Techniques for Borehole Logging of Geologic Materials', Information leaflet of Scintrex, Ltd., Concord, Ontario (Canada)
46. Senftle, F. E., Philbin, P. W. and Sarigianis, P. (1970). *Isotope Radiat. Technol.*, **7**, 411
47. Senftle, F. E., Duffey, D. and Wiggins, P. F. (1969). *J. Marine Tech. Soc.*, **3**, 9
48. Leipunskaya, D. I., Gauer, Z. E. and Flerov, G. N. (1959). *Soviet Atom. Energy*, **6**, 198
49. Abdullaev, A. A., Lobanov, E. M., Novikov, A. P., Romanov, M. M. and Khaidarov, A. A. (1960). *J. Anal. Chem. USSR*, **15**, 801
50. Lötzsch, W. (1964). *Kernenergie*, 776
51. Drynkin, V. I. *et al.* (1965). 'A Rapid Method for Determining the Content of Certain Elements in Ore Samples by Spectrometric Measurements of Neutron Capture γ-Radiation. Some Problems of the Use of Neutron Activation Analysis in Mineral Prospecting and Exploration', *Izd. Nedra*
52. Alaerts, L., Op de Beeck, J. P. and Hoste, J. (1973). *J. Radioanal. Chem.*, **15**, 601
53. De, A. K. and Meinke, W. W. (1958). *Anal. Chem.*, **30**, 1474
54. Cyprès, R., Noël, S. and Bettens, B. (1967). *Proc. 37th Int. Congr. Ind. Chem. Madrid*, **1**, 517
55. Lobanov, E. M., Romanov, O. M., Romanov, M. M. and Khaidarov, A. A. (1961). *J. Anal. Chem. USSR*, **16**, 23
56. Béress, M. (1966). *Radioisotope Instruments in Industry and Geophysics*, Vol. 1, 365 (Vienna: IAEA)
57. Strain, J. E. and Lyon, W. S. (1965). *Radiochemical Methods of Analysis*, 245 (Vienna: IAEA)
58. Tatar, Y. and Shiklosh, A. (1969). *Proc. Sci. Techn. Conf. Budapest*, *Instrumentation in Activation Analysis*, 54 (Moscow: Comecon)
59. Gureev, E. S., Islamov, T., Lobanov, E. M., Miranskii, I. A. and Khaidarov, A. A. (1967). *Izv. Akad. Nauk Uz. SSR, Fiz. Mat. Nauk*, **14**, 58
60. Mezhiborskaya, K. B. and Krasikova, M. I. (1970). *J. Anal. Chem. USSR*, **25**, 500
61. Plaksin, I. N., Slepchenko, I. F. and Starchik, L. P. (1961). *Dokl. Akad. Nauk SSSR*, **137**, 880
62. Ivanov, I. N., Markun, N. Y. and Shtan', A. S. (1970). *Tr. Vses. Nauch.-Issled. Inst. Radiat. Tekh*, 186
63. Tatar, Y. (1968). *Banyaszat* (*Hung.*), **101**, 287
64. Dugain, F. and Tatar, Y. (1970). *Annal. Instituti Geologici Publici Hungarici*, **LIV**, 375
65. Dugain, F. (1972). *Report ITE-61*, (Brussels: EEC Bureau Eurisotop)
66. Rhodes, J. R. (1969). *Isotope Radiat. Technol.*, **6**, 359
67. Van Grieken, R. and Hoste, J. (1972). *Annotated Bibliography on 14 MeV Neutron Activation Analysis*, Information booklet No. 65 (Brussels: EEC Bureau Eurisotop)
68. Nargolwalla, S. S. and Przybylowicz, E. P. (1973). *Activation Analysis with Neutron Generators*, *Chemical Analysis Series*, Vol. 39 (New York: Wiley-Interscience)
69. Morgan, J. W. and Ehmann, W. D. (1971). *Activation Analysis in Geochemistry and Cosmochemistry*, (A. O. Brunfelt and E. Steinnes, editors) 81 (Oslo: Universitetsforlaget)

70. Ehmann, W. D. and Morgan, J. W. (1971). *Proc. 2nd Lunar Sci. Conf., Cambridge, Mass.*, **2,** 1237
71. Eugster, M. P. (1969). *Handbook of Geochemistry* (K. H. Wedepohl, editor) Vol. II, Part I, Chapt. 8, E-1 (Berlin: Springer Verlag)
72. To-on, M. (1963). *USAEC Report TID-20472*
73. Knight, A., Watterson, J. I. W., Uken, E. A. and Faure, P. K. (1967). *Research Report No. 208* (Johannesburg: National Institute for Metallurgy)
74. Bartošek, J., Kašperec, L. and Skuton, B. (1970). *Geol. Prazk. (Czech.)*, **12,** 165
75. Weber, G. and Guillaume, M. (1970). *J. Radioanal. Chem.*, **5,** 379
76. Jeffery, P. G. and Bakes, J. M. (1970). *Proc. Soc. Anal. Chem.*, **7,** 71
77. Nargolwalla, S. S., Lundgren, F. A., Bettum, B., Thomas, P., Shideler, R., Ruegg, F. C. and Crambes, M. R. (1967). *NBS Tech. Note 428*, (Washington, D.C.: NBS)
78. Strain, J. E., Dyer, F. F. and Bate, L. C. (1968). *Anal. Div. Annu. Rep. ORNL-4343*, 75 (Tenn.: Oak Ridge National Laboratory)
79. Botvina, A. Y., Sokolskii, V. V. and Faizrakhamanova, N. R. (1968). *Tez. Dokl. 2-ogo Vses. Soveshch. Aktivat. Anal.*, 90
80. Dibbs, H. P. and McMahon, C. (1969). *Internal Report MS-69-79*, Dept. Energy, Mines and Resources, Mines Branch (Ottawa: Dept. Energy, Mines and Resources)
81. Wing, J. and Wahlgren, M. A. (1969). *J. Radioanal. Chem.*, **3,** 37
82. Gorski, L., Kusch, W. and Wojtkowska, J. (1964). *Talanta*, **11,** 1135
83. Hollingshead, R. G., Franz, H. W., and Martin, R. (1968). *Tech. Report 68-11* (Salt Lake City: Kennecot Research Center)
84. Pradzynski, A. (1969). *Nuclear Techniques and Mineral Resources*, 451 (Vienna: IAEA)
85. Navalikhin, L. V., Kireev, V. A., Talamin, Y. N. and Nikonov, G. S. (1971). *Soviet Atom. Energy*, **30,** 656
86. Wing, J. (1964). *Anal. Chem.*, **36,** 559
87. Vogt, J. R. and Ehmann, W. D. (1965). *Radiochim. Acta*, **4,** 24
88. Vogt, J. R. and Ehmann, W. D. (1965). *Geochim. Cosmochim. Acta*, **29,** 373
89. Ehmann, W. D. and Morgan, J. W. (1970). *Proc. Apollo 11 Lunar Sci. Conf.*, **2,** 1071
90. De Soete, D., Gijbels, R. and Hoste, J. (1972). *Neutron Activation Analysis, Chemical Analysis Series*, Vol. 34 (New York: Wiley-Interscience)
91. Bowen, H. J. M. and Gibbons, D. (1963). *Radioactivation Analysis* (Oxford: University Press)
92. Turkevich, A., Reed, G. W., Jr., Heydegger, H. R. and Collister, J. (1971). In *Proc. 2nd Lunar Sci. Conf., Cambridge, Mass.*, **2,** 1565
93. Greenland, L. P., Rowe, J. J. and Dinnin, J. I. (1971). *Geological Survey Research*, Chapt. B, Paper 750-B, 175 (Washington, D.C.: U.S. Geological Survey)
94. Gijbels, R. and Govaerts, A. (1973). *J. Radioanal. Chem.*, **16,** 7
95. Thoresen, P. (1964). *Acta Chem. Scand.*, **18,** 1054
96. Brunfelt, A. O. and Steinnes, E. (1966). *Geochim. Cosmochim. Acta*, **30,** 921
97. Kramer, H. H., Hampton, W. J., Molinski, V. J., Finn, J. J. and Wahl, W. H. (1968). *Nucl. Appl.*, **5,** 260
98. Wyttenbach, A. and Dulakas, H. (1968). *Chimia (Switz.)*, **22,** 484
99. Lamb, J. F., Prussin, S. G., Harris, J. A. and Hollander, J. M. (1966). *Anal. Chem.*, **38,** 813
100. Cobb, J. C. (1967). *Anal. Chem.*, **39,** 127
101. Gordon, G. E., Randle, K., Goles, G. G., Corliss, J. B., Beeson, M. H. and Oxley, S. S. (1968). *Geochim. Cosmochim. Acta*, **32,** 169
102. Gordon, G. E., Dran, J. C., Baedecker, P. A. and Anderson, C. F. L. (1969). *Proc. Symp. Gaithersburg, Modern Trends in Activation Analysis*, (J. De Voe, editor) NBS Spec. Publ. 312, 399 (Washington, D.C.: NBS)
103. Frána, J., Vobecký, M., Řanda, Z., Benada, J. and Kuncíř, J. (1971). *Radiochem. Radioanal. Lett.*, **8,** 97
104. Goles, G. G., Randle, K., Osawa, M., Schmitt, R. A., Wakita, H., Ehmann, W. D. and Morgan, J. W. (1970). *Proc. Apollo 11 Lunar Sci. Conf.*, **2,** 1165
105. Goles, G. G., Duncan, A., Lindstrom, D., Martin, M., Beyer, R., Osawa, M., Randle, K., Meek, L., Steinborn, T. and McKay, S. (1971). *Proc. 2nd Lunar Sci. Conf., Cambridge, Mass.*, **2,** 1063

106. Jérome, D. Y., Philippot, J. C. and Brichet, E. (1972). *Earth Planet. Sci. Lett.*, **13**, 436
107. Jérome, D. Y. and Philippot, J. C. (1973). *Geochim. Cosmochim. Acta*, **37**, 909
108. Kharkar, D. P. and Turekian, K. K. (1971). *Proc. 2nd Lunar Sci. Conf., Cambridge, Mass.*, *2*, 1301
109. Kunciř, J., Benada, J., Řanda, Z. and Vobecký, M. (1970). *J. Radioanal. Chem.*, **5**, 369
110. Melsom, S. (1970). *J. Radioanal. Chem.*, **4**, 355
111. Řanda, Z., Benada, J., Kunciř, J., Vobecký, M. and Frána, J. (1972). *J. Radioanal. Chem.*, **11**, 305
112. Schock, H. H. (1973). *Z. Anal. Chem.*, **263**, 100
113. Smales, A. A., Mapper, D., Webb, M. S. W., Webster, R. K., Wilson, J. D. and Hislop, J. S. (1971). *Proc. 2nd Lunar Sci. Conf., Cambridge, Mass.*, **2**, 1253
114. Vobecký, M., Frána, J., Řanda, Z., Benada, J., and Kunciř, J. (1971). *Radiochem. Radioanal. Lett.*, **6**, 237
115. Copeland, R. A., Frey, F. A. and Wones, D. R. (1971). *Earth Planet. Sci. Lett.*, **10**, 186
116. Craft, J. F. (1971). *Radiochem. Radioanal. Lett.*, **8**, 199
117. Das, H. A., De Koning, J. and Oosterom, M. G. (1971). *Radiochem. Radioanal. Lett.*, **6**, 203
118. Zaieev, E. I., Dnepronskii, I. S., Sotskov, Y. P., and Laptev, V. G. (1970). *Zh. Anal. Khim.*, **25**, 1922
119. Hertogen, J. and Gijbels, R. (1971). *Anal. Chim. Acta*, **56**, 61
120. Steinnes, E. (1971). *Anal. Chim. Acta*, **57**, 451
121. Das, H. A., Zonderhuis, J. and Rose, R. (1972). *J. Radioanal. Chem.*, **11**, 273
122. De Lange, P. W., Venter, J. H. and De Wet, W. J. (1969). *J. Radioanal. Chem.*, **2**, 219
123. Turkstra, J., Pretorius, P. J. and De Wet, W. J. (1970). *Anal. Chem.*, **42**, 835
124. Schroeder, G. L., Evans, R. D. and Ragaini, R. C. (1966). *Anal. Chem.*, **38**, 432
125. Vobecký, M., Řanda, Z., Benada, J. and Kunciř, J. (1970). *Česk. Akad. Ved. Res. Ustavu Jaderneho, Vyzkumu*, UJV-2319-CH
126. De Winter, J. (1973). *Thesis*, University of Ghent, Belgium
127. Vobecký, M., Frána, J., Bauer, J. Řanda, Z., Benada, J. and Kunciř, J. (1971). *Proc. 2nd Lunar Sci. Conf., Cambridge, Mass.*, **2**, 1291
128. Wölfle, R. and Herpers, U. (1970). *Mikrochim. Acta*, 933
129. Ehmann, W. D., McKown, D. M. and Morgan, J. W. (1971). *Activation Analysis in Geochemistry and Cosmochemistry* (A. O. Brunfelt and E. Steinnes, editors), 167 (Oslo: Universitetsforlaget)
130. Morrison, G. H., Gerard, J. T., Travesi, A., Currie, R. L., Peterson, S. F. and Potter, N. M. (1969). *Anal. Chem.*, **41**, 1633
131. Morrison, G. H., Gerard, J. T., Kashuba, A. T., Gangadharam, E. V., Rothenberg, A. M., Potter, N. M. and Miller, G. B. (1970). *Proc. Apollo 11 Lunar Sci. Conf.*, **2**, 1383
132. Morrison, G. H., Gerard, J. T., Potter, N. M., Gangadharam, E. V., Rothenberg, A. M. and Burdo, R. A. (1971). *Proc. 2nd Lunar Sci. Conf., Cambridge, Mass.*, **2**, 1169
133. Travesi, A., Palomares, J. and Adrada, J. (1971). Ref. 132, p. 1277
134. Laul, J. C., Case, D. R., Wechter, M., Schmidt-Bleek, F. and Lipschutz, M. E. (1970). *J. Radioanal. Chem.*, **4**, 241
135. Case, D. R., Laul, J. C., Lipschutz, M. E. and Schmidt-Bleek, F. (1969). *Radiochem. Radioanal. Lett.*, **1**, 143
136. Allen, R. O., Haskin, L. A., Anderson, M. R. and Muller, O. (1970). *J. Radioanal. Chem.*, **6**, 115
137. Graber, F. M., Lukens, H. R. and McKenzie, J. K. (1970). *J. Radioanal. Chem.*, **4**, 229
138. Denechaud, E. B., Helmke, P. A. and Haskin, L. A. (1970). *J. Radioanal. Chem.*, **6**, 97
139. Haskin, L. A., Helmke, P. A., Allen, R. D., Anderson, M. R., Korotev, R. L. and Zweifel, K. A. (1971). *Proc. 2nd Lunar Sci. Conf., Cambridge, Mass.*, **2**, 1307
140. Haskin, L. A., Allen, R. O., Helmke, P. A., Paster, T. P., Anderson, M. R., Korotev, R. L. and Zweifel, K. A. (1970). *Proc. Apollo 11 Lunar Sci. Conf.*, **2**, 1213
141. Potts, M. J., Early, T. O. and Herrmann, A. G. (1973). *Z. Anal. Chem.*, **263**, 97

142. Brunfelt, A. O. and Steinnes, E. (1973). *J. Radioanal. Chem.*, **13,** 11
143. Steinnes, E. (1971). *Activation Analysis in Geochemistry and Cosmochemistry*, (A. O. Brunfelt and E. Steinnes, editors) 113 (Oslo: Universitetsforlaget)
144. Brune, D. and Jirlow, K. (1964). *Nukleonik*, **6,** 242
145. Van der Linden, R. (1974). *Thesis*, University of Ghent, Belgium
146. Van der Linden, R., De Corte, F. and Hoste, J. (1973). In *Nuclear Data in Science and Technology*, Vol. 2, 241 (Vienna: IAEA)
147. Van der Linden, R., De Corte, F., Van den Winkel, P. and Hoste, J. (1972). *J. Radioanal. Chem.*, **11,** 133
148. Whitley, J. E., Bowden, P. and Angus, N. S. (1971). *Activation Analysis in Geochemistry and Cosmochemistry* (A. O. Brunfelt and E. Steinnes, editors) 55 (Oslo: Universitetsforlaget)
149. Henckelmann, R. (1971). *Radiochim. Acta*, **15,** 169
150. Henckelmann, R. and Born, H. J. (1973). *J. Radioanal. Chem.*, **16,** 473
151. Sine, N. M., Taylor, W. O., Webber, R. R. and Lewis, C. L. (1969). *Geochim. Cosmochim. Acta*, **33,** 121
152. Amiel, S. (1962). *Anal. Chem.*, **34,** 1683
153. Dyer, F. F., Emery, J. F. and Leddicotte, G. W. (1962). *USAEC Report ORNL-3342*
154. Gale, N. H. (1967). *Radioactive Dating and Methods of Low-level Counting*, 431 (Vienna: IAEA)
155. Bertine, K. K., Chan, L. H. and Turekian, K. K. (1970). *Geochim. Cosmochim. Acta*, **34,** 641
156. Carpenter, B. S. (1972). *Anal. Chem.*, **44,** 600
157. Fisher, D. E. (1970). *Anal. Chem.*, **42,** 414
158. Fisher, D. E. (1970). *Geochim. Cosmochim. Acta*, **34,** 630
159. Murali, V., Parekh, P. P. and Das Sankar, M. (1970). *Anal. Chim. Acta*, **50,** 71
160. Griffith, R. V. (1973). 'Results of the 1972 Survey on Track Registration', *USAEC Report UCRL-51362*
161. Cross, W. G. and Tommasino, L. (1969). *Proc. Symp. Clermont-Ferrand, Nuclear Track Registration in Insulating Solids and Applications*, **1,** 73
162. Johnson, D. R., Bagett, G. H. and Becker, K. (1970). *Health Physics*, **18,** 424
163. Agard, E. T. (1971). *Thesis*, Univ. of Toronto
164. Fleischer, R. L. and Price, P. B. (1964). *J. Geophys. Res.*, **69,** 331
165. Storzer, D. (1970). *Earth Planet. Sci. Lett.*, **8,** 55
166. Brunfelt, A. O. and Steinnes, E. (1971). *Talanta*, **18,** 1197
167. Brunfelt, A. O., Heier, K. S. and Steinnes, E. (1971). *Proc. 2nd Lunar Sci. Conf., Cambridge, Mass.*, **2,** 1281
168. Schmitt, R. A., Linn, T. A., Jr. and Wakita, H. (1970). *Radiochim. Acta*, **13,** 200
169. Wakita, H. and Schmitt, R. A. (1971). *Proc. 2nd Lunar Sci. Conf., Cambridge, Mass.*, **2,** 1231
170. Wakita, H., Rey, P. and Schmitt, R. A. (1971). Ref. 169, p. 1319
171. Parekh, P. P., Heimann, M. and Herr, W. (1972). *Radiochem. Radioanal. Lett.*, **12,** 153
172. Wänke, H., Wlotzka, F., Baddenhausen, H., Balacescu, A., Spettel, B., Teschke, F., Jagoutz, E., Kruse, H., Quijano-Rico, M. and Rieder, R. (1971). *Proc. 2nd Lunar Sci Conf., Cambridge, Mass.*, **2,** 1187
173. Filby, R. H., Haller, W. A. and Shah, K. R. (1970). *J. Radioanal. Chem.*, **5,** 277
174. Filby, R. H. and Haller, W. A. (1969). *Proc. Symp. Gaithersburg, Modern Trends in Activation Analysis* (J. De Voe, editor), NBS Spec. Publ. 312, 339 (Washington, D.C.: NBS)
175. Griffon, H. and Barbaud, J. (1951). *Ann. Pharm. Fr.*, **IX,** 35
176. Gordus, A. A. (1973). *J. Radioanal. Chem.*, **15,** 229
177. Cornelis, R. (1973). *J. Radioanal. Chem.*, **15,** 305
178. *General Atomic Reports 3664, 5566, 6152, 7041, 8013, 9807, 9822, 9829, 10276*
179. Perkons, A. K. and Jervis, R. E. (1966). *J. Forensic Sci.*, **11,** 50
180. Coleman, R. F., Cripps, F. H. and Stimson, A. (1965). *AWRE Report No. 0-78/65* (Aldermaston, UK: AWRE)
181. Coleman, R. F., Cripps, F. H., Stimson, A. and Scott, H. D. (1966). *AWRE Report No. 0-86/66* (Aldermaston, UK: AWRE)
182. Parker, J. B. (1966). *J. Forensic Sci. Soc.*, **6,** 33
183. Parker, J. B. (1967). *J. Forensic Sci. Soc.*, **7,** 134

184. Forslev, A. W. (1966). *J. Forensic Sci.*, **11**, 217
185. Quittner, P., Szabó, E., Perneczki, G. and Major, A. (1970). *J. Radioanal. Chem.*, **5**, 133
186. Lima, F. W., Shibata, H. and Atalla, L. T. (1964). *Proc. Symp. Radiochem. Methods of Analysis*, **I**, 119 (Vienna: IAEA)
187. Lima, F. W. (1966). *Proc. 1st Forensic Activation Analysis Conf., Texas* (*General Atomic Report, 8171*, 261)
188. Bate, L. C. (1965). *J. Forensic Sci.*, **10**, 60
189. Bate, L. C. and Dyer, F. F. (1966). *Proc. 1st Forensic Activation Analysis Conf., Texas* (*General Atomic Report, 8171*, 247)
190. Van den Berg, A. J., De Bruin, M. and Houtman, J. P. W. (1967). *Proc. Symp. Nucl. Activation Techn. Life Sci.* (*Bled*), 678 (Vienna: IAEA)
191. Van den Berg, A. J., De Goeij, J. J. M., Houtman, J. P. W. and Zegers, C. (1968). *Proc. Symp. Gaithersburg, Modern Trends in Activation Analysis*, NBS Spec. Publ. 312, 272 (Washington, D.C.: NBS (1969))
192. Cornelis, R. and Speecke, A. (1971). *J. Forensic Sci. Soc.*, **11**, 29
193. Obrusnik, I., Gislason, J., McMillan, D. K., D'Auria, J. and Potte, B. D. (1972). *J. Forensic Sci.*, **17**, 426
194. Obrusnik, I., Gislason, J., Maes, D., McMillan, D. K., D'Aura, J. and Potte, B. D. (1973). *J. Radioanal. Chem.*, **15**, 115
195. Cornelis, R. (1972). *Medicine, Science and the Law*, **12**, 188
196. Bush, H. D., Butterworth, A., Pearson, E. F. and Pounds, C. A. (1973). *J. Radioanal. Chem.*, **15**, 245
197. Vogt, J. R. and Hashii, D. M. (1973). *J. Radioanal. Chem.*, **15**, 355
198. Adams, F. and Hoste, J. (1973). *Forensic Sci.*, **2**, 29
199. Coleman, R. F. and Goode, G. C. (1973). *J. Radioanal. Chem.*, **15**, 367
200. Coleman, R. F. and Weston, N. T. (1968). *J. Forensic Sci. Soc.*, **8**, 32
201. Pro, M. J., Schlesinger, H. L. and Hoffman, C. M. (1966). *Proc. 1st Forensic Activation Analysis Conf., Texas* (*General Atomic Report, 8171*, 103)
202. Lima, F. W., Silva, C. M. and Guimarães, R., Jr. (1973). *J. Radioanal. Chem.*, **15**, 157
203. Jervis, R. E. (1967). *Proc. Symp. Nucl. Activation Techn. Life Sci.* (Amsterdam), SM91/67 (Vienna: IAEA)
204. Bate, L. C. (1973). *J. Radioanal. Chem.*, **15**, 193
205. Schlesinger, H. L. and Settle, D. M. (1971). *J. Forensic Sci.*, **16**, 309
206. Guy, R. D. and Pate, B. D. (1973). *J. Forensic Sci.*, **18**, 93; *J. Radioanal. Chem.*, **15**, 135
207. Lukens, H. R. and Guinn, V. P. (1971). *J. Forensic Sci.*, **16**, 301
208. Rötzscher, K., Mende, S., Flachowsky, J., Geisler, M. and Wehran, H. J. (1973). *J. Radioanal. Chem.*, **15**, 317
209. Weaver, J. N. and Bouman, W. B. (1971). *J. Forensic Sci.*, **16**, 521
210. Chan, R. K. H. (1972). *J. Forensic Sci.*, **16**, 93
211. Lyon, W. S., Ricci, E., Bate, L. C. and Dyer, F. F. (1973). *J. Radioanal. Chem.*, **15**, 399
212. Wyttenbach, A., Bajo, S. and Hug, E. (1973). *J. Radioanal. Chem.*, **15**, 9
213. Smith, H. (1973). *J. Radioanal. Chem.*, **15**, 71
214. Smith, H. (1964). *J. Forensic Sci. Soc.*, **4**, 192
215. Rodger, W. J. and Smith, H. (1967). *J. Forensic Sci. Soc.*, **7**, 86
216. Lenihan, J. M. A., Smith, H. and Harvey, W. (1972). *Proc. Symp. Nucl. Activation. Techn. Life Sci.* (*Bled*), SM157/19 (Vienna: IAEA)
217. Geisman, J. R., Carey, W. E., Gould, W. A. and Alban, E. K. (1969). *J. Food Sci.*, **34**, 295
218. Borra, G., Brandone, A., De Bernardi, G., Fornari, A., Orvini, E. and Rollier, M. A. (1973). *J. Radioanal. Chem.*, **15**, 51
219. Scott, H. D., Coleman, R. F. and Cripps, F. H. (1966). *AWRE Report No. 0-5/66* (Aldermaston, UK: AWRE)
220. Albu-Jaron, A. and Amiel, S. (1972). *J. Radioanal. Chem.*, **11**, 123
221. Pillay, K. K. S., Jester, W. A. and Fox, H. A., III (1973). *166th Nat. Meeting ACS, Chicago, Illinois*
222. Rudzitis, E., Kopina, M. and Wahlgren, M. (1973). *J. Forensic Sci.*, **18**, 93

223. Cowan, M. E., Purdon, P. L., Hoffman, C. M., Brunelle, R., Gerber, S. R. and Pro, M. (1973). *J. Radioanal. Chem.*, **15**, 203
224. Cornelis, R. and Timperman, J. (1974). *Medicine, Science and the Law*, **14**, 98
225. Gislason, J. and Pate, B. D. (1973). *J. Radioanal. Chem.*, **15**, 103
226. Schmitz, J. (1973). *J. Radioanal. Chem.*, **15**, 219
227. Krishnan, S. S. (1967). *J. Forensic Sci.*, **12**, 112
228. Krishnan, S. S. (1967). *J. Forensic Sci.*, **12**, 471
229. Krishnan, S. S. and Nichol, R. C. (1968). *J. Forensic Sci.*, **13**, 519
230. Krishnan, S. S., Gillespie, K. A. and Anderson, E. J. (1971). *J. Forensic Sci.*, **16**, 144
231. Menke, H., Leszczynski, Ch. and Weber, M. (1973). *J. Radioanal. Chem.*, **15**, 27
232. Pillay, K. K. S., Thomas, C. C., Jr., Hart, D. M., Didising, D. and Thomas, R. C. (1970). *Nucl. Appl. Techn.*, **8**, 73
233. Van Dalen, A., Das, H. A. and Zonderhuis, J. (1973). *J. Radioanal. Chem.*, **15**, 143
234. Houtman, J. P. W. and Turkstra, J. (1964). *Proc. Symp. Salzburg, Radiochemical Methods of Analysis* (Vienna: IAEA (1965))
235. Lux, F., Braunstein, L. and Strauss, R. (1968). *Proc. Symp. Gaithersburg, Modern Trends in Activation Analysis*, NBS Spec. Publ. 312, 216 (Washington, D.C.: NBS (1969))
236. Olin, J. S. and Sayre, E. V. (1968). *Proc. Symp. Gaithersburg, Modern Trends in Activation Analysis*, NBS Spec. Publ. 312, 246 (Washington, D.C.: NBS (1969))
237. Cotter, M. J., Meyers, P., Van Zelst, L. and Sayre, E. V. (1973). *J. Radioanal. Chem.*, **15**, 265
238. Karjala, D. S. (1971). *Calif. Law Rev.*, **59**, 997
239. Damsgaard, E. and Heydorn, K. (1972). *Proc. Symp. Nucl. Activation Techn. Life Sci. (Amsterdam)*, SM157/4 (Vienna: IAEA)
240. Sandstead, H. H. (1970) 'Uses of Activation Analysis in Studies of Mineral Element Metabolism in Man', *IAEA Rep. 122*, 51 (Vienna: IAEA)
241. Masironi, R. (1970). Ref. 240, p. 109
242. Reinhold, R. (1970). Ref. 240, p. 81
243. Kmet, J. (1970). Ref. 240, p. 131
244. Schroeder, H. A. and Nason, A. P. (1971). *Clin. Chem.*, **17**, 461
245. Schwarz, K. (1972). *Proc. Symp. Nucl. Activation Techn. Life Sci. (Bled)*, 3 (Vienna: IAEA)
246. Schiller, P., Cook, J. B., Kitzinger, A. and Woelfl, E. (1972). *Analyst*, **97**, 601
247. Whitley, J. E. and Muir, C. (1971). *Radiochem. Radioanal. Lett.*, **8**, 53
248. Swanson, A. A. and Truesdale, A. W. (1971). *Biochem. Biophys. Res. Commun.*, **45**, 1488
249. Fite, L. E., Wainerdi, R. E., Harrison, G. M. and Dogett, R. (1972). *Proc. Symp. Nucl. Activation Techn. Life Sci. (Bled)*, 488 (Vienna: IAEA)
250. Hislop, J. S. and Williams, D. R. (1972). *Analyst*, **97**, 78
251. Chattapadhyay, A. and Jervis, R. E. (1972). *Radiochem. Radioanal. Lett.*, **11**, 331
252. van der Mark, W. and Das, H. A. (1973). *J. Radioanal. Chem.*, **13**, 107
253. Ohno, S., Suzuki, M., Sasajima, K. and Iwata, S. (1970). *Analyst.* **95**, 210
254. Zurlo, N., Griffini, A. M. and Colombo, G. (1969). *Anal. Chim. Acta*, **47**, 203
255. Amos, M. D., Benett, P. A., Brodie, K. G., Lung, P. W. Y. and Matoušek, J. P. (1971). *Anal. Chem.*, **43**, 211
256. Hwang, J. Y., Ullucci, P. A., Smith, S. B., Jr. and Malenfant, A. L. (1971). *Anal. Chem.*, **43**, 1319
257. Hwang, J. Y., Ullucci, P. A. and Mokeler, C. J. (1973). *Anal. Chem.*, **45**, 795
258. Rose, G. A. and Wilden, E. G. (1973). *Analyst*, **98**, 243
259. Fry, B. W. and Taves, D. R. (1970). *J. Lab. Clin. Med.*, **75**, 1020
260. Tusl, J. (1970). *Clin. Chim. Acta*, **27**, 216
261. Neefus, J. D., Cholak, J. and Saltzman, B. E. (1970). *Amer. Ind. Hyg. Ass. J.*, **31**, 96
262. Versieck, J. M. J. and Speecke, A. B. H. (1972). *Proc. Symp. Nucl. Activation Techn. Life Sci. (Bled)*, 39 (Vienna: IAEA)
263. Versieck, J., Speecke, A., Hoste, J. and Barbier, F. (1973). *Clin. Chem.*, **19**, 472
264. Versieck, J., Speecke, A., Hoste, J. and Barbier, F. (1973). *Z. Klin. Chem. Klin. Biochem.*, **11**, 193
265. Lux, F. and Zeisler, R. (1972). *Z. Anal. Chem.*, **259**, 216
266. Lux, F. and Zeisler, R. (1972). *Z. Anal. Chem.*, **261**, 314

267. Coleman, R. F., Herrington, J. and Scales, J. T. (1973). *Brit. Med. J.*, **1,** 527
268. Gibbons, D., Hislop, J. S., Parker, A. and Webb, M. S. W. (1973). *Proc. 3rd Symp. Recent Develop. Neutron Activation Anal., Cambridge, UK*
269. Razeghi, M. and Paesa, B. (1973). *Radiochem. Radioanal. Lett.*, **13,** 95
270. Struempler, A. (1973). *Anal. Chem.*, **45,** 2251
271. Rottschafer, J. M., Boczkowski, R. J. and Mark, H. B., Jr. (1972). *Talanta*, **19,** 163
272. Levstek, M., Kosta, L., Dermely, M. and Byrne, A. R. (1972). *Proc. Symp. Nucl. Activation Techn. Life Sci. (Bled)*, 111 (Vienna: IAEA)
273. Damsgaard, E., Heydorn, K. and Rietz, B. (1972). Ref. 272, p. 119
274. Dubinskaya, N. A. and Pelekis, L. L. (1971). *J. Radioanal. Chem.*, **9,** 61
275. Van der Linden, R., De Corte, F. and Hoste, J. (1974). *Anal. Chim. Acta*, **71,** 263
276. Linekin, D. M. (1973). *JARI*, **24,** 343
277. Cornelis, R., Speecke, A. and Hoste, J. (1973). *Anal. Chim. Acta*, **68,** 1
278. Anderson, D. H., Morphy, J. J. and White, W. W. (1972). *Anal. Chem.*, **44,** 2094
279. Bowen, H. J. M. (1967). *Analyst*, **92,** 124
280. Bowen, H. J. M. (1969). *Advances in Activation Analysis* (J. M. A. Lenihan and S. J. Thompson, editors), 101 (London and New York: Academic Press)
281. Bowen, H. J. M. (1974). *J. Radioanal Chem.*, **19,** 215
282. Schramel, P. (1969). *J. Radioanal. Chem.*, **3,** 29
283. Fritze, K. and Robertson, R. (1971). *J. Radioanal. Chem.*, **7,** 213
284. Nadkarni, R. A. and Morrison, G. H. (1973). *Anal. Chem.*, **45,** 1957
285. (1971). *Certificate of Analysis SRM 1571* (Washington, D.C.: NBS)
286. Damsgaard, E. and Heydorn, K. (1973). *Proc. 3rd Symp. Recent Develop. Neutron Activation Anal., Cambridge, UK*
287. Heydorn, K. and Damsgaard, E. (1973). *Talanta*, **20,** 1
288. Morrison, G. H. and Potter, N. M. (1972). *Anal. Chem.*, **44,** 839
289. Brune, D., Düring, L., Larsson, B. and Lundqvist, H. (1973). *JARI*, **24,** 1
290. Demortier, G. F. A., Gillet, C. and Lefebvre, J. (1972). *Proc. Symp. Nucl. Activation Techn. Life Sci. (Bled)*, 23 (Vienna: IAEA)
291. Bornemisza-Pauspertl, P. (1973). *J. Radioanal. Chem.*, **14,** 189
292. Girardi, F. and Sabbioni, E. (1968). *J. Radioanal. Chem.*, **1,** 169
293. Ralston, H. K. and Sato, E. S. (1971). *Anal. Chem.*, **43,** 129
294. Török, G. and Diehl, J. F. (1971). *Radiochim. Acta*, **15,** 96
295. Török, G., Schelenz, R., Fischer, E. and Diehl, J. F. (1973). *Z. Anal. Chem.*, **263,** 110
296. Sabbioni, E., Pietra, R. and Girardi, F. (1970). *J. Radioanal. Chem.*, **4,** 289
297. Massart, D. L. (1970). *J. Radioanal. Chem.*, **4,** 265
298. Heurtebise, M. and Ross, W. J. (1972). *Anal. Chem.*, **44,** 596
299. Edgington, D. N. and Lucas, F. H., Jr. (1970). *J. Radioanal. Chem.*, **5,** 233
300. Plantin, L. O. (1972). *Proc. Symp. Nucl. Activation Techn. Life Sci. (Bled)*, 73 (Vienna: IAEA)
301. Schumacher, J. and Maier-Borst, W. (1972). *Kerntechnik*, **14,** 165
302. Tölg, G. (1972). *Talanta*, **19,** 1489
303. Samsahl, K. (1972). *Sci. Total Environ.*, **1,** 45
304. Samsahl, K. (1968), *Analyst*, **93,** 101
305. Thomas, A. D. and Smythe, L. E. (1973). *Talanta*, **20,** 469
306. Netsky, M. G., Harrison, W. W., Brown, M. and Benson, C. (1969). *Amer. J. Clin. Path.*, **51,** 358
307. La Fleur, P. D. (1973). *Anal. Chem.*, **45,** 1534
308. Weber, D. A. and Andrews, M. L. (1972). *J. Nucl. Med.*, **13,** 293
309. Boddy, K., East, B. W. and Robertson, I. (1970). *JARI*, **21,** 500
310. Cohn, S. H., Shukla, K. K., Dombrowski, C. S. and Fairchild, R. C. (1972). *Proc. Symp. Nucl. Activation Techn. Life Sci. (Amsterdam)*, 605 (Vienna: IAEA)
311. Boddy, K. and Glaros, D. (1973). *JARI*, **24,** 179
312. Battye, C. K., Knight, V., Marshall, T. O., Knight, A. and Godfrey, B. E. (1971). *Nucl. Instr. Meth.*, **92,** 595, 601
313. Boddy, K., Holloway, I. and Elliott, A. (1973). *JARI*, **24,** 428
314. Chamberlain, M. I., Fremlin, J. H., Holloway, I. and Peters, D. K. (1970). *JARI*, **21,** 725
315. Ettinger, K. V. (1973). *Proc. 3rd Symp. Recent Develop. Neutron Activation Anal., Cambridge, UK*

316. Al-Hiti, K., Al-Tikriti, A. and Ettinger, K. V. (1973). *Proc. 3rd Symp. Recent Develop. Neutron Activation Anal., Cambridge, UK*
317. Cohn, S. H. and Dombrowski, C. S. (1971). *IEEE Trans. Nucl. Sci.*, **NS-18,** 73
318. Chettle, D. R., Harvey, T. C. and Fremlin, J. H. (1973). *Proc. 3rd Symp. Recent Develop. Neutron Activation Anal., Cambridge, UK*
319. Anderson, J., Battye, C. K., Osborn, S. B., Tomlinson, R. W. S., Fry, F. A. and Newton, D. (1972). *Proc. Symp. Nucl. Activation Techn. Life Sci.* (*Bled*), 571 (Vienna: IAEA)
320. Nelp, W. B., Denney, J. D., Murano, R., Hinn, G. M. and Chestnut, C. H. (1972). Ref. 319, p. 627
321. Hoch, F. L., Kinas, R. A. and Jones, J. D. (1971). *Anal. Biochem.*, **40,** 86
322. Cornelis, R. and Hoste, J. (1973). *J. Radioanal. Chem.*, **13,** 419
323. Leu, M. L., Strickland, G. T., Bechner, W. M., Chen, T. S. M., Wang, C. C. and Yeh, S. I. (1970). *J. Lab. Clin. Med.*, **76,** 432
324. Henke, G., Möllmann, H. and Althoff, W. (1971). *Klin. Wschr.*, **49,** 284
325. Battistone, G. C., Lewri, E. and Lofberg, R. (1970). *Clin. Chim. Acta*, **30,** 429
326. Schicha, H., Feinendegen, L. E., Kasperek, K., Klein, H. J. and Siller, V. (1970). *Beitr. Path. Bd.*, **141,** 227; (1972). **146,** 55
327. Cotzias, G. C., Miller, S. T. and Edwards, J. (1966). *J. Lab. Clin. Med.*, **67,** 831
328. Fernandez, A. A., Sobel, C. and Jacobs, J. L. (1963). *Anal. Chem.*, **35,** 1721
329. Heydorn, K. and Nørgärd, K. (1973). *Talanta*, **20,** 835
330. Speecke, A. (1973). *Proc. 3rd Symp. Recent Develop. Neutron Activation Anal., Cambridge, UK*
331. Liebscher, K. and Smith, H. (1968). *Arch. Environ. Health*, **17,** 881; Lenihan, J. M. A. and Smith, H. (1967). *Proc. Symp. Nucl. Activation Anal. Techn. Life Sci.* (*Amsterdam*), 601 (Vienna: IAEA)
332. Hagenfeldt, R., Plantin, L. O. and Diczfalusy, E. (1973). *Acta Endocrinol.*, **72,** 115
333. Versieck, J., Speecke, A., Hoste, J. and Barbier, F. (1974). *Acta Endocrinol.*, **76,** 783
334. De Rudder, J. and Speecke, A. (1973). Unpublished results
335. Christian, G. D. (1969). *Anal. Chem.*, **41,** 24A
336. Hansen, L. C., Scribner, W. G., Gilbert, W. T. and Sievens, R. E. (1972). *Anal. Chem.*, **43,** 349
337. Urone, P. F. and Anders, H. K. (1950). *Anal. Chem.*, **22,** 1317
338. Koch, H. J., Smith, R. E., Shimp, N. F. and Connor, J. (1956). *Cancer*, **9,** 499
339. Monacelli, R., Tanaka, H. and Yoe, J. H. (1956). *Clin. Chim. Acta*, **1,** 577
340. Miller, D. O. and Yoe, J. H. (1959) *Clin. Chim. Acta*, **4,** 378
341. Paixao, L. M. and Yoe, J. H. (1959). *Clin. Chim. Acta*, **4,** 507
342. Herring, W. B., Leavell, B. S., Paixao, L. M. and Yoe, J. H. (1960). *Amer. J. Clin. Nutrit.*, **8,** 846
343. Gofman, J. W., de Lalla, O., Johnson, G., Krich, E. L., Lowe, O., Martin, W., Piluso, D. L., Tandy, R. K., Upham, F., Weitzel, R. and Wilbur, D. (1961). *USAEC Report UCRL-9897*
344. Schroeder, H. A., Balassa, J. J. and Tipton, I. H. (1962). *J. Chron. Dis.*, **15,** 941
345. Bowen, H. J. M. (1964). *Analyst*, **89,** 658
346. Morrison, G. H. (1967). *Proc. Symp. Nucl. Activation Anal. Techn. Life Sci.* (*Amsterdam*), 211 (Vienna: IAEA)
347. Van Kooten, W. J., Mali, J. W. H., De Goeij, J. J. M. and Houtman, J. P. W. (1967). Ref. 346, p. 567
348. Feldman, F. J., Knoblock, E. C. and Purdy, W. C. (1967). *Anal. Chim. Acta*, **38,** 489
349. Mertz, D. P., Koschnick, R., Wilk, G. and Pfeilsticker, K. (1968). *Z. Klin. Chem. Klin. Biochem.*, **6,** 171
350. Fritze, K. and Robertson, R. (1968). *J. Radioanal. Chem.*, **1,** 463
351. (1969). *Internist Observer*, **7,** 4
352. Meloni, S., Brandone, A. and Maxia, V. (1969). *JARI*, **20,** 757
353. Behne, D. (1969). *J. Radioanal. Chem.*, **3,** 17
354. Savory, J., Mushak, P., Sunderman, F. W., Estes, R. H. and Roszel, N. O. (1970). *Anal. Chem.*, **42,** 294
355. Welz, B. and Wiedering, E. (1970). *Z. Anal. Chem.*, **252,** 111
356. Hambidge, K. M. (1971). *Anal. Chem.*, **43,** 103
357. Niedermeier, W., Griggs, J. H. and Johnson, R. S. (1971). *Appl. Spectrosc.*, **25,** 53

358. (J. R. Vogt, editor) (1971). *University of Missouri, Nucl. Sci. Group Tech. Report*, 74
359. Savory, J., Glenn, M. T. and Ahlstrom, J. A. (1972). *J. Chromatogr. Sci.*, **10**, 247
360. Kasperek, K., Schicha, H., Siller, V., Feinendegen, L. E. and Höch, A. (1972). *Proc. Symp. Nucl. Activation Techn. Life Sci. (Bled)*, 517 (Vienna: IAEA)
361. Behne, D. and Diehl, F. (1972). Ref. 360, p. 407
362. Wolf, W. R., Taylor, M. L., Hughes, B. M., Tiernan, T. D. and Sievers, R. E. (1972). *Anal. Chem.*, **44**, 616
363. Maxia, V., Meloni, S., Rollier, M. A., Brandone, A., Patwardhan, V. N., Waslien, C. I. and El Shami, S. (1972). *Proc. Symp. Nucl. Activation Techn. Life Sci. (Bled)*, 527 (Vienna: IAEA)
364. Agterdenbos, J., Van Broeckhoven, L., Jütte, B. A. H. G. and Schuring, J. (1972). *Talanta*, **19**, 341
365. Davidson, I. W. F. and Secrest, W. L. (1972). *Anal. Chem.*, **44**, 1809
366. Tessari, G. and Torsi, G. (1972). *Talanta*, **19**, 1059
367. Davidson, I. W. F. and Burt, R. L. (1973). *Amer. J. Obstet. Gynecol.*, **116**, 607
368. Versieck, J., Speecke, A., Hoste, J. and Barbier, F. (1974). *Clin. Chem.*, **20**, 1141
369. Van Grieken, R. and Hoste, J. (1972). *Annotated Bibliography on 14 MeV Neutron Activation Analysis*, Eurisotop Office Inform. Booklet 65 (Brussels: Eurisotop Office)
370. Adams, F., Op de Beeck, J., Van den Winkel, P., Gijbels, R., De Soete, D. and Hoste, J. (1971). *CRC Crit. Rev. Anal. Chem.*, **1**, 455
371. Huysmans, K., Gijbels, R. and Hoste, J. (1973). *Talanta*, **20**, 843
372. Lievens, P., Speecke, A. and Hoste, J. (1974). *Anal. Chim. Acta*, **67**, 269
373. Vandecasteele, C., Speecke, A. and Hoste, J. (1972). *The Determination of Oxygen in Non-Ferrous Metals by 14-MeV Neutron Activation Analysis*, Eurisotop Office Inform. Booklet 68 (Brussels: Eurisotop Office)
374. Gijbels, R., Speecke, A. and Hoste, J. (1968). *Anal. Chim. Acta*, **43**, 183
375. Vandecasteele, C., Van Grieken, R., Gijbels, R. and Speecke, A. (1973). *Anal. Chim. Acta*, **65**, 1
376. Nikolaenko, O. and Shtan, A. (1969). *Radiation Technology*, 175 (Moscow: Atomizdat)
377. Nargolwalla, S., Crambes, M. and De Voe, J. (1968). *Anal. Chem.*, **40**, 667
378. Vandecasteele, C., Van Grieken, R., Gijbels, R. and Speecke, A. (1973). *Anal. Chim. Acta*, **64**, 187
379. Nargolwalla, S., Crambes, M. and Suddueth, J. (1970). *Anal. Chim. Acta*, **49**, 425
380. Leal Horta, J. and Dos Santos, A. (1973). *Anal. Chim. Acta*, **63**, 359
381. Brune, D. and Jirlow, K. (1969). *J. Radioanal. Chem.*, **2**, 49
382. Storm, E. and Israel, H. (1967). 'Photon Cross-sections from 0.001 to 100 MeV for Elements 1 through 100,' Report LA-3753
383. Van Grieken, R., Gijbels, R., Speecke, A. and Hoste, J. (1968). *Anal. Chim. Acta*, **43**, 199
384. Avery, A., Bendall, D., Butler, Y. and Spinney, K. (1960). 'Methods of Calculation for Use in the Design of Shields for Power Reactors', *AERE Report R-3216* (Harwell, UK: AERE)
385. Allen, R., Carter, R. and Taylor, H. (1963). *Fast Neutron Physics*, (J. Marion and J. Fowler, editors) Part II, 1429 (New York: Interscience)
386. Hughes, D. and Schwartz, R. (1958). 'Neutron Cross-Sections', *USAEC Report BNL-325*, 2nd Ed.
387. Evans, R. (1967). *The Atomic Nucleus*, 628 (New York: McGraw-Hill)
388. Van Grieken, R., Speecke, A. and Hoste, J. (1972). *J. Radioanal. Chem.*, **10**, 95
389. Grosjaen, C. and Bossaert, W. (1965). *Table of Absolute Detection Efficiencies of Cylindrical Scintillation Gamma-Ray Detectors* (Ghent: Ghent University)
390. Wolicki, E. (1972). 'Nuclear and Ion Beam Techniques for Surface and Near-Surface Analysis', *NRL Report No. 7477* (Washington, D.C.: NRL)
391. Albert, P. (1969). *Le Vide*, **141**, 149
392. Chevarier, N., Giroux, J., Tran, M. and Tousset, J. (1967). *Bull. Soc. Chim. Fr.*, 2893
393. Adams, F., Op de Beeck, J., Van den Winkel, P., Gijbels, R., De Soete, D. and Hoste, J. (1971). *CRC Crit. Rev. Anal. Chem.*, **1**, 455
394. Ricci, E. and Hahn, R. (1965). *Anal. Chem.*, **37**, 742
395. Ricci, E. and Hahn, R. (1967). *Anal. Chem.*, **39**, 794
396. Kuin, P. and Reynders, J. (1972). *Proc. Conf. Saclay, Modern Trends in Activation Analysis*; (1973). *J. Radioanal. Chem.*, **16**, 403

397. Kim, C. (1969). *Radiochem. Radioanal. Lett.*, **2,** 53
398. Kim, C. (1971). *Anal. Chim. Acta*, **54,** 407
399. Nozaki, T., Yatsurugi, Y. and Akiyama, N. (1970). *J. Radioanal. Chem.*, **4,** 87
400. Nozaki, T., Yatsurugi, Y., Akiyama, N. and Imai, I. (1968). *Proc. Conf. Gaithersburg, Modern Trends in Activation Analysis*, Vol. 2, 842 (Washington, D.C.: NBS Spec. Publ. 312 (1969))
401. Alexandrova, G., Demidov, A., Kotelnikov, G., Plashakova, G., Sukhov, G., Choporov, D. and Schmanenkova, G. (1967). *Soviet Atom. Energy*, **23,** 797
402. Debrun, J. and Barrandon, J. (1972). *CNRS Int. Colloquim on Activation Analysis*; (1973). *J. Radioanal. Chem.*, **17,** 291
403. Kuin, P. (1967). *Proc. 2nd Int. Conf. on Practical Aspects of Activation Analysis with Charged Particles*, 31 (Brussels: Euratom, (1968))
404. Chen, N. (1970). *Radiochem. Radioanal. Lett.*, **4,** 375
405. Rook, H. and Schweikert, E. (1969). *Anal. Chem.*, **41,** 959
406. Rook, H., Schweikert, E. and Wainerdi, R. (1968). *Proc. Symp. Gaithersburg, Modern Trends in Activation Analysis*, Vol. 2, 768 (Washington, D.C.: NBS Spec. Publ. 312 (1969))
407. Williamson, C., Boujot, J. and Picard, J. (1966). *CEA Report 3042*
408. Tran, M. and Tousset, J. (1968). *Proc. Symp. Gaithersburg, Modern Trends in Activation Analysis*, Vol. 2, 754 (Washington, D.C.: NBS Spec. Publ. 312 (1969))
409. Vandecasteele, C., Adams, F. and Hoste, J. (1974). *Anal. Chim. Acta*, **71,** 67
410. Dabney, S., Swindle, D., Beck, J., Francis, G. and Schweikert, E. (1972). *Proc. Conf. Saclay, Modern Trends in Activation Analysis*, Paper M69; (1973). *J. Radioanal. Chem.*, **16,** 375
411. Riddle, D. and Schweikert, E. (1972). *Proc. Conf. Saclay, Modern Trends in Activation Analysis*, Paper M74; *J. Radioanal. Chem.*, **16,** 413
412. Engelmann, C., Gosset, J., Loeuillet, M., Marschal, A., Ossart, P. and Boisier, M. (1968). *Proc. Symp. Gaithersburg, Modern Trends in Activation Analysis*, Vol. 2, 819 (Washington, D.C.: NBS Spec. Publ. 312 (1969))
413. Giroux, J., Talvat, M., Thomas, J. and Tousset, J. (1971). *Bull. Soc. Chim. Fr.*, 706
414. Schweikert, E. and Rook, H. (1970). *Anal. Chem.*, **42,** 1525
415. Gross, C., Gaetano, G., Tucker, T. and Baker, J. (1972). *J. Electrochem. Soc.*, **119,** 926
416. Giroux, J., Thomas, J., Tran, M. and Tousset, J. (1970). *The Uses of Cyclotrons in Chemistry, Metallurgy and Biology* (C. Amphlett, editor), 319 (London: Butterworth)
417. Giroux, J., Talvat, M., Thomas, J. and Tousset, J. (1970). *J. Radioanal. Chem.*, **6,** 423
418. Vialatte, B. and Revel, G. (1972). *Anal. Chem.*, **44,** 1077
419. Vialatte, B. (1972). *CNRS Int. Colloquium on Activation Analysis*, Paper C24; (1973). *J. Radioanal. Chem.*, **17,** 301
420. Giroux, J. and Thomas, J. (1972). *CNRS Int. Colloquium on Activation Analysis*; (1974). *J. Radioanal. Chem.*, **19,** 129
421. Engelmann, C. (1971). *J. Radioanal. Chem.*, **7,** 281
422. Krasnov, N., Dmitriyev, P., Dmitriyeva, S., Konstantinov, I. and Molin, G. (1970). *Applications of Cyclotrons in Chemistry, Metallurgy and Biology*, (C. Amphlett, editor), 341 (London: Butterworths)
423. Sippel, R. and Glover, E. (1960). *Nucl. Instr. Meth.*, **9,** 37
424. Meyers, P. (1969). *Archeometry*, **11,** 67
425. Mayer, A. (1972). *Solid State Tech.*, April, 38
426. Kraft, G. (1972). *The Technologic and Economic Importance of the Determination of Oxygen in Non-Ferrous Metals*, Euratom Inform. Booklet 59 (Brussels: Eurisotop Office)
427. Rook, H., Schweikert, E. and Wainerdi, R. (1968). *Anal. Chem.*, **40,** 1194
428. Martin, J. and Haas, E. (1972). *Z. Anal. Chem.*, **259,** 97
429. Martin, J. and Haas, E. (1971). *Radiochem. Radioanal. Lett.*, **7,** 121
430. Martin, J., Schnöller, M. and Haas, E. (1972). *Z. Anal. Chem.*, **259,** 105
431. Engelmann, C. (1967). *Int. J. Appl. Radiat. Isotopes*, **18,** 569
432. Schuster, E. and Wohlleben, K. (1969). *Z. Anal. Chem.*, **245,** 239
433. Vialatte, B. (1971). *J. Radioanal. Chem.*, **8,** 269
434. Nozaki, T., Makide, Y., Yatsurugi, Y., Endo, Y. and Akiyama, N. (1972). *Bull. Chem. Soc. Jap.*, **45,** 2776

435. Marschal, A., Gosset, J. and Engelmann, C. (1971). *J. Radioanal. Chem.*, **8,** 243
436. Benmalek, M., Chermette, H., Martelet, C., Sandino, D. and Tousset, J. (1972). *Proc. Conf. Saclay, Modern Trends in Activation Analysis*; Univ. Lyon: LYCEN 7249
437. Faure, L., Boissier, M. and Tousset, J. (1972). *J. Radioanal. Chem.*, **10,** 213
438. Faure, L., Giroux, J. and Tousset, J. (1972). *J. Radioanal. Chem.*, **10,** 223
439. Engelmann, C., Gosset, J. and Rigaud, J. (1970). *Radiochem. Radioanal. Lett.*, **5,** 319
440. Vialatte, B., Barrandon, J., Alexandrov, S., Bourrelly, I., Cleyrergue, C., Deschamps, N. and Jaffrezic, H. (1970). *Radiochem. Radioanal. Lett.*, **5,** 59
441. Blondiaux, G. and Vialatte, B. (1971). *Radiochem. Radioanal. Lett.*, **8,** 251
442. Nozaki, T., Okuo, T., Akutzu, H. and Furukawa, M. (1966). *Bull. Chem. Soc. Jap.*, **39,** 2685
443. Nozaki, T., Yatsurugi, Y., Akiyama, N., Endo, Y. and Makide, Y. (1972). *CNRS Int. Colloquium on Activation Analysis*, Paper C44; (1974). *J. Radioanal. Chem.*, **19,** 109
444. Rommel, H. (1966). *Anal. Chim. Acta*, **34,** 427
445. Engelmann, C., Gosset, J. and Rigaud, J. (1970). *Radiochem. Radioanal. Lett.*, **5,** 319
446. Gebauhr, W. and Martin, J. (1973). *The Determination of Carbon and Nitrogen in Silicon and Germanium*, Eurisotop Inform. Booklet 87 (Brussels: Eurisotop Office)
447. Engelmann, C. and Marschal, A. (1971). *Radiochem. Radioanal. Lett.*, **6,** 189
448. Vandecasteele, C., Adams, F. and Hoste, J. (1974). *Anal. Chim. Acta*, **72,** 269
449. Engelmann, C. (1973). *The Use of Activation Analysis by Gamma-Photons and Charged Particles to determine the Oxygen Content in Non-Ferrous Metals*, Eurisotop Office Inform. Booklet 74 (Brussels: Eurisotop Office)
450. Revel, G. and Albert, P. (1968). *J. Nucl. Mater.*, **25,** 87
451. Saito, K., Nozaki, T., Tanaka, S., Furukawa, M. and Cheng, H. (1963). *Int. J. Appl. Radiat. Isotopes*, **14,** 357
452. Schweikert, E., McGinley, J., Francis, G. and Swindle, D. (1972). *CNRS Int. Colloquium on Activation Analysis*; (1974). *J. Radioanal. Chem.*, **19,** 89
453. Zadvornyi, A., Gorenko, A., Skakun, N. and Klyucharev, A. (1970). *J. Anal. Chem. USSR*, **25,** 292
454. Engelmann, C. (1970). *J. Radioanal. Chem.*, **6,** 227
455. Debrun, J., Barrandon, J. and Albert, P. (1969). *Bull. Soc. Chim. Fr.*, 1011
456. Lee, D., Stauffacker, C. and Markowitch, S. (1970). *Anal. Chem.*, **42,** 994
457. Sokolskii, V., Novgorodov, A., Kazachevskii, I., Artemev, O., Kazakov, V. and Trushlyakov, F. (1971). *Aktiv. Anal. Mater. Vses Soveshch*, 2nd Ed., 206, 1968
458. Lee, D., Lamb, J. and Markowitch, S. (1971). *Anal. Chem.*, **43,** 542
459. Debrun, J., Riddle, D. and Schweikert, E. (1972). *Anal. Chem.*, **44,** 1386
460. Pretorius, R. and Schweikert, E. (1971). *J. Radioanal. Chem.*, **7,** 319
461. Schweikert, E. (1970). *Trans. Amer. Nucl. Soc.*, **13,** 58
462. Debrun, J., Barrandon, J. and Albert, P. (1970). *Ann. Chim.*, **5,** 357
463. McGinley, J. and Schweikert, E. (1972). *Proc. Conf. Saclay, Modern Trends in Activation Analysis*; (1973). *J. Radioanal. Chem.*, **16,** 385
464. Kohn, A., Barranolon, J., Delrun, J., Vallaolon, M. and Vialatte, B. (1974). *Anal. Chem.*, **46,** 1737
465. Vandecasteele, C., Adams, F. and Hoste, J. (1975). *Anal. Chim. Acta*, **76,** 27
466. Vandecasteele, C. and Hoste, J. (1975). *Anal. Chim. Acta*, **79,** 302
467. Vandecasteele, C. and Hoste, J. (1975). *J. Radioanal. Chem.*, **27,** 465
468. Krivan, V., Swindle, D. and Schweikert, E. (1974). *Anal. Chem.*, **46,** 1626
469. Barrandon, J., Benaben, P., Debrun, J. and Valladon, M. (1974). *Anal. Chim. Acta*, **73,** 39
470. Benaben, P., Barrandon, J. and Debrun, J. (1975). *Anal. Chim. Acta*, **78,** 129

3
The Application of Computer Techniques to Instrumental Neutron Activation Analysis

J. P. OP DE BEECK and J. HOSTE
Rijksuniversiteit Gent

3.1 INTRODUCTION

Increasing use is being made in activation analysis of instrumental or non-destructive techniques, where γ-spectrometry with the aid of semiconductor detectors has replaced most of the measuring methods used previously. In view of the instrumental complexity, the required speed and the volume of data resulting from such a counting technique, the application of a computer for data acquisition and data reduction has become a necessity.

This description and discussion of the application of computer techniques to activation analysis will be limited to those cases where the computer does more than merely replace another calculating device. The field of discussion will be further narrowed by excluding methods that have not been appreciably improved or expanded during at least the past five years. These methods include the analysis of complex decay curves[1-4] for counting devices that have no energy-resolving ability and the use of least squares techniques to unfold complex γ-spectra[2,5-7] from detectors with limited resolution such as NaI(Tl) scintillation crystals.

Thorough reviews concerning data reduction methods and computer calculations in activation analysis have been published for NaI(Tl) and Ge(Li) detector type spectra by Yule[8,9], Quittner *et al.*[10], Op de Beeck *et al.*[11] and Turkstra *et al.*[12].

The application of computers to γ-spectrometry can be separated into four main sections. The first deals with the acquisition of the data, when one merely uses the very high speed of the computer in its real-time application domain. The second makes use of the extensive arithmetic and logic programming facilities of the computer in order to operate on the data at the end of the measurement, and to perform what is now generally termed data reduction. A third application is the qualitative analysis of γ-spectra, requiring a library file on magnetic tape or disc with hundreds to thousands of γ-energies that have to be scanned in order to identify the peak energies resulting from the data reduction of a spectrum.

The fourth application deals with complete activation analysis programs that routinely perform all calculations, starting from the original spectra of samples and standards down to the final completely corrected concentration values.

3.2 DATA ACQUISITION

Computerised data acquisition can be performed at several levels of sophistication. It is usually not realised that commercially available γ-spectrometers built around pulse height analysis (PHA) multi-channel analysers (MCA) contain complete computer systems. Indeed, the MCA consists of a digital processor, a computer memory and input and output facilities. In this case, however, the only input facility is an analogue-to-digital converter (ADC), and the processor is hardware programmed once and for all mainly to increase the content of the memory locations whose addresses are specified by the ADC. As output facilities one usually has many options available that are compatible with most general purpose (GP) computers, such as cartridge or industry compatible magnetic tape, punched paper tape and teleprinter–puncher combinations. As a rule, commercial MCAs have a built-in oscilloscope display unit as a standard option.

Theoretically, in order to convert an MCA into a GP computer, it would be sufficient to add a few more input devices and to expand the processor for general programming purposes. In practice, however, the opposite is done: a mini- or medium-sized computer is provided with an ADC at one of its input channels, and used as a MCA, under control of a user-provided program[13-17]. Typically, mini-computers have a memory of 8 to 12 K words (16 bits); medium computers have memories from 16 to 32 K. Several computer manufacturing companies now offer complete γ-spectrometry PHA systems built around their standard small computers.

The advantage of computer-based systems lies in the fact that in between data acquisition periods the computer can be used to perform relatively simple data reduction procedures. Often the manufacturers provide the necessary software for both data acquisition and reduction. In general, for activation analysis the latter should be examined critically and its quantitative performance checked very carefully. In practice, users will quickly saturate and exhaust the data reduction capacity of such small computer systems and turn to larger and faster computational facilities.

At a higher level of sophistication we can consider the so-called 'fifty-fifty' systems. In this configuration, a hardware-programmed processor and a GP processor share the same memory in perfect symbiosis. While the first is busy with data acquisition, using part of the memory, the second performs some calculations on previously collected data. Alternatively, in between data acquisition periods, the GP processor can directly access the memory section where the data have been stored by the other precessor, without making use of peripheral memory devices.

The highest level of sophistication in data acquisition consists of one or more commercial MCAs connected on-line with standard GP computers by means of the parallel data output available on every MCA. This configuration allows direct memory-to-memory transfers and is as a rule selected if a computer is located in the immediate vicinity of the MCA or if several MCAs are present and it is economically feasible to replace a special peripheral device for every MCA by a single small computer.

An example of the last case[18] is schematically shown in Figure 3.1. Up to eight MCAs are connected to a single PDP-11-20 computer with only a memory of 4 K. The computer's task is limited to the constant polling of the MCAs in order to detect whether one is ready to transfer data. Simultaneous transfer requests from several MCAs are kept in a queue. Book-keeping and labelling of the spectra is carried out, and the spectra, together with a label

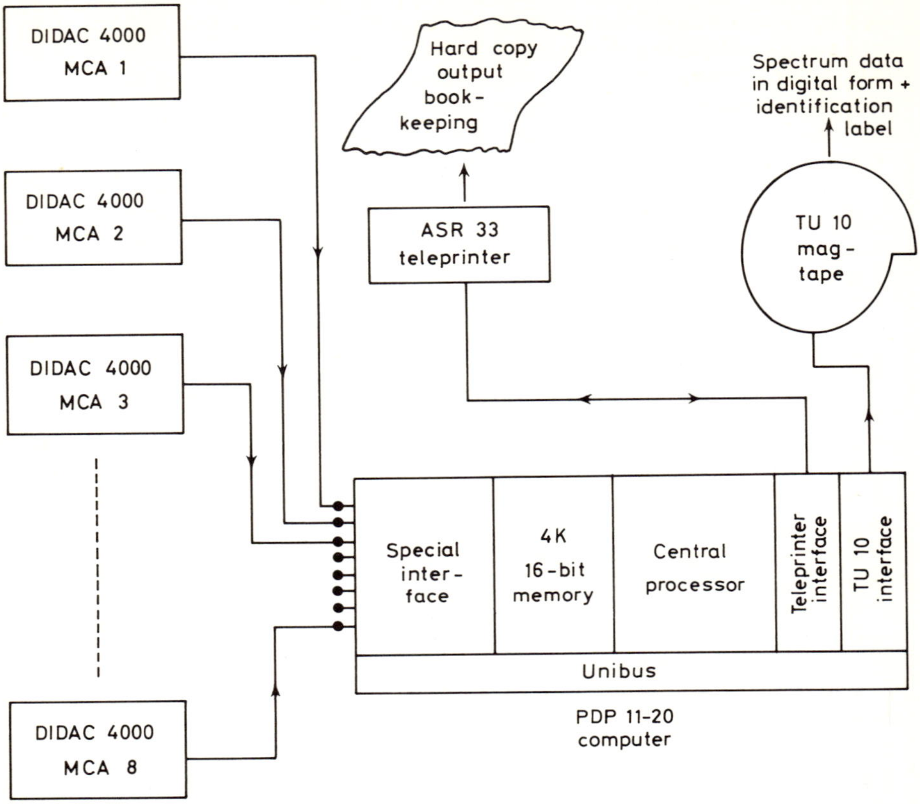

Figure 3.1 Block scheme of a moderately complex data acquisition system

defining the MCA and the sequence number of the spectrum for that MCA, are written on industry-compatible magnetic tape.

Often, while laboratories together with their counting as well as computer facilities are growing, complex situations of data acquisition may occur[19]. Combinations of, and variations on, one or more of the methods described above may co-exist under particular circumstances.

A final word should be said about the medium on which the spectra are stored, usually in binary coded decimal (BCD), before being transferred to a computer. It should be noted that BCD is the best coding format for hard-copy peripherals, but that pure binary is a much more efficient means for storage of data if they have to be computer-processed. For Ge(Li) detector type spectra, punch cards are not very convenient because the punching time is relatively long and most small-to-medium computers do not have a card-reader input. Paper punch-type is very popular for small computers, but a typical teleprinter–puncher combination with a speed of 10 digits per second takes 45 minutes to punch a complete spectrum at the MCA side, and an equal amount of time to read the data back into the computer. Magnetic tape is the only convenient way to temporarily store and retrieve 4000-channel spectra. Fast card and paper tape punchers and readers are available, but their price is an appreciable fraction of the total price of a small computer or an MCA. The same is true for industry-compatible magnetic tape units. Inexpensive cartridge-type tape units and discs have recently been introduced for small computers and in the future will probably offer the best compromise between speed and price.

3.3 DATA REDUCTION

3.3.1 Scope and concepts

Data reduction in a broad sense is the process of replacing a certain amount of data by a smaller amount, in a useful or profitable way. The latter requires that the information content of the original data is preserved as much as possible and, often, even increased. In a more narrow and specific sense, data reduction has been described as a method to distinguish between useful and useless data in γ-spectra[20].

In its general conception, data reduction covers almost all aspects of data

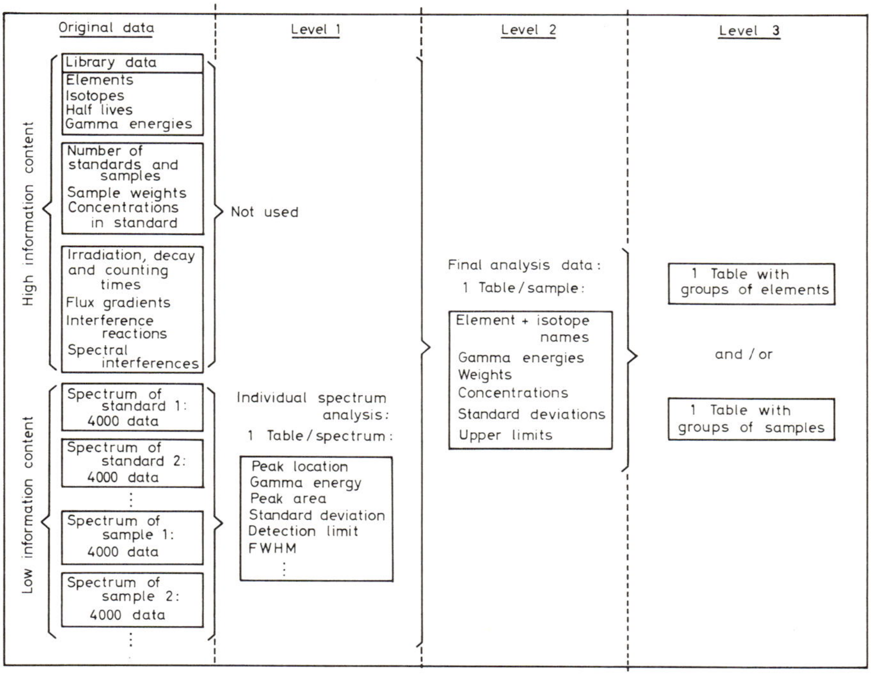

Figure 3.2 Detailed data reduction sequence with three hierarchical levels

handling, as shown in the scheme represented by Figure 3.2. All data reduction has to start from original data. These consist partly of numerical data with a high information content of a specific nature, such as the element and isotope library, sample weights, virtual concentrations due to nuclear interference reactions, flux gradient corrections, etc. The main part of the original data consists of numerical data with a very low information content, e.g. the counts in individual channels of an MCA. Because of the mismatch in information content between the two kinds of data, their direct comparison is not meaningful. The information content of the second kind of data has to be increased by the level 1 data reduction step, which is the data reduction concept in its specific sense mentioned above[20].

The amount of original data at the base of the results of the highest data reduction level is very large. If Ge(Li) detector-type spectra are considered from several standards and correspondingly more samples (typically in a ratio between 1:1 to 10:1 or more), many thousands of numbers have to be handled. Furthermore, the number of elementary mathematical operations required is a multiple of the amount of numbers involved. A detailed description of the first data reduction level, the individual spectrum analysis, will be given later.

After the first level, all data, partly original and partly from the first reduction, can be combined in order to produce numerical data with a high information content, shown on level 2 in Figure 3.2. The data reduction methods used at this level are all straightforward and well known. They consist mainly of several correction procedures due to differences in irradiation, decay and counting times, interferences by (n,p), (n,α) or (n,γ,β^-,n,γ) reactions, spectral interferences, etc. These corrections, together with the comparison of standard and sample data with the library data and with each other, constitute the classical base of the activation analysis method as such, and can be found in every good textbook covering this subject[21]. It should be noted how the useful information content of the level 2 data has increased, whereas all information about the detailed spectral shape has been lost. It is important to realise that data reduction intensifies the information content of a specific nature at the expense of all detailed information about the direct physical measurements. Therefore a good data reduction procedure, if carried out by computer, should save as much as possible of the information lost at each level on a hard-copy peripheral device for subsequent use.

In general, data reduction in activation analysis is considered to be finished after level 2. However, in view of the fact that activation analysis is particularly suited for the simultaneous determination of many elements that in natural substances occur only in trace amounts, an additional reduction step may be considered. As trace elements do not contribute to the overall chemical nature of substances, they offer a particular method for identification. Alternatively, small differences in trace element concentration in biological material may indicate important physiological factors.

For a large number of samples of a more or less similar kind, the many individual elemental concentration values can be reduced to a single numerical value describing the relationships between different chemical elements or between different samples. Such methods have been called fingerprinting, pattern recognition or similarity analysis.

3.3.2 Individual spectrum analysis

The principal aspects and features of a γ-spectrum taken with a Ge(Li) detector are illustrated in Figure 3.3a. The characteristics of the shape of isolated peaks and Compton edges, together with definitions of important concepts, are outlined in Figure 3.3b.

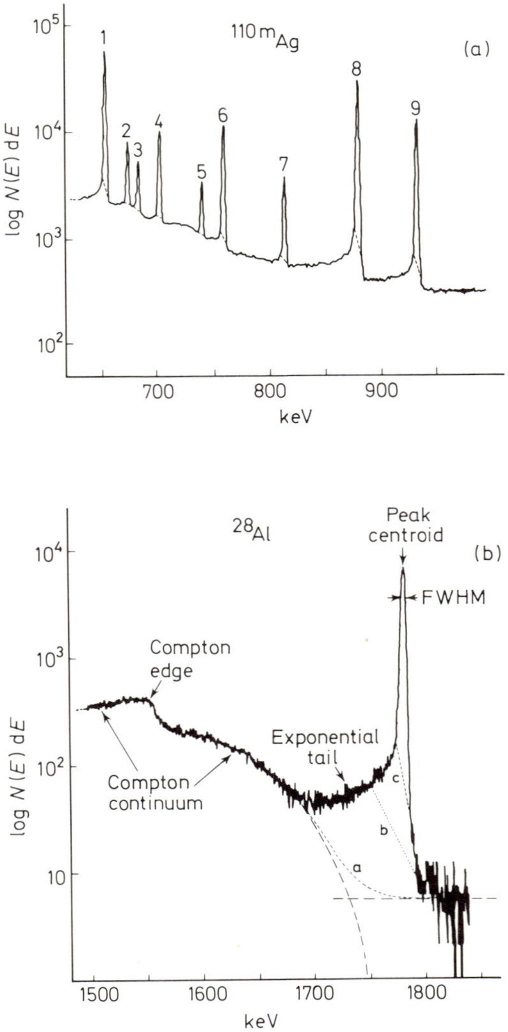

Figure 3.3 (a) Aspect of typical Ge(Li) spectrum peaks with baselines determined by the same number of channels for each peak; (b) detailed peak shape for a monoenergetic isotope, with definition of important concepts; baselines a, b, and c are explained in the text

3.3.2.1 Preliminary data treatment

The measured data are usually presented to the computer in BCD format, and have to be converted to pure binary. Furthermore, the data have to be checked for transcription errors from the temporary storage medium. Neither of these data editing problems is trivial. The first requires a machine-level language program, whereas for the second no satisfactory general solution exists.

Subsequent data treatment is usually carried out with convolution techniques. The method has been extensively described by Savitzky and Golay[22]. This technique, sometimes referred to as digital filtering, consists of deriving a new set of data from the measured spectral data by means of the following equation:

$$Y(i) = a_0 \times y(i) + \sum_{k=1}^{m} a_k \times [y(i+k) \pm y(i-k)] \qquad (3.1)$$

in which $y(i)$ = original number of counts measured in channel i of the MCA and $a_0, a_1 \ldots, a_m$ are a set of parameters defining the desired convolution effect. Depending on the value of the parameters $a_0, a_1, \ldots, a_m$ and the sign in the second term of the right-hand side of equation (3.1), a smoothed version of the original spectrum or any order of derivative can be obtained.

Smoothing is widely used in preliminary data treatment in order to reduce the noise resulting from statistical fluctuations on the number of counts in the individual channels of the MCA. Although very effective for NaI(Tl) detector-type spectra, smoothing methods cannot be applied to Ge(Li) detector-type spectra without changing the shape of some of the features of the spectra, especially of the more intense peaks. Yule[23, 24] has extensively discussed the effect of smoothing on a spectrum, and has suggested several smoothing techniques in order to avoid peak shape distortions. He concluded that good smoothing should be adapted to, and vary with, the local features in a spectrum. Furthermore, he claims that a single uniform smoothing of the whole spectrum, with carefully selected convolution parameters, only introduces a small distortion in overlapping peaks.

The influence of smoothing on peaks in spectra from high-resolution detectors as well as more general spectral features has also been described by Tominaga *et al.*[25] and by Groszwendt[26]. Tominaga *et al.*[25] suggested Gaussian-type filter functions instead of polynomials[22], because they induce less peak shape distortion for the same smoothing effect. Op de Beeck[27] has shown that the amount of distortion varies according to the number of channels contributing to a peak, and also to the ratio of the peak height to the underlying continuum. The smoothed data can, therefore, not be fitted to a standard peak shape, and furthermore the effective energy resolution observed in the smoothed data may be markedly decreased, which makes it more difficult to analyse interfering peaks.

Some of the authors mentioned above[25-27] made use of the Fourier transform to study the smoothing effect, because this transformation is especially suited to study linear systems for which convolution is the numerical counterpart. Inouye[28] has shown that the Fourier transform provides an ideal method for smoothing, combining the highest smoothing effect with the lowest peak distortion. An entire spectrum is transformed into frequency space, multiplied by an adequately chosen cut-off type filter function, and back-transformed into channel space. Unfortunately, the computation time involved in this method is prohibitive.

As a general conclusion on spectrum smoothing, it appears that in spite of its very successful applications to NaI(Tl) detector-type spectra, it should be handled with care for high-resolution detector-type spectra. If the spectral shape is not to be touched, only a very soft smoothing is possible, which is not very effective in removing statistical fluctuations. A stronger smoothing may be helpful to improve the shape of peak tops of moderately small peaks as a part of a peak detection method. For very small peaks that are scarcely visible among the statistical fluctuations of the continuous parts of the spectra, little or no improvement can be expected.

The peak-distorting effects of smoothing can be used advantageously in order to produce a so-called base spectrum from a normal spectrum. A base spectrum coincides with the original spectrum in all features except the peaks. Subtraction of a base spectrum from its original spectrum produces a third spectrum which in the ideal case contains only net peaks. Both Ralston and Wilcox[29] and Groszwendt[30] have described methods based on repetitive averaging of the original spectral data in order to obtain a close approximation to the ideal base spectrum. Fujii *et al.*[31] have generated a base spectrum by fitting a distribution through all 'true' minima of the spectrum. Minima not belonging to the base spectrum were rejected by testing the slope of the straight line connecting adjacent minima. These methods will be further discussed in the paragraph dealing with methods for baseline generation and peak area calculation.

3.3.2.2 Peak detection methods

Total absorption peaks (TAP) are presented in a Ge(Li)-type γ-spectrum (Figure 3.3) as narrow, sharp, nearly triangular shaped spikes, superimposed on a slowly varying continuous so-called background distribution. The latter is made up of a relatively small contribution from the true detector background and mainly of the added contributions of the Compton continuum of one or several high-energy γ-ray peaks.

The typical features of such a TAP lend themselves very well to an easy identification in a numerical fashion. This has led to direct detection of peaks from the original or smoothed data by means of simple heuristical tests. Dooley *et al.*[32] used count increases in adjacent smoothed channel groups to detect the presence of a TAP.

With $n(i)$ representing the number of counts recorded in channel i of an MCA, the condition

$$n(i-2) < n(i) - P[n(i)]^{\frac{1}{2}} > n(i+2) \tag{3.2}$$

with P equal to a numerical factor chosen by the analyst, was used by Guzzi *et al.*[33] to detect the top of a TAP. Dauk[34] used the condition

$$[n(i-1) < n(i)] \wedge [n(i+1) < n(i+2)] \tag{3.3}$$

to detect the low-energy slope of a peak, together with the condition

$$n(i-1) \geqslant n(i) \tag{3.4}$$

to locate the peak top. Slavic and Bingulac[35] used the following computer *IF* function:

$$IF\,(i) = \begin{cases} -1 \text{ for } n(i) \leqslant n(i-1) \\ +1 \text{ for } n(i) > n(i-1) \end{cases} \tag{3.5}$$

This function goes through a characteristic sequence −1, +1, −1 for a peak and the valley preceding it usually on the low-energy side. A similar, but slightly more complicated, method was used by Borchardt *et al.*[36].

The previous methods all rely on algebraic inequalities which will only work well with favourable statistics. If the noise content of the peaks becomes appreciable, more sophisticated peak detection methods are necessary. By far the most popular method uses the smoothed first and/or higher derivatives calculated from the original data according to a simplified least-squares method[22] of equation (3.1). Yule[8,9,37,38] has extensively examined and propagated these methods. In Figure 3.4a–d a peak is represented by the

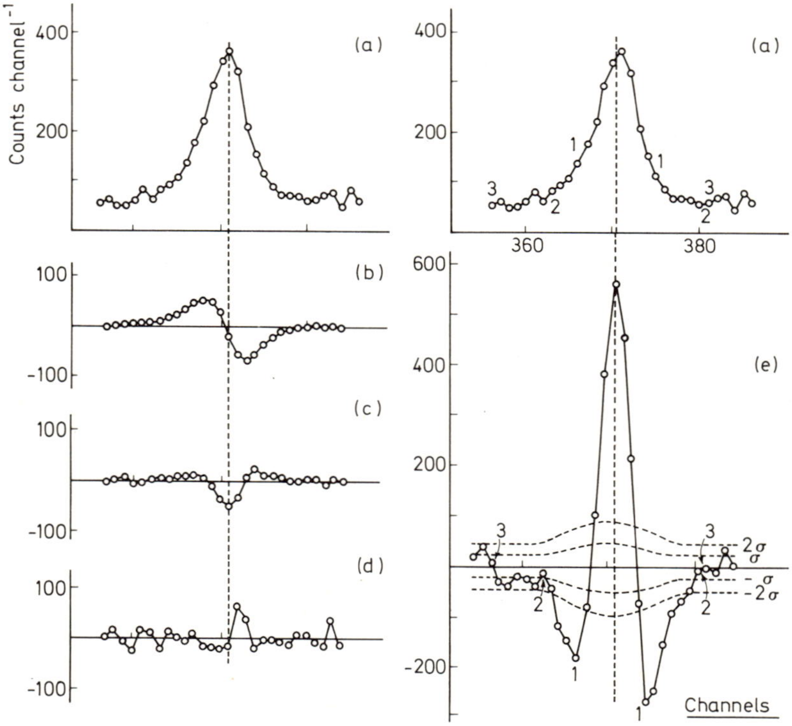

Figure 3.4 (a) Typical peak in a Ge(Li) spectrum; (b) smoothed first derivative; (c) smoothed second derivative; (d) smoothed third derivative; (e) square wave convoluted spectrum with indication of possible peak boundaries

original measured data, as well as the first, second and third smoothed derivatives calculated according to Savitzky and Golay[22]. The first derivative changes sign with a negative slope at the top of a TAP as well as at any other broad or narrow maximum present in the spectrum. Therefore, a few other tests concerning the local shape of the spectrum associated with the detected maximum have to be made in order to be sure of the presence of a TAP. In addition, a statistical significance test on the peak height is required. This method for peak detection is reliable and was used by Yule[39], Bowman[40] and Op de Beeck[19].

In order to avoid the additional tests, higher-order derivatives have been tried. The second derivative, for example, exhibits a deep negative minimum corresponding to the top of a TAP, which makes it possible to discriminate between peaks and other less pronounced maxima in a spectrum. The same information can be obtained from the slope of the first derivative where it changes sign, but the characteristic symmetrical shape of the second derivative is also useful to test the symmetry of the peaks, especially in the case of overlapping peaks. The second derivative method has been used by Verheijke[41] and Nyman[42]. Combinations of several derivatives have been used, mainly as an aid to resolve peak multiplet structures[38, 43, 44]. However, the generation of several sets of secondary spectra containing higher-order derivative data, useful as they may be, is often prohibitive with respect to the required computer memory and execution time. Furthermore, it is shown in Figure 3.4b–d that the smoothed higher derivatives become more and more influenced by statistical fluctuations; for moderately small peaks, the third derivative is effectively no longer useful.

Mariscotti[45] has suggested the calculation of difference spectra instead of derivatives and he stressed the importance of the smoothed second difference spectrum as a peak and multiplet detecting method. Difference spectra are computationally more efficient since the multiplications associated with equation (3.1) are replaced by simple subtractions. Unfortunately the intensity of a second difference spectrum is of the same order of magnitude as its own statistical fluctuations. Therefore several averaging runs on the whole difference spectrum are required, which annihilates the gain in computation speed.

Black[46] has suggested the use of cross-correlation methods used in geophysics and communications to detect a signal of specified shape in random noise. The cross-correlation is defined by

$$C(i) = \lim_{T \to \infty} (2T)^{-1} \int_{-T}^{+T} f(t) \times g(t+i)\, \mathrm{d}t \tag{3.6}$$

in which i = channel number; $f(t)$ = functional shape of a TAP; and $g(t+i)$ = the measured spectrum.

Because of the discrete nature of the spectrum data, the convolution integral (3.6) is reduced to the simple numerical convolution of equation (3.1):

$$C(i) = \sum_{t=-m}^{m} f(t) \times g(i+t) \tag{3.7}$$

Because a real spectrum contains intense slowly varying continua in addition to peaks and random noise, equation (3.7) has to be modified to

$$C(i) = \sum_{t=-m}^{+m} f(t)[g(i+t) - A_i] \tag{3.8}$$

where

$$A_i = \frac{1}{2m+1} \sum_{t=-m}^{m} g(i+t) \tag{3.9}$$

A_i is only a very crude estimate of the continuum present in the specified channel of the spectrum. The cross-correlation spectrum exhibits very intense positive peaks centred at the centroids of the TAP, but no other useful information can be extracted from it. It is, therefore, computationally inefficient.

Overlapping peaks (multiplets) may offer special difficulties for peak detection methods. Distinction can be made between 'open' multiplets, and 'closed' multiplets. The first type allows the direct observation of the components indicated by valleys between, or shoulders on, the individual peaks. If the distance between two peaks becomes very small, 'closed' doublets are observed that can only be recognised by a slight peak shape distortion or an excessive FWHM (full width at half maximum) value. As cited above, multiplets with components of sufficient intensity can often be analysed by higher-order derivatives[38, 43, 44]. A general solution including closed and low-intensity multiplets may probably be obtained by Inouye's resolution improvement filter[47]. This method again makes use of the Fourier transform and its inverse, after multiplication by a suitable filter function. This filter has a deconvolution effect and actually removes part of the peak broadening occurring in the detector and in the electronics of the γ-spectrometer, resulting in an improved energy resolution. Apart from its mathematical complexity, the method has the undesirable effect of generating ringing oscillations at the base of the peaks. Sterlinski[48] has suggested the use of simple convolution parameters in conjunction with equation (3.1) in order to obtain the same improvement in resolution, but the parasitic oscillations are in this case even stronger.

It is appropriate at this point to realise that apart from the methods that detect TAPs directly, all other methods which generate a new spectrum from the original data, either by direct numerical convolution, Fourier transformations or cross-correlation methods, can be brought back to the application of equation (3.1) with a suitable set of parameters a_k. It should, therefore, be possible to define the most profitable set of parameters combining the advantages of the different methods. 'Profitable' may be defined here by the following set of rules for the spectrum derived with the ideal parameters from the original data:

(a) It should have a non-zero value only in the immediate vicinity of a peak.
(b) It should contain information about the peak centroid, peak boundaries and multiplet structure.
(c) It should be proportional to the ratio of peak height to underlying continuum.

(d) It should be simple to calculate, short and easy to program and fast in execution.

The closest approach to this ideal derived spectrum is the SWC (square wave convoluted) spectrum suggested by Op de Beeck[27]. The result of its application to a real TAP is shown in Figure 3.4e and should be compared with the weak signals for the numerically computed derivatives[22] of the same peak (Figures 3.4b–d). It is obtained by convoluting the original data with a symmetric one-period square wave of eight channels width and varying between the values -1 and $+1$. Its main features are a strong positive peak (analogous to Black's cross-correlation peaks) corresponding to the top of a TAP, a very characteristic shape similar to the inverted second derivative or smoothed second difference, an essentially zero signal in the absence of peaks, and a computationally simple and fast approach because only additions and subtractions are required. In practice the values of the SWC spectrum are put equal to zero if they are smaller than twice their standard deviation. The latter value is also indicated in Figure 3.4e. As the positive peaks in the SWC spectrum have a much smaller FWHM than the peaks in the original spectrum, the resolution of the components in multiplets is considerably improved. Closed doublets may even be transformed into open structures.

A very simple peak detection can be applied after subtraction of the base spectrum from its original spectrum[29-31], as was explained in the section on preliminary data treatment. Since only net peaks remain, a simple direct peak detection can be applied. However, the base spectra methods practically available produce remnants of other maxima present in the spectra apart of the true TAPs, such that additional peak shape tests are required[12].

Finally, once a peak has been detected, its exact location, within a fraction of a channel unit, has to be calculated. This can be done by fitting a parabola through a few channels around the top of the peak, making use of a simplified method based on the convolution coefficients for obtaining the smoothed first and second derivative[22]. Alternatively, the exact centroid (C) can be obtained, less accurately, from the weighted first moment for a few channels (n) around the top of the peak:

$$C = \frac{\sum_{i}^{i+n-1} i \times y(i)}{\sum_{i}^{i+n-1} y(i)} \tag{3.10}$$

where i = channel number and $y(i)$ = counts in channel i.

The exact location of a TAP should of course only be calculated from the peak data after correction for the continuum underlying the peak. Methods of performing the latter operation are discussed in the next two sections.

3.3.2.3 Precision and accuracy of a peak area calculation method

If one is interested in the absolute intensity of the γ-ray lines in a spectrum, the exact integral over the peak area needs to be known. For activation analysis, however, a quantity exactly proportional to this peak area is often sufficient because only the ratio of peaks, corresponding to the same γ-energy in spectra from samples and standards, enters into the calculations.

In Figure 3.3 a few important definitions concerning the characteristics of a TAP have been indicated. To these, the concepts of peak boundaries, base area and baseline have to be added, as illustrated by Figure 3.5a.

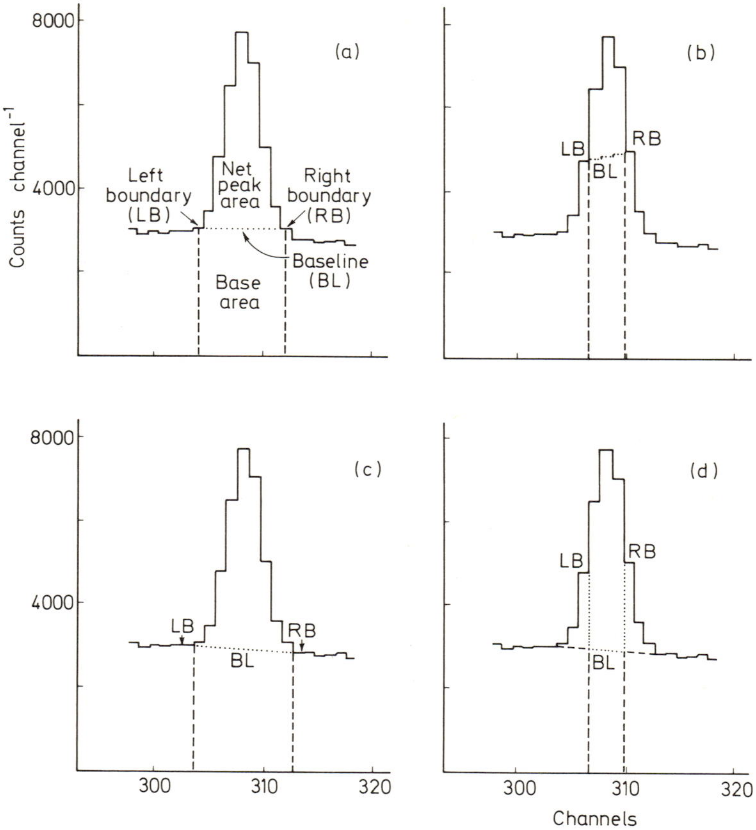

Figure 3.5 (a) Illustration of the formal definition of peak boundaries, peak area and base area; (b), (c) and (d) are different definitions of peak boundaries and baseline as explained in the text

As a rule, peaks are superimposed on continuous distributions from the Compton effect of higher-energy γ-rays. In order to subtract this continuum, only a small number of channels at the location of the peak need to be considered. The channels defining this small section of the spectrum are called the peak boundaries. The calculated line indicating the separation between the peak area and the underlying continuum is called the baseline. The integral of the continuum under the peak is called the base area (or sometimes background).

The definitions of peak boundaries and baseline are of crucial importance in spectrum analysis, because they alone will ultimately establish the largest part of the precision and accuracy of the data reduction method as a whole. Indeed, subtraction of the base area from the original spectrum within the interval of the peak boundaries results in a numerical value that is in the end a measure of the amount of a specified element present in the sample from which the γ-spectrum was taken. The precision and accuracy of this numerical value, therefore, largely determines the precision and accuracy of the final analytical result.

Most methods of baseline estimation have been tailored around relatively intense well-defined peaks with a peak-to-base ratio of 1 or higher, superimposed on a very flat continuum, and with no other peaks in the immediate vicinity. It is often tacitly suggested that this represents the average situation. In practical activation analysis work, however, the opposite is true. Peaks are mostly of low intensity, ill-defined and with low peak-to-base ratios. Many peaks are situated on definitely non-linear true baselines and/or are very close to other peaks. The number of overlapping peaks is, as a rule, not small, nor is the degree of overlap usually negligible, in spite of some authors stating that 'overlapping peaks are rather scarce'[41] or 'overlapping of photopeaks . . . is at present not a serious problem and is only occasionally encountered'[12].

The basic problem in designing a general prescription to define the peak boundaries or the baseline can be appreciated by looking at the details of the peak slopes brought about very distinctively in the semi-log presentation of Figures 3.3a and 3.3b. A number of obvious conclusions can be made immediately. The TAPs are not symmetrical. The high-energy side approximates to a Gaussian function, but the low-energy side exhibits a marked 'exponential tail' (Figure 3.3b). The importance of this tail decreases virtually as the peak-to-base (PTB) ratio decreases. The true baseline under a peak may deviate appreciably from a straight line if the peak is not situated on an important continuum (Figure 3.3b), if it is located in the vicinity of the Compton edge of an important higher energy peak, or if it is superimposed on the low-energy tail of an adjacent peak. The true peak area of a TAP cannot be determined exactly because, theoretically, part of it is smeared over a large part of the whole spectrum. Fortunately, a large fraction of the area ($>90\%$, depending on the quality of the detector) can be found in a small number of channels corresponding to an interval of 3–4 FWHM. It is this large fraction that practically defines the total net peak area. Although not visible in Figure 3.3, the FWHM of the peaks increases with increasing γ-energy. This effect is relatively small for large-volume Ge(Li) detectors (20–60 cm^3), but may be very important for very small ones such as used for low-energy photon detectors (LEPD). In this case the FWHM may increase by a factor of three in the energy range from 20 to 400 keV. It is clear that any method for peak boundary and baseline determination will have to cope with serious sources of systematic error.

The accuracy of a method is determined by its invariance against systematic changes that involve peak width and shape. Systematic errors may result if calculated peak areas corresponding to peaks of the same γ-energy but with different shape or width are compared in separate spectra (e.g. samples and standards). Differences in peak width can be traced back in part to different counting times for the spectra, such that slow electronic drifts will influence the spectra differently. Differences in peak shape originate from differences in base area intensity and baseline shape combined with the relatively small number of channels occupied by a peak in a Ge(Li) type spectrum, such that the width of a channel is an appreciable fraction of the total peak width. Peak shape distortions, including an increase in FWHM, may also be induced by high counting rates, giving undesirable electronic effects in the amplifier section of the spectrometer, or alternatively by inadequate preliminary data treatment. A final and obvious source of systematic error resulting in both an increasing peak width as well as a peak shape distortioning is spectral interference or peak overlap.

In the literature[12, 49, 50], the concept 'precision of a peak area calculation method' is sometimes encountered. By itself, however, the precision of any calculation method is extremely good. For example, the repetitive application of any peak area calculation method to the same spectrum results in exactly the same numerical value for a specified peak. In γ-spectrometry, precision problems stem mainly from the Poisson statistics governing the count rates in individual channels. It may be assumed that the fluctuations in count rate for any two channels have practically very little covariance. Furthermore, it is the nature of Poisson statistics that they are relatively more important the smaller the number of counts in a channel. The precision of the result of a peak area calculation is thus determined by the precision of the measurement of the spectrum. The method of calculation is involved in the precision of the result only in as far as error propagation calculations are applied to the mathematical operations that are used in the calculation. This means that methods of peak area calculation vary in the degree of use of the available precision, but they can never be the source of additional variance. Practically, a method of peak area calculation will give a more precise result compared with another method for the same peak if it minimises the use of channels with low count rate (at the peak boundaries) in the mathematical operations necessary to obtain the result. It should be noted finally that the precision of the result of the peak area calculation will subsequently be called 'precision of the peak area calculation method'. Experimentally, this precision can only be obtained by applying the method to a series of spectra taken from the same source under identical circumstances and conditions such that the spectra differ only by statistical fluctuations with exclusion of all differences in intensity or peak shape.

Comparative studies, both from a theoretical and an experimental point of view, for different baseline estimation procedures have already been made by Turkstra *et al.*[12], Baedecker[49] and Hertogen *et al.*[50]. In the second paper the peaks, to which the various methods under test are applied, were counted in the presence of increasing amounts of a high-energy γ-emitter. In the first and the third papers the test peaks were measured separately and spectra with increasing intensity of a high-energy γ-emitter were added to them in the memory of the MCA or the computer. In this way, the peak broadening as well as the pulse pile-up losses due to high counting rates were avoided, and separated from the accuracy properties of the baseline estimation methods. In the second paper an additional experiment was carried out with a series of spectra taken from the same isotope mixture during the same counting time, in order to test the different methods for their behaviour when only counting statistics are involved. As far as precision is concerned, only Baedecker[49] made the correct experiments, allowing the deduction of relevant conclusions. The two other papers have tried to derive conclusions on the general precision of peak area calculation methods based on a series of single measurements, with inherently a very different precision, without appropriate weighting, whereas simultaneously systematic errors of unknown intensity and sign were involved. All three papers failed to separate precision and accuracy in the systematic error study carried out by varying the peak-to-base ratio. It is clear that as this ratio is decreased by increasing the base, both systematic errors and statistical fluctuations increase simultaneously, so that the second may partly or completely mask the first, independent of the ratio considered. This probably explains the contradictory conclusions reached by various authors for certain methods. Only in Baedecker's paper were the systematic errors so important as to not be largely obscured by the precision.

3.3.2.4 Peak boundary and baseline determination

The methods for peak boundary and baseline definition can be considered as belonging to one of the following general approaches:

(a) *A priori* definition of the baseline.

(b) Baseline estimated from the originally measured or smoothed data in sections of the spectrum adjacent to the TAP.

(c) Baseline estimated from a secondary spectrum derived from the original.

In every case, the calculated baseline may be given either by a straight line or by another function resulting in a curved baseline, in an attempt to obtain a better approximation to the true baseline and thus avoid systematic errors. Furthermore, the calculated baseline may be placed as close as possible to the true baseline in order to obtain the closest approach to the true net peak area (TPA = total peak area method) and to be able to compare peak intensities for different γ-energies (e.g. for comparator methods in activation analysis). Alternatively, the peak boundaries, baseline, or both, may be deliberately placed above the true baseline so that only a fraction of the total net peak area is obtained (PPA = partial peak area method) in order to improve the precision and accuracy of the peak area calculation.

(a) A priori *baseline methods*—The typical method for an *a priori* baseline estimation is the original peak area calculation of Covell[51]. The peak boundary positions are chosen far above the true base of the peaks, so as to enclose a fixed number of channels, and the baseline is a linear function connecting the two peak boundaries, as illustrated in Figure 3.5b. This is obviously a PPA method, the constant fraction of the peak area being determined by the constant number of channels between the peak boundaries. Typically, the latter are chosen at one-third to one-half of the peak height.

The method was originally designed for application to NaI(Tl) detector-type spectra. These broad peaks, covering many channels, tolerate the use of this method, especially since the peaks are either close to each other or the curvature of the true baseline cannot be estimated from the contents of the channels adjacent to the peaks. Therefore, a 'high' baseline minimises systematic errors due to true baseline curvature and moderately overlapping peaks. For Ge(Li)-type spectra, however, the method suffers from severe precision and accuracy problems because the boundary channels are chosen on the steep slopes of the peak sides, exactly at the points that are most sensitive to peak distortion and broadening effects of all kinds.

The original peak area calculation according to Covell[51] is given by

$$A = \sum_{i=-k}^{k} n(i) - (k + \tfrac{1}{2})[n(k) + n(-k)] \tag{3.11}$$

in which i = channel number relative to the centroid of the peak; $n(i)$ = number of counts in channel i; and $2k + 1$ = number of channels including the peak boundaries.

The variance on A is formally given by

$$V_A = \sum_{i=-k}^{k} n(i) + (k + \tfrac{1}{2})^2[n(k) + n(-k)] \tag{3.12}$$

from which it is clear that the boundary channels, which have inherently the lowest precision, have the highest weight in the precision calculation for the peak area. Equation (3.12) is not quite correct, however. Indeed, the boundary channels $n(k)$ and $n(-k)$ are included in the positive as well as the negative part in equation (3.11), which means that they are subtracted from themselves. This operation has a variance of zero because of the perfect covariance, and should not be included in equation (3.12). Equations (3.11) and (3.12) have to be replaced by

$$A = \sum_{i=-(k-1)}^{k-1} n(i) - (k - \tfrac{1}{2})[n(k) + n(-k)] \tag{3.13}$$

and

$$V_A = \sum_{i=-(k-1)}^{k-1} n(i) + (k - \tfrac{1}{2})^2[n(k) + n(-k)] \tag{3.14}$$

For NaI(Tl)-type peaks, the error on the variance is negligible, but for Ge(Li) peaks it is important because k is a very low number (2–4).

Sterlinski[52,53] has shown that if A is considered as a function of k, the summation of A values for decreasing k values results in a quantity S that is proportional to the peak area and is defined by

$$S = k \times n(0) + \sum_{i=1}^{k-1} (k - 2i + \tfrac{1}{2})[n(-i) + n(i)] - (k - \tfrac{1}{2})[n(k) + n(-k)]$$

or

$$S = k \times n(0) + \sum_{i=1}^{k/2} (k - 2i + \tfrac{1}{2})[n(-i) + n(i)]$$

$$- \sum_{i=(k+2)/2}^{k-1} (2i - k - \tfrac{1}{2})[n(-i) + n(i)] - (k - \tfrac{1}{2})[n(k) + n(-k)] \tag{3.15}$$

It is obvious that in the variance of S, the boundary channels have a much smaller relative weight than in equation (3.14). As a result, the relative variance for equation (3.15) can be better than for equation (3.13). However, as the negative part in equation (3.13) is expanded with additional terms in equation (3.15), the expected gain in precision will not always be important.

Heydorn and Lada[54] have shown that the precision of A, calculated by equation (3.13), goes through a maximum for varying values of k. The ratio $V_A^{\frac{1}{2}}/A$ goes through a broad minimum at a value of k approximately equal to the FWHM of the peak. This corresponds to an interval of two FWHM, or to peak boundaries that are lying low and close to the true baseline. The specific advantages of the Covell method are in that case lost. For Ge(Li)-type spectra such a baseline comes so close to the true baseline that practically a TPA method is being used.

If V_s represents the variance of S, then the ratio $V_s^{\frac{1}{2}}/S$ also goes through a minimum as a function of k, but this minimum is not pronounced and occurs at a k value of 2–3 FWHM units. In Figure 3.6a the experimental relative standard deviations $V_A^{\frac{1}{2}}/A$ [equation (3.13)] and $V_s^{\frac{1}{2}}/S$ [equation (3.15)] are plotted as a function of k for the peak shown in Figure 3.5a. It can be concluded from this figure that only when equations (3.13) and (3.15) are applied as a TPA method does the latter offer a better precision, and then only for low peak-to-base ratios. Baedecker[49] observed a slightly better overall precision for Covell's method compared with Sterlinski's equation for his pure counting statistics test.

For a decreasing peak-to-base ratio, Baedecker[49], Hertogen *et al.*[50] and Turkstra *et al.*[12] for the Covell method, and Baedecker[49] for the Sterlinski method, found experimentally that these methods when compared with others show the largest systematic errors.

(b) *Methods with baseline obtained from the original spectrum*—It follows from the previous discussion that the simplest TPA method is merely equation (3.13) with peak boundaries placed low enough to ensure that the resulting baseline lies close to the true baseline, as shown in Figure 3.5c. However, instead of fixing the value of k *a priori*, it is usually determined *in situ*, depending on the shape and intensity of the peak and the local features of the spectrum. This technique, which has been called the classical method by Yule[8], was first applied by Westermark and Sjöstrand[55] to a NaI(Tl) spectrum for a low energy peak. The method is much more useful in Ge(Li) spectrometry, where a good estimate for the true baseline can easily be obtained from as little as one channel in the regions adjacent to both sides of the peak. The several TPA methods described in the literature differ only in the way the boundary channels or boundary regions are selected.

Intuitively, one would define the peak boundaries as those channels where the contribution of the peak becomes negligible in comparison with the counting statistics on the spectrum continuum, because this is where the peak contribution virtually disappears. Actually these channels are usually closely approximated by the points of highest curvature at the lower end of the peak slopes.

In Figure 3.3a such peak boundaries were determined for one of the smaller peaks (peak no. 5) and boundaries were established for the other peaks corresponding to the same number of channels in the interval between the

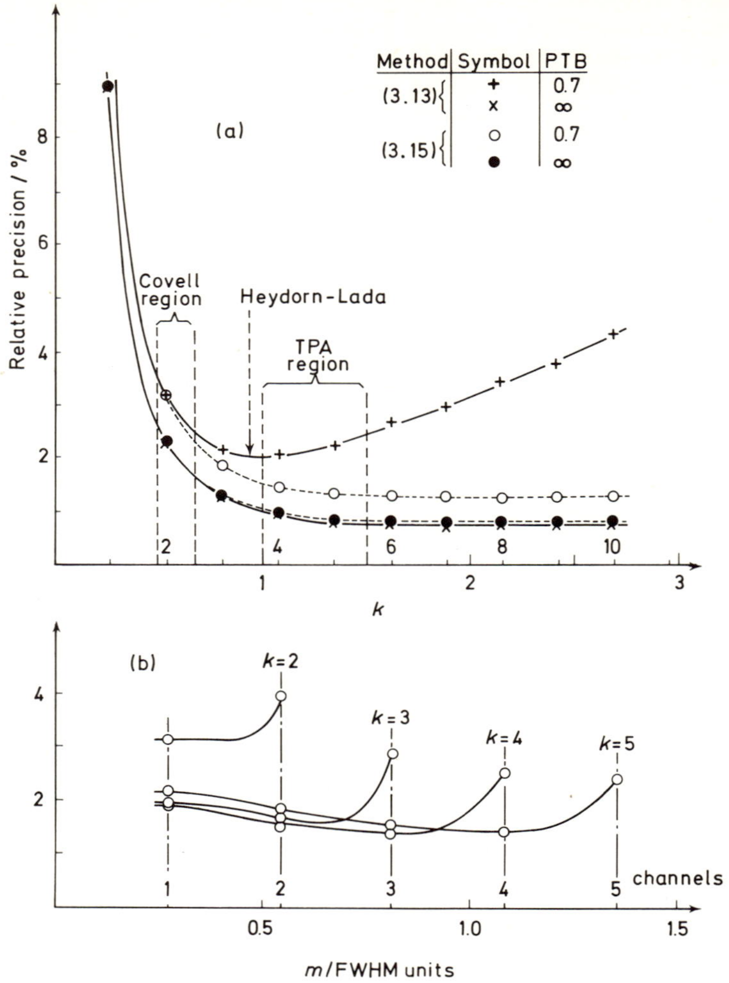

Figure 3.6 Relative precision of peak area calculation methods: (a) with baseline going through the peak boundaries; (b) with peak boundaries above the baseline. The parameters k and m are explained in the text

boundaries. It can be seen that for the larger peaks this results practically in a baseline considerably above the spectrum continuum. It may be concluded from this that TPA methods defined by peak boundaries based on highest curvature, or on lines tangential to the spectral regions immediately adjacent to the peak, suffer from a systematic error. As the peak-to-base ratio becomes smaller, a decreasing fraction of the peak area is obtained. On the other hand, these TPA methods which adapt their boundaries to the local shape of the peak are very insensitive against all kinds of peak distorting and broadening effects.

Figure 3.3b shows how the difficulties in locating the peak boundaries increase dramatically as the peak-to-base ratio becomes very large. Both the true baseline (baseline a) as well as the line connecting the points of highest curvature (baseline b) correspond to too large a number of channels, owing to the low-energy tail of the peak. However, a more realistic baseline (baseline c) does not correspond to characteristic or easily identified channels.

It can be concluded that, with TPA methods that obtain peak boundaries directly from the spectrum, the imitation of the human observation by eye should generally be avoided. As a rule the smaller peaks need wider boundaries and the more intense peaks narrower ones than the observer tends to allocate in a graphically presented spectrum. In Figure 3.6a it can be seen that TPA methods with a baseline close to the true baseline have on average a better precision than methods with a high baseline. Nevertheless, the interval between the peak boundaries for TPA methods should be kept as small as possible, in order to screen the area calculation against the influence of true baseline curvature and interference by adjacent peaks.

Wasson[56] has suggested an interesting PPA method which is illustrated by Figure 5.5d. The straight baseline is constructed as close as possible to the true baseline, but the peak boundaries are chosen far above it such that the wings of the peak are effectively truncated. Thus, those parts of the peaks are discarded that are most vulnerable to low precision and systematic error on the baseline positioning. Baedecker[49] found experimentally that the precision and accuracy of this method ranked among the best of the methods he had compared. The PPA value according to Wasson's method is calculated by the equations:

$$A = \sum_{i=-(k-m)}^{k-m} n(i) - (k - m + \tfrac{1}{2})[n(k) + n(-k)] \qquad (3.16)$$

in which $1 \leqslant m \leqslant k$. The variance of A is given by

$$V_A = \sum_{i=-(k-m)}^{k-m} n(i) + (k - m + \tfrac{1}{2})^2[n(k) + n(-k)] \qquad (3.17)$$

In equation (3.17) the relative weight of the baseline boundary channels in the variance has been decreased compared with the TPA methods [equation (3.14)]. For $m = 1$, equations (3.16) and (3.17) are reduced to equations (3.13) and (3.14). In Figure 3.6b the relative precision $V_A^{\frac{1}{2}}/A$ is shown for the peak presented in Figure 3.5a. such that the results can be directly compared with those of Figure 3.6a. Apparently for a k value within the TPA region, an optimum value of m lies around $0.8k$. For such a low effective number of channels $(1 + 0.4k)$ in the peak integration interval, accuracy problems may arise in the case of peak broadening.

For the determination of the peak boundaries Guzzi *et al.*[33] scan the spectrum, starting from the highest channel number in the region where a peak is located, and search for pairs of adjacent channels, at both sides of the peak, for which the difference in content is smaller than the standard deviation on one of the channels. For this method, the number of channels between the peak boundaries is strongly dependent on the intensity of the spectrum and the PTB ratio. For increasing counting statistics and decreasing PTB ratio, a systematically smaller number of channels and a correspondingly smaller fraction of the true peak area will be obtained. The method may be considered as a TPA technique. A straight line through the peak boundaries defines the baseline, and the precision on the resulting peak area is given by equation (3.14).

Op de Beeck[19, 57] has used a method for locating boundaries by searching for a true (non-statistical) minimum at both sides of the peak. If such a minimum is not found within a range of 1.5 FWHM on either side of the peak, a statistical significance test, analogous to the one of Guzzi *et al.*[33], is used within the interval between 1.5 and 2.5 FWHM on either side of the peak. If this test also fails, the peak boundary is fixed at 2.5 FWHM from the peak top. The baseline is a linear function connecting the peak boundaries and, therefore, the precision is given by equation (3.14). The method suffers from two systematic errors with opposing effects. The first one, giving rise to negative errors, arises because a smaller fraction of the total peak area is obtained as the PTB ratio decreases. The second one, giving rise to positive errors, results from the selection of a statistical minimum coinciding with a true minimum as a peak boundary. This results in underestimating the base area. The second error is only important for relatively large statistical fluctuations. For small peaks situated on a large continuum, the negative error will predominate.

Choy and Schmitt[58] and Quittner[59] have defined peak boundaries by using least squares methods to fit polynomials through several data points in the regions just adjacent to the peak. Quittner used second-degree polynomials on both sides of the peak, whereas Choy and Schmitt used a quadratic function of the channel number for the count rate on the left of the peak, and for the logarithm of the count rate on the right side. Quittner defines a baseline as a cubic polynomial that has the same slope and value as the quadratic polynomials at the central points of the left and right fitting interval. Baedecker[49] and Hertogen *et al.*[50] have found that Quittner's method provides a very precise TPA result, mainly because a very large number of channels is used for estimating the base area. The accuracy is also good on condition that some 20 channels on either side of the peak are not influenced by other peaks, and that the continuum under the peak is only very slightly curved. For a seriously curved true baseline the method fails badly. It is, however, very successful in the case of very low PTB ratios, especially if statistical fluctuations are important. For larger PTB ratios, Quittner's cubic baseline results in peak areas practically identical with those obtained from a straight line drawn through his peak boundaries. Choy and Schmitt[58] define a straight baseline through the minima of the parabolas fitted to both peak sides. Although their method has not been systematically studied or compared with others, it can be assumed that it has the same limitations as Quittner's method. Precision and accuracy will be slightly less favourable because of the straight baseline.

Tominaga *et al.*[25] have suggested fitting a single polynomial through groups of channels adjacent to both sides of the peak. Hertogen *et al.*[50] have fitted a single parabola directly through three or five points to the left and the right of the peak. Considering the small number of points, the gap between them and the influence of statistical fluctuations, this method will practically result in the equivalent of a straight line through the average of the three or five points at both sides of the peak.

It may be concluded for the polynomial fitting methods that they should not be used in a routine general spectrum analysis program because of their rather severe limitations and computational complexity. Furthermore, the fitting regions and the number of channels in these regions have to be defined *a priori*. As shown in Figure 3.3b, the curvature of the spectrum in the regions immediately adjacent to a peak is not always representative of the true baseline and it can be seriously doubted whether any polynomial, no matter how obtained, is a more useful approximation to the true baseline than a straight line.

A base spectrum, as defined and discussed previously, in the sections on preliminary data treatment and peak detection, would be the ideal method for providing a curved baseline simultaneously for all peaks in the spectrum. Peak boundary selection prior to the baseline construction is not necessary, and the base spectrum is supposed to approach the true baseline under the peaks very closely. In practice, however, a few important drawbacks should be mentioned. The averaging method suggested by Groszwendt[26, 30] and Ralston and Wilcox[29] makes it very difficult to calculate the precision on the generated base spectrum. Indeed, a rapidly increasing covariance between adjacent channels in the base spectrum will occur as a result of the repetitive averaging steps. It will, therefore, be practically impossible to obtain the proper precision estimate on the final peak area result by using the standard error propagation equations. Fujii's method[31] for generating a base spectrum through all true minima in the original spectrum has not been described in detail. It can be expected, however, that locating all true minima in any spectrum is a formidable task and cannot be completely free of failure.

All base spectrum methods have difficulties in forcing the base spectrum through pronounced maxima of the spectrum that do not correspond with TAPs. At Compton edges, for example, the base spectra will be systematically too low, as was found by Turkstra *et al.*[12] for the Ralston and Wilcox method, which led them to the conclusion that the method is not suitable for general use. Furthermore, in regions of the spectrum with high peak density or for very intense peaks on a low base, the average of even a large part of the spectrum is seriously affected by the presence of the peaks and, therefore, it is no longer a good measure of the base spectrum. As a result, the averaging methods provide a base spectrum that is systematically too high. This was proven experimentally by Hertogen *et al.*[50] who concluded, nevertheless, that the Ralston and Wilcox method competes well with others.

As a general conclusion, it may be pointed out that methods which use the original or smoothed spectrum to determine the peak boundaries and the baseline suffer from two major difficulties. In the first place, as the PTB decreases it becomes increasingly easier virtually to recognise peak boundaries and to fit a straight line or a polynomial through the continuum under the peak, but the resulting peak area shows an increasing systematic negative error. On the other hand, for high PTB values it becomes difficult to indicate characteristic channels as peak boundaries corresponding to a sufficiently narrow interval for peak area integration. The latter is especially true if one requires the same fraction of the true total peak area in order to compare peaks of the same γ-energy but with very different PTB ratios.

(*c*) *Methods with baseline obtained from a derived spectrum*—Yule[37,39] has pointed out that the peak boundaries may be advantageously located in the smoothed first derivative spectrum. Indeed, for an isolated peak on a flat continuum, the first derivative goes from positive to zero at the left peak boundary and from negative to zero at the right peak boundary (see Figure 3.4b). In the presence of statistical fluctuations, however, the first derivative will fluctuate around zero. These fluctuations will start at a shorter distance from the peak centroid as the counting statistics become relatively more important. This results in a systematic negative error on the peak area for a decreasing PTB ratio. Furthermore, if the peak is located on a continuum with a substantial slope, the first derivative will not reach zero in the immediate vicinity of the peak, which results in a boundary too far removed from the peak. Bowman[40] has used Yule's method to locate peak boundaries and valleys between slightly overlapping peaks with the aid of the first derivative.

Higher derivatives have not been used up to now to locate peak boundaries, although the second derivative or difference spectrum has very interesting properties in this respect. It exhibits extrema for all the points with highest curvature and these include the peak top as well as two points at the lower end of the peak sides at approximately one FWHM from the peak centroid. Unfortunately these extrema, except for the peak top, are easily blurred by counting statistics (see Figure 3.4c).

The SWC spectrum suggested by Op de Beeck[27] has all the advantages of a smoothed second derivative and is not so sensitive to counting statistics (see Figure 3.4e). It is, therefore, very well suited for peak boundary location. Peak boundaries may be selected at the channels coinciding with the negative minima (Figure 3.4e: 1–1), or at the channels where the negative SWC spectrum becomes smaller in absolute value than its own standard deviation (Figure 3.4e: 2–2), or even further separated at the channels where the negative SWC spectrum becomes zero or positive (Figure 3.4e: 3–3). The deep negative minima are not recommended as peak boundaries for TPA methods because the corresponding baseline lies too far above the continuum. Baselines defined by the other points are usually very close to each other. An interesting combination is the use of the SWC spectrum to detect peaks and locate their centroids and boundaries, followed by Wasson's peak area calculation[56] with 2–2 or 3–3 (Figure 3.4e) as a baseline and 1–1 as peak integration limits.

Until now, convolution transformed spectra have not been explored much for peak boundary location, but the interesting properties of the SWC spectrum point to a promising direction for further research.

(d) *Peak fitting methods*—The methods described in the previous three sections, which will be called direct methods, have all been conceived only for single peak structures or very slightly overlapping peaks. Peaks are said to overlap if there is at least one channel to which both peaks contribute substantially. In that case, a meaningful peak boundary for the estimation of the baseline between the peaks cannot be obtained. It is then necessary to define peak boundaries that include all components of the complex peak structure. Although some of the methods described in the previous sections for locating peak boundaries could be used for this purpose, one has to object against the concept of a single baseline under two or more peaks simultaneously. Indeed, the curvature of the base spectrum over such a large number of channels is no longer negligible; this rules out a straight baseline. Furthermore, the curvature of the spectrum in the regions adjacent to the multiplet cannot be used as an estimate for the true baseline curvature under the overlapping peaks.

A second important problem is the fact that even if a meaningful baseline can be defined and a good estimate of the base area subtracted, the resulting net counts in the channels were the peaks overlap have to be divided between the different components. In the literature, procedures are described to solve this problem with the aid of peak fitting methods based on the least squares principle. Most authors have modified the Gaussian functions that are to be fitted to the TAPs of a spectrum, in such a way as to represent the experimental shape of a TAP accurately. On the other hand, practically in all cases an over-simplified shape for the continuum under the peaks has been used.

Freiburg *et al.*[44] and East[50] used a linear function to represent the base spectrum. Dauk[34], Mariscotti[45] and Varnell and Trischuk[61] suggested a quadratic polynomial to fit the base spectrum, whereas Slavic and Bingulac[35] used a linear function, but have provided the possibility for quadratic or cubic polynomials in order to obtain a better fit. Considering that a complex peak structure is obtained by a superposition of single peak structures like the one shown in Figure 3.3b, it is clear that the underlying continuum consists of a series of rounded steps that can only be relatively well represented by a high-order polynomial. In this respect, the interesting approach of Gunnink and Niday[62] should be mentioned. Under the complex peak structure they construct a series of steps coinciding with a first estimate of the centroids of the different components, whereas the height of the steps is proportional to the first estimate of the intensity of the different components. Subsequently, the steps are smoothed to obtain a rounded step function which is probably the best estimate of the true baseline that can be obtained.

As far as peak models are concerned, Mariscotti[45], East[60] and Slavic and Bingulac[35] used simple unmodified Gaussian functions with a FWHM which varied according to the experimentally determined width of the TAPs as a function of energy. Dauk[34] used the correction method, for the left side of a TAP, suggested by Routti and Prussin[63], who conceived corrections at both the low- and high-energy sides of the peaks. Varnell and Trischuk[61] designed correction functions for pulse pile-up effects at both sides of the peak and an additional correction for the low-energy tail of the peak, resulting in a total of seven parameters that have to be estimated before the fitting of a single peak can be started. Clearly, from Figure 3.3b it can be concluded that a correction for the low-energy tail of the peak is a real necessity as this represents an important deviation from the Gaussian function. Further refinements of the peak model may be of little practical importance, especially in view of the simple approximation of the continuum under the peak.

The ultimate in peak fitting methods has been described by Heath *et al.*[64, 65]. In their approach, which was originally designed for NaI(Tl)-type spectra, but later adapted to Ge(Li)-type spectra, the total spectrum shape including TAP and Compton continuum for mono-energetic isotopes was determined experimentally as a function of γ-energy. As a result, the total response function of the detector was obtained in terms of a series of parameters. With these, it is possible to fit a complex γ-spectrum in its entirety with a sum of response functions for the various γ-energies that contribute to the spectrum. This method proved to be very successful for scintillation-type spectra, but turned out to be somewhat more difficult for Ge(Li) detectors for practical reasons. Unfortunately the method is too complex and lengthy to be used for activation analysis work.

All peak fitting methods are relatively more complicated and time-consuming than direct methods. Therefore they should be avoided for analysis of single peaks, especially if a small to medium size computer is to be used. However, for the analysis of multiplets, peak fitting is the only worthwhile solution at the moment. Therefore, if one decides to analyse overlapping peak structures, the peak parameters for single peaks should be obtained by a peak fitting method as well, in order to minimise the influence of systematic errors which may be of quite different natures for the fitting and direct methods.

Peak fitting methods suffer from several practical drawbacks, especially for the analysis of multiplets. As the number of components and an estimate of their intensity and location have to be provided before the fitting, some preliminary detailed information about the multiplets has to be acquired. Furthermore, the required peak model is not simple and contains several additional parameters that have to be estimated before the fitting. These shape parameters are themselves functions of more fundamental parameters, such as γ-energy, as well as of purely experimental parameters that are not always easily controlled or standardised: factors determining the quality of the detector; stability of the electronic counting chain as a function of count rate and temperature; geometry and scattering properties of the source being counted as well as the environment of the detector, etc.

Finally, all fitting methods depend on the least-squares concept and minimise the χ^2 value represented by a hypersurface in a space with as many dimensions as parameters in the combined peak and continuum model. Owing to the non-linearity of the regression equations, iterative calculation procedures are required to locate a minimum on the χ^2 surface. There is always a finite probability, especially if the initial estimate of the parameters is not very good, that small spurious χ^2 minima may end the iteration so that the best fit may never be reached.

3.3.2.5 Determination of important peak parameters

In the section on peak detection methods we have already discussed how an accurate peak location can be calculated in channel units after subtraction of the continuum from the spectrum under the peak. Several methods of carrying out such a subtraction and obtaining an estimate of the net peak area or a value proportional to it, in a more or less precise and accurate way, have been described in the sections on peak boundary and baseline determination. At least another three important peak parameters have to be determined.

The peak location has to be converted from channel units to energy units in order to carry out an identification of the isotopes present in the spectrum. The FWHM of the peak has to be obtained in order to discriminate between true single peaks and closed doublets. Finally, a detection limit for the measurement of each TAP superimposed on a continuum is necessary to provide a measure of the sensitivity of the analysis.

(a) *Energy calibration*—The conversion of peak locations from channel units to energy units is carried out by calibration of the γ-spectrometer as a whole with the aid of isotopes having γ-transitions of accurately known energy. Theoretically there is a linear relationship between the energy transferred by a γ-photon to the detector and the pulse height obtained from the detector. However, in the amplification, and mainly in the analogue-to-digital conversion steps, a certain amount of non-linearity is introduced that is sufficiently important to make the direct identification of isotopes extremely difficult. The relation between γ-energy and channel number has to be calibrated with the aid of a polynomial with an order depending on the quality of the electronics. The required calibration equation is given by

$$E = A_0 + A_1C + A_2C^2 + A_3C^3 + \ldots \qquad (3.18)$$

in which E = energy and C = channel number. The parameters A_0, A_1, A_2 . . . can be determined by least squares fitting the experimentally determined channel numbers from a series of standard γ-sources to the well known energies of these standards. However, since the channel numbers are subject to statistical errors, this results in the following polynomial:

$$C = B_0 + B_1E + B_2E^2 + \ldots \qquad (3.19)$$

The parameters A_0, A_1, A_2, . . . are then obtained by least squares fitting the standard energies to the channel numbers calculated with equation (3.19), which results in the polynomial given by equation (3.18). Starting from order one, the correct polynomial order is obtained by checking some measure for goodness of fit (χ^2, average of squared residuals, etc.) for an insignificant decrease corresponding to the increase of the order of the polynomial.

The calibration procedure has to be repeated more or less frequently depending on the stability of the electronics. However, if the non-linearity is not pronounced and the electronic shift is small, a simple linear correction applied to equation (3.18) is often sufficient:

$$E = F(A_0 + A_1C) + A_2C^2 + \ldots + \mathrm{d}E \qquad (3.20)$$

The correction parameters F and $\mathrm{d}E$ are obtained from the measurement of two standard γ-energies only. If E_1 and E_2 represent the energies of the standard γ-lines (preferably a high- and a low-energy line with $E_2 > E_1$) and if x_1 and x_2 are the observed channel numbers corresponding to E_1 and E_2 at the time of the calibration, whereas x_1' and x_2' are the observed channel numbers after a shift has occurred, then the parameters F and $\mathrm{d}E$ are given by

$$F = \frac{x_2 - x_1}{x_2' - x_1'} \qquad (3.21)$$

and

$$\mathrm{d}E = E_1\left(1 - \frac{x_2 - x_1}{x_2' - x_1'}\right) + (E_2 - E_1)\left(\frac{x_1 - x_1'}{x_2' - x_1'}\right) \qquad (3.22)$$

(b) *Peak width*—The determination of the width of a peak expressed as the FWHM is not easy, especially if counting statistics are relatively important. Furthermore, taking into account that there exist several peak broadening effects it should only be decided that a peak is in fact a closed doublet if the observed FWHM is considerably larger than expected. In order to make this decision the FWHM has to be known as a function of energy for a given detector and its associated electronics. Theoretically the following relationship exists between resolution and energy:

$$(\text{FWHM})^2 = kf\varepsilon E + \rho^2 \tag{3.23}$$

where k = proportionality constant; f = Fano factor; ε = energy required to produce one electron–hole pair (2.98 eV for Ge); E = energy of the gamma photon; and ρ = contribution to the resolution by electronic factors (noise from detector leakage current, preamplifier, etc.).

The FWHM will become practically constant below a certain energy, depending on the quality of the detector and associated electronics. Experimentally, the FWHM is determined for a series of standard peaks by fitting the experimental data to the following equation:

$$(\text{FWHM})^2 = AE + B \tag{3.24}$$

The parameters A and B are obtained from the least squares fitting procedure. The determination of the FWHM for the individual peaks can be done with the aid of the SWC spectrum[27] because the total width of the positive part of the central peak in this spectrum has a simple linear relationship to the FWHM of the original peak. This relationship can be determined experimentally as a function of γ-energy. For calibration purposes the FWHM values can be obtained with better accuracy by the method of Zimmerman[66]. For any three equidistant points of a Gaussian function $y(x)$, separated by unity on the abscissa, the following equation holds:

$$\sigma^2 = 2(x - x_0)/\log_e[y(x - 1)/y(x + 1)] \tag{3.25}$$

where x_0 = abscissa of the peak centroid; and σ = width of the Gaussian function (FWHM = 2.355σ).

(c) *Detection limit*—The detection limit required for the estimation of the sensitivity of the analytical method is an *a priori* detection limit predicting the probability that the method will be able to detect a TAP with a given intensity. Altshuler and Pasternack[67] have pointed out that such a detection limit results from the interaction of two kinds of errors. The first type in spectrometry is to decide that a peak is present when there is none; this is to mistake a statistical fluctuation for a true TAP. The second type is to decide that a TAP is absent, when actually a small peak is present, but masked by the variance on the spectrum continuum. Currie[68] has defined a detection limit based on these two types of error for radioactivity measurements of the kind in which a background can be measured in the absence of the signal of interest. This is never possible in spectrometry, however, where the 'background' is the sum of several Compton continua of which some are due to higher-energy peaks of the same isotope as the peak being considered. As a result, most peak detection methods will be successful when Currie's equations predict a failure, if the latter are applied (unjustified) to the base area estimated from the spectral regions adjacent to the peak. The *a priori* detection limit is actually given by the ability of the peak detection method to recognise a peak shape among the random statistical fluctuations on a spectrum continuum. The peak shape can be roughly reduced to two parameters: peak height or intensity and peak width. These parameters have an opposing effect on the ability to recognise a peak. Indeed, for a given constant peak area it will be increasingly easy to distinguish the peak from the narrow statistical fluctuations as its width increases. On the other hand, the correspondingly decreasing peak height will make it more difficult to recognise the peak by its intensity.

Clearly the theoretical definition of a meaningful *a priori* detection limit is in general very difficult if not impossible. Head[69] has suggested that errors of the first kind can be determined experimentally, for a given method, from Compton continua in real spectra as a function of the average continuum height. However, as he was obliged to keep the peak width constant, his approach is of little practical importance for routine spectrum analysis.

If we restrict ourselves to errors of the first kind, for a given location in a specified measured spectrum, an *a posteriori* detection limit can be conceived which corresponds to the largest TAP that can be present but completely masked by the variance on the local spectrum continuum at that specified location. This *a posteriori* detection limit is important since it constitutes the basis for calculation of upper limits for the concentrations of those elements that give no detectable TAP in the sample spectra.

As in the case of an *a priori* detection limit, the same implications concerning the peak shape have to be dealt with. However, if we assume that we have a peak detection method that has a strong ability to discriminate against all spectral structures that do not resemble a peak very closely, then the decision whether a peak is present or not is determined only by the intensity of the supposed peak in relation to the variance of the local continuum. Furthermore, we can then consider the largest number of spurious peaks that can possibly be present in a spectrum as being given by the maximum number of statistical maxima with approximately the same total width (= 2 FWHM) as a true peak:

$$M \approx \frac{N}{2\ \mathrm{FWHM}} - n \tag{3.26}$$

where N = total number of channels in the spectrum; FWHM = average width of a true peak in the spectrum; and n = total number of true peaks, present in the spectrum.

For 4000 channels, 40 true peaks, 0.5 keV per channel and 2 keV FWHM, M amounts to approximately 460 spurious peaks.

Let the signal of interest, used to test the presence of a peak, be given by

$$S_{A,B} = f(A', B') \tag{3.27}$$

in which A = net peak area; B = integral of Compton continuum under the peak; $A' = aA$; $B' = bB$; and a and b = proportionality constants defined by the peak detection method being used. S can be the peak area, the peak height, the intensity of the central minimum in the second derivative spectrum, the central maximum in the SWC or cross-correlation spectrum, or any other parameter related to the peak intensity. The expected value of S in the absence of a peak has to be zero. Therefore, $S_{0,B}$ is distributed around zero as shown in Figure 3.7, curve 1. We can now select arbitrarily any positive

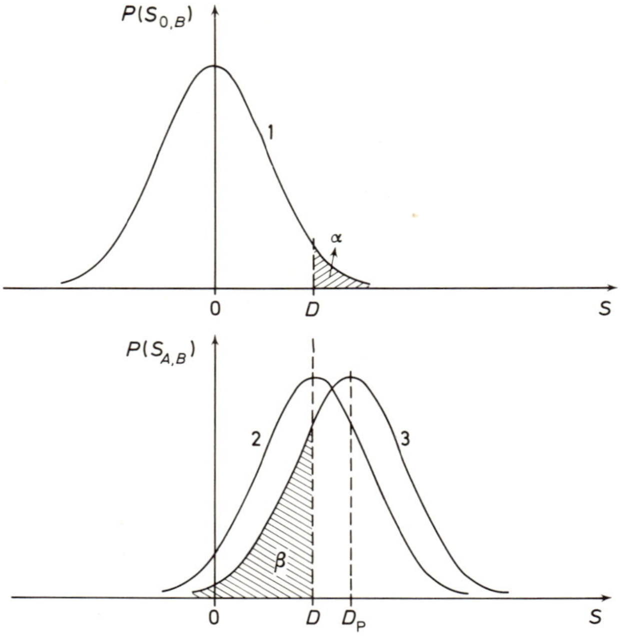

Figure 3.7 Definition of *a priori* and *a posteriori* detection limits

non-zero value of S, e.g. D on Figure 3.7, to serve as a detection limit. A fraction a exists for each choice which is the probability of making an error of the first kind, so that aM spurious peaks will be accepted on average. A peak giving a signal $S_{A,B}$ equal to D on average gives a distribution of S values around D, as shown in Figure 3.7, curve 2. As a consequence, peaks on the detection limit will on average be missed in 50% of the cases. As this error of the second kind is too large for an *a priori* detection limit, the latter is always higher than the *a posteriori* limit so that only a fraction β, corresponding to the limit D_P in Figure 3.7, of the peaks will not be found. A value of 0.05 is recommended for β.

In practical spectrometry we want aM to be small and as a result values for a of the order of 0.005 or less are recommended. This value for a fixes the *a posteriori* limit D by the relation

$$D = k_a \times \sigma_{S_{0,B}} \tag{3.28}$$

in which $\sigma_{S_{0,B}}$ = standard deviation of the distribution of $S_{0,B}$; and k_a = constant corresponding to a probability = $(1 - 2a)$ for the standardised normal distribution. The *a priori* detection limit is then given by

$$D_P = D + k_\beta \times \sigma_{S_{A,B}} \tag{3.29}$$

with $\sigma_{S_{A,B}}$ being estimated from the distribution of S values corresponding to peaks that give rise to an average value for $S = D_P$; D is apparently a function of B only and can, therefore, be calculated by applying the error propagation equations to equation (3.27) with $A' = 0$. If, for example, we use the SWC spectrum to detect peaks, then

$$S_{A,B} = (aA + bB) - (a'A + bB) \tag{3.30}$$

$$\sigma_{S_{0,B}} = (2bB)^{\frac{1}{2}} \tag{3.31}$$

and

$$D = k_a(2bB)^{\frac{1}{2}} \tag{3.32}$$

The set peak area A_D corresponding to a SWC value of D is then given by setting $S_{A_D,B}$ equal to D:

$$A_D = \frac{k_a}{a - a'}(2bB)^{\frac{1}{2}} \tag{3.33}$$

On the other hand, A_D may be directly determined experimentally by using an estimate B' of B, determined from channels adjacent to the peak, so that:

$$\sigma_{A_D} = (A_D + B + \sigma_{B'}^2)^{\frac{1}{2}} \tag{3.34}$$

In Table 3.1 the values of D, A_D and σ_{A_D} (assuming that $\sigma_{B'}^2 = B$) are given for a series of a and k_α values, for a square wave convolute spanning 8

Table 3.1 Detection limit and corresponding total peak area and its expected standard deviation, together with the average number of spurious peaks, as a function of the probability for errors of the first kind

$a/\%$	k_a	D/counts	A_D/counts	$\delta_{A_D}/\%$	*Number of spurious peaks per 100 channels*
0.2	2.880	204	611	20.5	0.023
1	2.327	165	494	25.2	0.115
2	2.054	145	435	28.5	0.230
5	1.645	116	349	35.4	0.575
10	1.282	91	272	45.5	1.15
20	0.841	59	178	69.2	2.30

channels and a Compton continuum value of 500 counts per channel. The expected number of spurious peaks per 100 channels is also given, assuming that the Compton height is constant over 100 channels. Obviously very small peaks can be detected, on a 50% second type error basis with a minimum of spurious peaks (first type error), on condition that the peak shape can be recognised with absolute certainty. In practice this is, of course, not true and a larger number of spurious peaks will be observed for a given a, since M is larger than predicted by equation (3.26). Therefore a has to be further decreased empirically, giving a rise in D until aM becomes acceptable.

3.3.3 Calculation of analytical results

The operations performed at level 2 in the data reduction sequence (Figure 3.2) consist mainly of a simple comparison of a sample–standard spectrum pair. This comparison involves the ratios of peak areas (or detection limits) for peaks corresponding to the same γ-energy. Taking into account the weights of the samples and of the elements of interest in the standards, final analytical results can be obtained in terms of weights and concentrations. Before the peak areas resulting from the individual spectrum analysis can be compared, they have to undergo a series of simple and straightforward corrections.

If any of the irradiation, decay or counting periods is different for samples and standards, corrections have to be applied in accordance with the individual half-lives of the isotopes. If any of the isotopes present decays appreciably within the counting period, and the dead time of the analyser used is not negligible, dead time losses occur that cannot be compensated for by the normal life time operation of the analyser. As a result, special dead time corrections have to be applied[70-72]. If appreciable flux gradients exist at the irradiation site, flux monitors can be irradiated together with standards and samples, and flux correction factors derived from them. Often, the same isotope is produced from chemically different elements. Corrections for these interference reactions can be estimated by calculation, or more accurately determined experimentally. If the individual spectrum analysis does not provide accurate peak areas for overlapping peaks, corrections can be carried out, based on accurately known or measured γ-ray abundance ratios, if the interfering peaks belong to isotopes having more than one peak in their γ-spectrum. Finally, corrections for neutron shadow effects in either or both samples and standards may be necessary. However, as these effects cannot be calculated or measured very accurately, it is recommended that they should be estimated in advance and that their occurrence during the irradiation should be prevented.

It should be noted that some of the correction factors such as those for dead time, interference reactions, flux gradients, etc. have to be ready at the level 2 data reduction step. They have to be deduced from time or activity measurements, which have to go through a level 1 data reduction step in parallel with the individual spectrum analyses.

3.3.4 Similarity analysis

At level 3 of the data reduction scheme (Figure 3.2), several methods are available to derive qualitative and quantitative conclusions concerning the chemical similarity between many samples or the analogous behaviour of a large number of chemical elements in many samples. All these methods are mathematically complex and include the manipulation of large data matrices, which necessitates the use of a computer.

The development of instrumental multi-element analysis techniques, among which activation analysis based on Ge(Li) spectrometry plays an important role, puts an enormous amount of final analytical data at the disposal of the analytical chemist within a very short time. The automation of all steps of the analysis, including partial or complete real-time computer control, frees the analyst from the burden of physically participating in the act of analysis, and allows him more time to speculate on the use of methods for interpreting his results. The design and evaluation of such methods should be performed by the analytical chemist, since he is ultimately the best judge of the accuracy and precision of his work.

In the literature, methods can be found to group either elements or samples. For the elements, linear correlation coefficients can be used to describe simple linear relationships between pairs of elements in a group of samples. Care should be taken, however, to verify that all concentration values, for a single element, used in the calculation of a linear correlation coefficient belong to a single normally distributed population. If the latter is not true, meaningless coefficients leading to wrong conclusions will be obtained. Subsequently, factorial analysis can be used to detect larger groups of elements that are linearly correlated. These methods have already been applied to geochemical problems[73-75].

The grouping of samples in terms of their chemical similarity is more complex. First of all a distinction must be made between those elements that define the chemical nature of the sample (major and minor constituents), and the trace elements which contribute little or nothing to the chemical character of the sample. In the latter case the term 'fingerprinting' can be justified.

The series of trace element concentrations in a sample can be considered as constituting a specific pattern characterising a sample only as far as chemical genesis and evolution are concerned. The similarity analysis is then reduced to a problem of pattern recognition which can be solved with the aid of computerised learning machines[76, 77]. Operating either directly on activities in channels of NaI(Tl) spectra[77] or on trace element concentrations[76], complex spectra of archaeological artefacts can be separated into distinct groups, on condition that a sample population for each group is available to assist the machine in its learning phase to make the correct decisions.

Instead of concentrations, the ratios of concentrations for the same elements between different samples[78], or the ratio of ratios of concentrations of different trace elements within a sample[79], can be considered. The variance on these ratios[78] or the fraction of the ratios falling within experimentally determined limits[79] can then be used as criteria for grouping the samples.

A third alternative is to represent the chemical similarity between two samples by a function (similarity index) of all concentration differences for the trace elements determined in the samples. Taxonomical classification methods[80-82] can then be used, resulting in hierarchical groupings. The latter can be represented by a topological tree, the projection of which in a plane immediately gives the dissimilarity structure and hierarchy of the groups at a glance. An example derived from trace element analysis of coal samples[83] is given in Figure 3.8. This graphical representation is called a dendrogram and can be considered as a convenient and comprehensive means of summarising the results of a dissimilarity analysis. As indicated in Figure 3.8, the samples are located at the top of the branches. The ordering of the

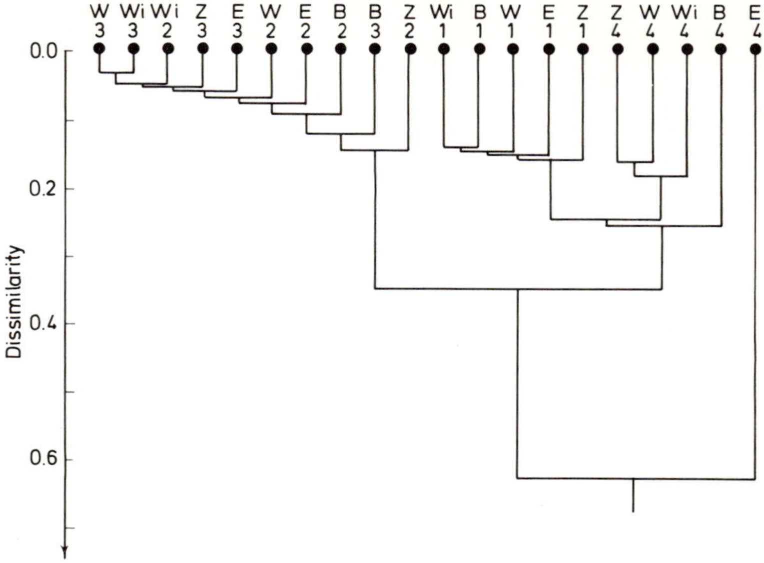

Figure 3.8 Classification dendrogram summarising the similarity among 20 different coal samples. All mining locations are situated in Belgium: B, Beringen; E, Eisden; W, Waterschei; Wi, Winterslag; Z, Zolder. The numerical codes refer to the kind of coal: 1, coal dust intended for use in electrical power plants; 2, fine coal size 0/10 intended for coke production; 3, same as 2 but size 10/80 for heating of houses; 4, 'mixtures' intended for use in electrical power plants

samples is determined by the projection of the tree carried out so as to minimise the overlap (crossing) of branches. Therefore the horizontal parts of the branches have no quantitative meaning. The length of the vertical part of the branches is a measure of the overall chemical dissimilarity, as indicated by the scale at the left, for the samples (or groups of samples) connected by these branches. The nodes of the branches are representative for two or more similar samples and are comparable with single samples or other nodes at a higher hierarchy in the classification. In Figure 3.8 it is clear that grouping occurs as a function of the kind of coal rather than of the locality in which it was produced. The homogeneity of the groups, their mutual dissimilarity and the occurrence of spurious samples is obvious.

3.4 QUALITATIVE ANALYSIS

In Section 3.3.2.5 we discussed how peak locations in channel units can be accurately converted to energy values. These energy values are highly characteristic of the isotope that emitted the γ-rays which gave rise to the peak. In principle, therefore, the energy values can be used to identify the isotopes contributing to the spectrum being considered. In practice, however, a series of difficulties arise which make the identification procedure far from trivial.

In spite of the precision with which the location of a peak can be determined in channel units, and the accuracy with which the latter can be converted to energy units, the identification in general cannot be performed unambiguously because of the high density of γ-energies per unit energy in the 70–1500 keV interval where most unstable isotopes emit. Typically, as shown in Table 3.2, the average energy density may be as high as 2 keV^{-1}. The picture is

Table 3.2 Gamma energy density as a function of energy for a large number of isotopes

Energy interval /keV	*Number of γ-energies*	*Energy interval* /keV	*Number of γ-energies*
0–100	134	1200–1300	87
100–200	237	1300–1400	67
200–300	195	1400–1500	56
300–400	177	1500–1600	45
400–500	162	1600–1700	34
500–600	138	1700–1800	33
600–700	152	1800–1900	36
700–800	132	1900–2000	20
800–900	139	2000–2100	19
900–1000	103	2100–2200	12
1000–1100	106	2200–2300	12
1100–1200	89	2300–2400	9

further obscured by the limited resolution of present-day detectors, the occurrence of double and single escape peaks, and the masking effect of Compton distributions for low intensity peaks.

By speculating on the half-life of the isotope, in relation to the irradiation, decay and counting time of the samples, some isotopes can be eliminated with high probability in order to facilitate the peak-to-isotope assignment.

The simplest isotope identification method requires an already rather elaborate logical decision scheme. The problem can be split up into monoenergetic isotopes and those that emit γ-rays with two or more characteristic energies. In complex spectra, the monoenergetic isotopes offer the largest difficulties. If a peak of approximately the required energy has not been found, the isotope may still be present, but its peak may be masked by a neighbouring more intense peak or Compton distribution. If such a peak is found, one has to check whether one is not dealing with a single or double escape peak or a peak belonging to another isotope emitting a γ-ray of nearly the same energy.

For isotopes emitting γ-rays with different energies, the previous remarks on monoenergetic isotopes are equally valid. However, as the number of energies per isotope increases, it becomes more and more probable that at least one peak can unambiguously be assigned. Furthermore, if more than one peak can be assigned to the same isotope, the identification probability increases sharply, especially if a check on the correct peak intensity ratio is made.

Special programs based on sophisticated decision schemes have already been published[84, 85]. Programs for quantitative gamma spectrum analysis incorporating more or less elaborate peak assignment routines have also been described[62, 86].

In practice, however, activation analysts seldom require a computer-assisted isotope identification. Usually sufficient information about the samples is available to allow a fast and easy peak assignment by hand, whereas routine analyses of large numbers of similar samples need no qualitative analysis at all. Arguments against systematic isotope identification for every spectrum analysed are:

(a) The relatively large size of the necessary computer routine.

(b) The extent of the library with isotope data (some 550 isotopes including their half-lives, and some 2000 γ-energies with relative intensities).

(c) The appreciable computer time lost in library searching.

A few authors[41, 87] have included a check on the half-life in order to confirm or help the identification of isotopes. This practice is not recommended for general use, since it requires that each sample is counted several times, and even then the half-life values are obtained with only poor precision.

3.5 TOTAL ACTIVATION ANALYSIS PROGRAMS

A complete activation analysis program would in theory include all steps and methods described in Sections 3.3 and 3.4 above, except probably for the similarity analysis discussed in Section 3.3.4. In practice, however, such comprehensive programs can only be run efficiently in very large and fast computers[62, 87]. If the latter are not available, several simplified approaches are possible that still fulfil most of the needs of the activation analyst.

First of all, the complete qualitative analysis can be taken out of the program and replaced by simpler routines that work with a very limited isotope library, containing only the isotopes for which the user requires final analyses results[19, 36]. In this way, such results can be obtained within a reasonable time from medium sized computers (e.g. 32K words memory) or even smaller ones (16K word memory or less), provided that in the latter case overlay structures or virtual memory are available[19].

A further reduction in program size and execution time can be obtained by not performing a complete spectrum analysis for all peaks present in sample and standard spectra. The program instead instructs the computer to look only at those locations in each spectrum where peaks are expected corresponding to the energies listed in the limited isotope library[88]. In this way, final analytical results can be obtained quickly with small to medium sized computers.

Finally, the required computational work could be further diminished by replacing the classical relative analysis method, using a standard for every element to be determined, by a comparator method[89]. Although such methods were suggested several years ago, they have been abandoned mainly because of the lack of reliable physical data for most nuclides of interest. Recently, however, the quality of these data has been much improved so that the use of comparator methods has become more feasible than before. Unfortunately the gain in computational efficiency has to be compensated by rigorous control of irradiation and counting conditions and an accurate monitoring of the detector response.

References

1. Gardner, D. G., Gardner, J. C., Laush, G. and Meinke, W. W. (1959). *J. Chem. Phys.*, **31,** 978
2. O'Kelley, G. D. (1968). *Proc. Symp. Gatlinburg, Application of Computers to Nuclear and Radiochemistry* (UAEC, NAS-NS-3107, 1963)
3. Munzel, H. (1964). *Proc. Symp. Salzburg, Radiochemical Methods of Analysis II*, 141 (Vienna: IAEA)
4. (1966). *NBS Technical Note*, No. 404, (J. R. De Voe, editor) 187
5. Parr, R. M. and Lucas, H. F. (1964). *IEEE Trans. Nucl. Sci.*, NS-11, 349
6. Schonfeld, E., Kibbey, A. H. and Davis, W., Jr. (1966). *Nucl. Instr. Methods*, **45,** 1
7. Gavron, A. (1969). *Nucl. Instr. Methods*, **67,** 245
8. Yule, H. P. (1969). *Proc. Int. Conf. Gaithersburg, Modern Trends in Activation Analysis, II*, 1155 (NBS Spec. Publ. 312) (Washington D.C.: NBS)
9. Yule, H. P. (1971). *Proc. Symp. Kjeller, Activation Analysis in Geochemistry and Cosmochemistry*, 145 (A. O. Brunfelt and E. Steinnes, editors) (Oslo: Universitetsforlaget)
10. Quittner, P. and Wainerdi, R. E. (1970). *Atomic Energy Rev.*, **8,** 361
11. Op de Beeck, J. P. (1971). *Computer Applications in Instrumental and Radiochemical Activation Analysis*, (Ohio: CRC Press)
12. Turkstra, J., Van Rensburg, M. C. J. and De Wet, W. J. (1970). *PEL 204* (Pretoria: Atomic Energy Board)
13. Thompson, C. J. (1969). *Proc. Int. Conf. Gaithersburg, Modern Trends in Activation Analysis, II*, 1121 (NBS Spec. Publ. 312) (Washington D.C.: NBS)
14. Pierce, T. B., Webster, R. K., Hallett, R. and Mapper, D. (1969). *Proc. Int. Conf. Gaithersburg, Modern Trends in Activation Analysis, II*, 1116 (NBS Spec. Publ. 312) (Washington D.C.: NBS)
15. De Bruin, M. and Korthoven, P. J. (1972). *Anal. Chem.*, **44,** 2382
16. Lenkszus, F. R. and Strauss, M. G. (1969). *Nucl. Instr. Methods*, **72,** 188
17. Kellie, J. D. and Crawford, G. I. (1969). *Nucl. Instr. Methods*, **72,** 329
18. Op de Beeck, J. P. (1972). Unpublished work
19. Op de Beeck, J. P. (1972). *J. Radioanal. Chem.*, **11,** 283
20. Adams, F., Op de Beeck, J. P., Van den Winkel, P., Gijbels, R., De Soete, D. and Hoste, J. (1971). *CRC Critical Rev. Anal. Chem.*, **1,** 455
21. De Soete, D., Gijbels, R. and Hoste, J. (1972). *Neutron Activation Analysis, Monographs in Analytical Chemistry and Applications*, Vol. 34 (P. J. Elving and I. M. Kolthoff, editors) (New York: Wiley-Interscience)
22. Savitzky, A. and Golay, M. J. E. (1964). *Anal. Chem.*, **36,** 1627
23. Yule, H. P. (1967). *Nucl. Instr. Methods.*, **54,** 61
24. Yule, H. P. (1972). *Anal. Chem.*, **44,** 1245
25. Tominaga, H., Dojyo, M. and Tanaka, M. (1972). *Nucl. Instr. Methods.*, **98,** 69
26. Groszwendt, B. (1971). *PTB-Mitteilungen*, **2,** 82
27. Op de Beeck, J. P. (1973). *Proc. Symp. Cambridge, Recent Developments in Neutron Activation Analysis*, in press
28. Inouye, T., Harper, T. and Rasmussen, N. C. (1969). *Nucl. Instr. Methods*, **67,** 125
29. Ralston, H. R. and Wilcox, G. E. (1969). *Proc. Int. Conf. Gaithersburg, Modern Trends in Activation Analysis, II*, 1238 (NBS Spec. Publ. 312) (Washington D.C.: NBS)
30. Groszwendt, B. (1971). *Nucl. Instr. Methods.*, **93,** 461
31. Fujii, I., Inouye, T., Muto, H. and Onodera, K. (1969). *Analyst*, **94,** 189
32. Dooley, J. A., Gorrell, J. H., Polishuk, P. and Young, M. (1969). *Proc. Int. Conf. Gaithersburg, Modern Trends in Activation Analysis, II*, 1090 (NBS Spec. Publ. 312) (Washington D.C.: NBS)
33. Guzzi, G., Pauly, J., Girardi, F. and Dorpema, B. (1967). *CETIS-EUR 3469e* (Ispra, Varese: Euratom)
34. Dauk, J. (1971). *QA-3-71* (Cologne: Inst. Nucl. Chem., University)
35. Slavic, I. A. and Bingulac, S. P. (1970). *Nucl. Instr. Methods.*, **84,** 261
36. Borchardt, G. A., Hoagland, G. W. and Schmitt, R. A. (1970). *J. Radioanal. Chem.*, **6,** 241
37. Yule, H. P. (1968). *Anal. Chem.*, **40,** 1481

38. Yule, H. P. (1969). *Proc. Int. Conf. Gaithersburg, Modern Trends in Activation Analysis, II*, 1256 (NBS Spec. Publ. 312) (Washington D.C.: NBS)
39. Yule, H. P. (1966). *Anal. Chem.*, **38,** 103
40. Bowman, W. W. (1971). *Nucl. Instr. Methods.*, **96,** 135
41. Verheijke, M. L. (1972). *J. Radioanal. Chem.*, **10,** 299
42. Nyman, B. (1973). *Nucl. Instr. Methods*, **108,** 237
43. Morrey, J. R. (1968). *Anal. Chem.*, **40,** 905
44. Freiburg, C., Erdtmann, G. and Petri, H. (1971). *Proc. Symp. Kjeller, Activation Analysis in Geochemistry and Cosmochemistry*, 167, (A. O. Brunfelt and E. Steinnes, editors) (Oslo: Universitetsforlaget)
45. Mariscotti, M. A. (1967). *Nucl. Instr. Methods*, **50,** 309
46. Black, W. W. (1969). *Nucl. Instr. Methods*, **71,** 317
47. Inouye, T. (1972). *Nucl. Instr. Methods*, **104,** 541
48. Sterlinski, S. (1972). Personal communication
49. Baedecker, P. A. (1971). *Anal. Chem.*, **43,** 405
50. Hertogen, J., De Donder, J. and Gijbels, R. (1973). *Nucl. Instr. Methods*, **115,** 197
51. Covell, D. F. (1959). *Anal. Chem.*, **31,** 1785
52. Sterlinski, S. (1968). *Anal. Chem.*, **40,** 1995
53. Sterlinski, S. (1970). *Anal. Chem.*, **42,** 151
54. Heydorn, K. and Lada, W. (1972). *Anal. Chem.*, **44,** 2313
55. Westermark, T. and Sjöstrand, B. (1960). *Int. J. Appl. Radiation Isotopes*, **9,** 1
56. Wasson, J. T. (1971). Cited in Baedecker, P. A. (1971). *Anal. Chem.*, **43,** 405
57. Op de Beeck, J. P., Baedecker, P. A. and Siegel, J. (1967). *MIT-905-108* (Cambridge, Mass.: MIT)
58. Choy, S. C. and Schmitt, R. A. (1965). *Nature*, **205,** 758
59. Quittner, P. (1969). *Nucl. Instr. Methods*, **76,** 115
60. East, L. V. (1971). *Proc. DECUS Fall Meeting, San Francisco*, 141 (Maynard: Digital Eq. Corp.)
61. Varnell, L. and Trischuk, J. (1969). *Nucl. Instr. Methods*, **76,** 109
62. Gunnink, R. and Niday, J. B. (1972). *UCRL-51061* (California: Lawrence Livermore Lab.)
63. Routti, J. T. and Proussin, S. G. (1969). *Nucl. Instr. Methods*, **72,** 125
64. Heath, R. L. (1966). *Nucl. Instr. Methods*, **43,** 209
65. Helmer, R. G., Heath, R. L., Schmittroth, L. A., Wayne, G. A. and Wagner, L. M. (1967). *Nucl. Instr. Methods*, **47,** 305
66. Zimmerman, W. (1961). *Rev. Sci. Instr.*, **32,** 1063
67. Altshuler, B. and Pasternack, B. S. (1963). *Health Phys.*, **9,** 292
68. Currie, L. A. (1968). *Anal. Chem.*, **40,** 586
69. Head, J. N. (1972). *Nucl. Instr. Methods*, **98,** 419
70. Wiernik, M. (1971). *Nucl. Instr. Methods*, **95,** 13
71. Bartosek, J., Adams, F. and Hoste, J. (1972). *Nucl. Instr. Methods*, **103,** 45
72. De Wispelaere, C., Op de Beeck, J. P. and Hoste, J. (1973). *Anal. Chem.*, **45,** 547
73. Parks, J. M. (1966). *J. Geol.*, **74,** 703
74. McCammon, R. B. (1968). *Amer. Assoc. Petrol. Geol. Bull.*, **52,** 12178
75. Saager, R. and Esselaar, P. A. (1969). *Economic Geol.*, **64,** 445
76. Kowalski, B. R. and Schatzki, T. F. (1972). *Anal. Chem.*, **44,** 2176
77. Wangen, L. E. and Isenhour, T. L. (1970). *Anal. Chem.*, **42,** 737
78. Borchardt, G. A., Harvard, M. E. and Schmitt, R. A. (1971). *Quaternary Res.*, **1,** 247
79. Anders, O. U. (1972). *Anal. Chem.*, **44,** 1930
80. Jardine, N. and Sibson, R. (1971). *Mathematical Taxonomy* (London: Wiley)
81. Sokal, R. R. and Sneath, P. H. A. (1963). *Principles of Numerical Taxonomy* (London: Freeman)
82. Op de Beeck, J. P., Putman, J., Gijbels, R., De Donder, J. and Hoste, J. (1973). *Proc. Int. Conf. Rome–Venice, Application of Nuclear Methods in the Field of Works of Art* (Rome: Accademia Nazionale dei Lincei)
83. Block, C. (1973). Personal communication
84. Adams, F. and Dams, R. (1969). *Radiochem. Radioanal. Lett.*, **1,** 163
85. Dooley, J. A., Gorrell, J. N., Thompson, J. M. and Hoffmann, E. (1968). *Proc. Int. Conf. Gaithersburg, Modern Trends in Activation Analysis II*, 1148 (NBS Spec. Publ. 312) (Washington D.C.: NBS)

86. Watterson, L., Watterson, I. I. W. and Rasmussen, S. E. (1972). *NIM-1476* (Johannesburg: Nat. Inst. Metallurgy)
87. Anders, O. U. (1969). *Anal. Chem.*, **41,** 428
88. Op de Beeck, J. P. and De Donder, J. (1973). Unpublished work
89. De Corte, F., Speecke, A. and Hoste, J. (1969). *J. Radioanal. Chem.*, **3,** 205

4
Analytical Techniques for Air Pollution Studies

P. W. WEST and V. DHARMARAJAN
Louisiana State University

4.1 INTRODUCTION

This review of analytical methods used for the study of air pollution includes background information on established methods, together with discussions of recent advances and trends. The field of air pollution monitoring has undergone a rapid and dramatic change during the past two decades. During this period, a transition from a limited number of crude analytical measurements to a wide variety of highly sophisticated methods has taken place.

The study of air quality is a very challenging and important problem. The ambient air is a complex mixture of gases, liquids and solids, and consists of a variety of elements, molecules and complex species. These constituents very often co-exist because of the great dilution of the system, in spite of the fact that in more concentrated systems they would not be compatible. The magnitude of the analytical problems can be appreciated when it is realised that the individual pollutants are usually present in concentrations of parts per million to parts per billion or less, and measurements must often be made on microgram or nanogram amounts of material.

It is often overlooked that the analytical process must include the selection of a representative sampling site, together with proper procedures for collecting the sample, preserving it, processing it, and finally measuring significant species selected within the sample mixture. There is usually the likelihood that samples undergo change during the sampling process. It is therefore important that the analyst is aware of the complex nature of air samples and the likelihood that incompatible species in it will react during the collection and concentration of samples. In this connection, it is appropriate to call attention to the fact that most monitoring equipment on the market is not specific. Also, monitors tend to give a false sense of security because they usually produce vast amounts of data. The basic need for the evaluation of air quality is reliability.

In this review, a division is made between gaseous and particulate pollutants. Up to this time, most emphasis has been placed on gaseous pollutants, because these include most of the acute toxins. It may be unfortunate that the presence of acute toxins has attracted so much attention because only in rare cases do the concentrations of such agents reach lethal levels. In most cases, slight exposure to such agents as SO_2, chlorine or H_2S do not seem to have lasting effects, even when the exposure is continuous. It is quite possible that future interests will be directed to a much greater extent to the cummulative and additive toxins, which although less dramatic, may be far more insidious. These hazards appear more often among the particulate pollutants than in the gaseous groups. This review considers to some extent the belief that more emphasis should be placed on identifying and measuring particulate species. It is not sufficient to evaluate dusts, for example, simply as a nuisance. The mass of 'fall-out' or 'suspended dust' and the measurement of a 'dust stain index' give little indication of possible health hazards. Particulate samples should be analysed to determine the nature of materials (particularly carcinogens) actually present.

Space does not permit detailed discussion of sampling schedules and techniques, nor is it possible to consider in detail such important matters as reference methods, equivalent methods, standards, collaborative testing and calibration programmes. Also, organic pollutants could not be considered. What is presented, therefore, should be regarded as a guide and used in the knowledge that valid data come only from the intelligent application of appropriate techniques.

4.2 GASEOUS INORGANIC POLLUTANTS

4.2.1 Sulphur dioxide

Sulphur dioxide is one of the most widely distributed and important air pollutants. It is formed when substances containing sulphur such as coal, fuel oil, gasoline or sulphide minerals are combusted. Sulphur dioxide itself is an irritant, but its effect is greatly enhanced when it is oxidised to SO_3 and combined with moisture to form sulphuric acid, a process which takes place naturally and is the ultimate fate of combusted sulphur. There are several methods reported for the determination of SO_2 in ambient air, the most important of which are discussed below.

4.2.1.1 West–Gaeke method

(a) *Principle*—The method[1] is based on the formation of the stable dichlorosulphitomercurate(II) complex, when SO_2 is absorbed from air and a solution of 0.1 M sodium tetrachloromercurate(II) is used as an absorber. The complex once formed is resistant to oxidation by the other constituents in air such as oxygen, ozone, oxides of nitrogen, etc.[2]. The non-volatile dichlorosulphitomercurate(II) complex reacts with acid-bleached *p*-rosaniline and formaldehyde to form an intense red–purple *p*-rosaniline–methyl sulphuric acid complex[3]. The absorbance of the solution, measured spectrophotometrically at 560 nm, is proportional to the amount of SO_2 absorbed.

(b) *Range*—The method is applicable to a wide range of concentrations of SO_2 from 0.01 to 0.40 p.p.m. (or 25–1060 $\mu g\ m^{-3}$).

(c) *Interference*—This method is not subject to interference from other pollutants present in usual concentrations. Excessive amounts of oxides of nitrogen, ozone and heavy metals may interfere but their effects can be eliminated. Interference due to oxides of nitrogen is eliminated by addition of sulphamic acid[4,5], and that due to heavy metals by the addition of EDTA (ethylenediaminetetra-acetic acid, disodium salt)[6]. Ozone interference can be eliminated by a time-delay technique[6].

(d) *Precision and accuracy*—A relative standard deviation of 4.6% at the 95% confidence level has been reported for the analytical procedure using standard samples[5].

(e) *Sampling and determination*—A typical sampling train set-up is shown in Figure 4.1[7]. The size of the impinger and the volume of the absorbing solution used will depend on the duration of sampling. 10 ml of the absorbing solution is recommended for 1 h sampling; 24 h sampling is usually carried out with 50 ml of absorber. The desired flow rate is maintained using the needle valve and the rotameter and sampling time is measured accurately to establish the volume of air sampled.

After collection, the tetrachloromercurate(II) solution is transferred to the laboratory where *p*-rosaniline and formaldehyde solutions are added. The

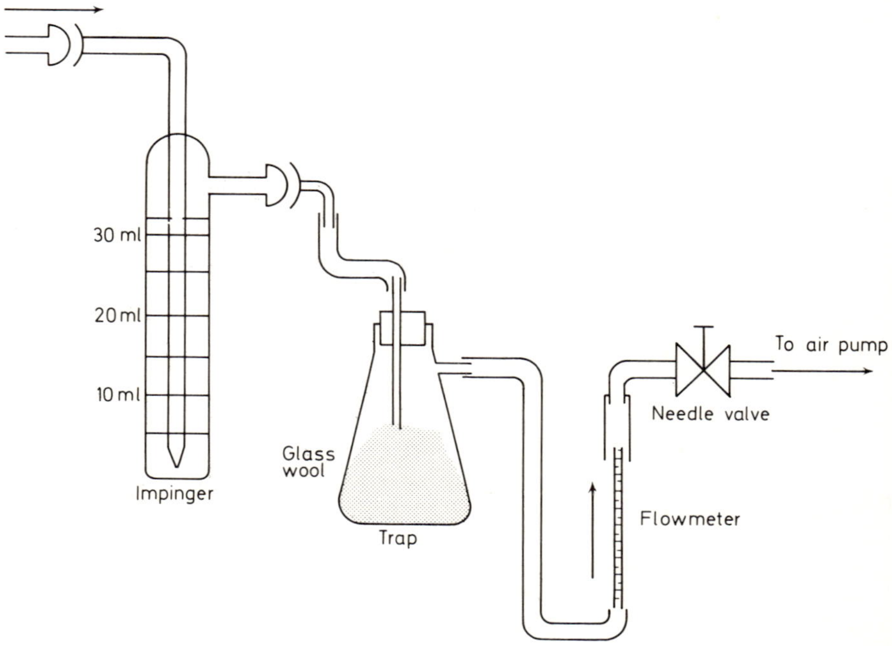

Figure 4.1 Sampling train for the determination of SO_2 by the West–Gaeke procedure

solution is allowed to sit for 30 min for full colour development and the absorbance is then measured at 560 nm. A blank run should always be made using the same reagents. The amount of SO_2 found is determined by reference to a calibration curve.

(f) *Calibration using standard sulphite solution*—Standard solutions of the dichlorosulphitomercurate(II) complex, containing known amounts of SO_2, can easily be prepared by adding appropriate amounts of a standard sodium sulphite solution to a tetrachloromercurate(II) solution. The sulphite solution is prepared by dissolving sodium sulphite in deionised water and standardising it by iodimetric titration. The colour is then developed in the standards as described before, the absorbances are measured for the various standards and the values used to prepare a calibration curve.

(g) *Calibration using SO_2 permeation tubes*—Atmospheres containing accurately known amounts of sulphur dioxide at levels of interest can be prepared using permeation tubes[8-10]. The permeation tube consists of a small tube made of FEP Teflon in which a small amount of liquid is sealed. The sulphur dioxide permeates through the walls of the tube at a steady rate, thereby providing a source of nanogram to microgram amounts of SO_2. The SO_2 permeating from the tube is transported by an inert carrier gas to a mixing chamber where it is accurately diluted to the levels of interest with SO_2-free air. Standard samples are then taken as described above. The samples are analysed as described previously and a calibration curve is drawn.

(h) *Expression of results*—It would be very desirable to have one uniform system of expressing the concentration of gases in air, but unfortunately use of both p.p.m. and $\mu g\ m^{-3}$ are common in the literature. Their definition and relation to one another is given below.

$$SO_2\ (\text{p.p.m.}) = \frac{\text{microlitres of } SO_2 \text{ in sample}}{\text{volume of air sampled in litres}}$$

$$SO_2\ (\mu g\ m^{-3}) = \frac{\text{weight of } SO_2 \text{ (in micrograms) in sample}}{\text{volume of air sampled in cubic metres at s.t.p.}}$$

$$SO_2\ (\text{p.p.m.}) = SO_2\ (\mu g\ m^{-3}) \times 3.82 \times 10^{-4}$$

4.2.1.2 *Coulometric method*[11,12]

(a) *Principle*—This method, which is generally used for continuous monitoring, is based on the reducing properties of SO_2. Air to be analysed is continuously drawn through a cell containing acidified bromide solution. By electrolysis of the Br^- solution, a known amount of Br_2 is generated. The Br_2 thus generated oxidises the SO_2 absorbed, causing depletion of Br_2, which is measured by an oxidation–reduction sensor electrode. The system then reacts so that current passes through the generating electrodes to maintain a fixed concentration of Br_2. The amount of current required to maintain the desired concentration in the cell is proportional to the amount of SO_2 introduced. The current is measured accurately with a current integrator which is directly calibrated in terms of p.p.m. of SO_2. Some modifications in the circuitry have been proposed to improve the sensitivity[13,14].

(b) *Range*—The range of SO_2 detectable by this method is 0.1–10 p.p.m. with a sensitivity of 0.1 p.p.m. A 90% response to a change in concentration requires *ca.* 30 s.

(c) *Interferences*—Obviously the presence of other reducing agents such as H_2S, thiols and organic sulphides will interfere, yielding higher results, while oxidising agents such as Cl_2, Br_2, NO_2, or O_3 will interfere to yield low results for SO_2.

(d) *Precision and accuracy*—The reproducibility of such instruments is within 2% of the full-scale deflection for a 24 h period, and the expected accuracy is between 5 and 10%.

(e) *Sampling and determination*—A typical layout of the components of a coulometric analyser, shown in Figure 4.2, consists of the sampling probe,

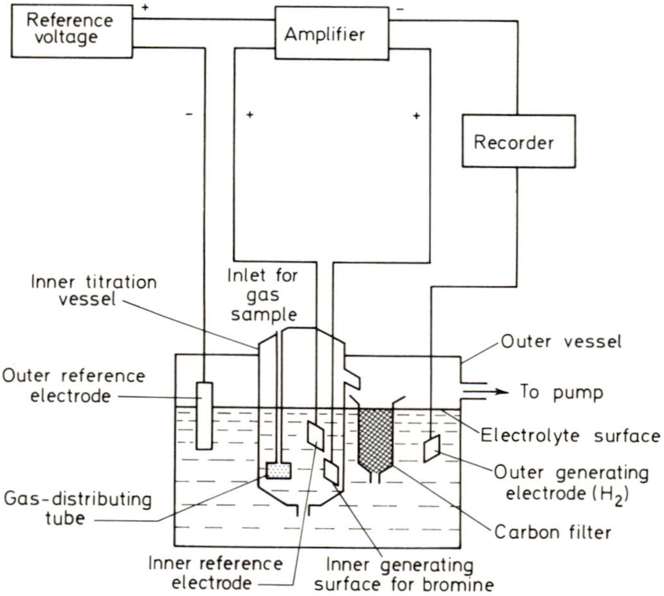

Figure 4.2 Coulometric cell for SO_2 determination

Br^- absorption cell and electrode systems for generating and sensing Br_2. An initial zero is established by maintaining a continuous stream of air, free of reactive materials (obtained by incorporating a scrubber containing activated charcoal and soda lime on the sampling line), and also by keeping an optimum Br_2 concentration in the electrolysis cell. Once the zero is adjusted, samples can be drawn at the same rate for a continuous record of the SO_2 concentration. In commercial instruments, the zeroing is done automatically and periodically.

(f) *Calibration*—The best method for the calibration of this technique is the use of permeation tubes. Calibration by gas dilution techniques can be used but the results are less accurate. For the details of gas dilution techniques, the reader is referred to Ref. 11.

4.2.1.3 The SO_2 *permeation method*[15]

This method is valuable for determining the average concentration of SO_2 over long periods of time such as hours, a day or even several days. It is especially useful for large air-pollution surveys, and where low cost is a factor. The method is essentially the same as the West–Gaeke method except in sampling. The sampling is ingeniously accomplished by using the ability of SO_2 to permeate through synthetic membranes. The permeation device used for sampling consists of a small Pyrex tube (o.d. = 41 mm), to one end of which is cemented a silicone membrane. Tetrachloromercurate(II) solution is added to the cell which is then closed with a rubber stopper having a capillary insert, which serves to equalise any changes in atmospheric pressure. A diagram of the permeation cell is given in Figure 4.3. The cell, wrapped

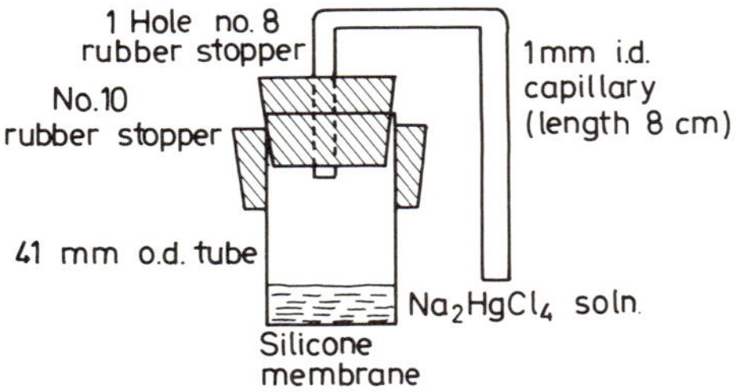

Figure 4.3 Permeation cell for SO_2

with black tape to protect the collected samples from sunlight, is placed inside a shelter which has air vents to provide free air flow and the assembly is then placed at any desired sample location. Sulphur dioxide permeates through the membrane at a constant rate, dependent only on the ambient concentration of SO_2. When it comes into contact with the tetrachloromercurate solution, it is fixed as the dichlorosulphitomercurate(II) complex. At the end of the desired sampling time the solution is brought back to the laboratory, transferred to a beaker, *p*-rosaniline and formaldehyde are added, and the absorbance of the colour developed is measured. The concentration of SO_2 is calculated by referring to a calibration curve.

Since the permeation rate constant of the individual membrane varies, it is important to calibrate each cell, using the standard techniques of calibration described previously. The same cell can be used again and again. This method has the unique advantage of being simple, inexpensive, yet reliable, and is as sensitive as the other methods. It is ideal for air pollution surveys because it requires no electrical power.

4.2.1.4 Miscellaneous methods

In the past few years a tremendous amount of effort and money have been spent in developing automated analysers for monitoring ambient levels of SO_2 and also stationary source and vehicular emission monitors[16,17]. This has resulted in an array of instruments based on various physical and chemical properties of SO_2. In the United States alone there are well over 60 commercially available types of monitor, involving 15 distinctly different principles of operation, with prices ranging from \$800 to \$20 000[17]. Except for summarising these methods in Tables 4.1 and 4.2, even a brief review of these methods is beyond the scope of this chapter. However, interested readers may refer to Ref. 13 for the details.

Table 4.1 Ambient air monitors for gaseous inorganic pollutants*

Principle of operation	*General measuring range*†	*General response times*†	*Cost range*†/\$	*Gases monitored*
Conductimetric	0.004–5.0	20 s–4 min	1500–5000	SO_2, NO, NO_x, H_2S, NH_3, Cl_2, CO_2
Colorimetric	0.002–4.0	2–7 min	2000–4000	SO_2, H_2S, NO, NO_2, NO_x
Coulometric	0.002–10	2–5 min	2000–4000	O_3, SO_2, H_2S
Electrochemical transducer	0.01–10	10 s–2 min	1000–2000	NO_2, NO_x, CO, H_2S, SO_2
Flame photometric detector	0.01–10	10–30 s	3000–5000	Total sulphur and phosphorus compounds
Gas chromatography with flame photometric detector	0.01–10	4–5 min	5000–7000	Sulphur and phosphorus compounds
Correlation spectroscopy	1–1000	1–32 s	20 000	SO_2, NO_2
Second-derivative spectroscopy	0.005–10	24 s	12 500	SO_2, NO, NO_2, O_3, NH_3, etc.
Condensation nuclei	0.001–5.0	45 s	5500	SO_2, NH_3, Hg, CO, NO_x

* Adapted from Ref. 17

† Measuring range, response times and cost vary from manufacturer to manufacturer

4.2.2 Hydrogen sulphide[11]

Hydrogen sulphide is one of the most commonly encountered pollutants, produced in nature by the enzymic or bacterial decay of sulphur-containing animal and vegetable organic matter. In industry it is found in the waste gases produced in the manufacture of paper, pulp, rayon, staple fibre, rubber, carbon disulphide, sulphur dyes, etc. It is also emitted during petroleum refining and coal-tar production. Of the several methods available for monitoring H_2S, the following are important.

Table 4.2 Stationary source monitors for gaseous inorganic pollutants*

Principle of operation	*Measuring range*†/p.p.m.	*Response time*†	*Cost range*†/$	*Gases monitored*
Conductimetric	1.4–8000	2 min	5600	SO_2, CO_2
Colorimetric	0.01 p.p.m.–100%	2 min	4500	SO_2, H_2S
Coulometric	0.05–1000	5 min	4000–5000	SO_2, H_2S
Electrochemical transducer	0.01–10 000	5 s–3 min	1000–3500	SO_2, NO_2, NO_x, CO, H_2S
Flame photometric detectors	25–10 000	12 s	8500	Total sulphur and phosphorus compounds
Non-dispersive absorption spectroscopy	10 p.p.m.–100%	0.5–10 s	1500–4000	SO_2, CO, NO, NO_2, etc.
Dispersive absorption spectroscopy	0.1–3000	30 s	4000	SO_2, CO, NO, NO_2, etc.
Correlation spectroscopy	10–5000	20 s	8000	SO_2
Second-derivative spectroscopy	0.1 p.p.m.–100%	20 s	9000	SO_2, NO, NO_2, O_3, NH_3, etc.

* Adapted from Ref. 17
† Measuring range, response times and cost vary from manufacturer to manufacturer

4.2.2.1 Methylene blue method[19-21]

(a) *Principle*—The method is based on absorbing H_2S in a suitable medium containing an alkaline suspension of cadmium hydroxide[19]. The absorbed H_2S reacts with *p*-aminodimethylaniline in the presence of $FeCl_3$ in a strongly acidic medium to form the water-soluble dyestuff methylene blue (1). The absorbance of methylene blue, measured at 670 nm, is proportional to the concentration of H_2S. The stoichiometric reaction in which methylene blue is formed is:

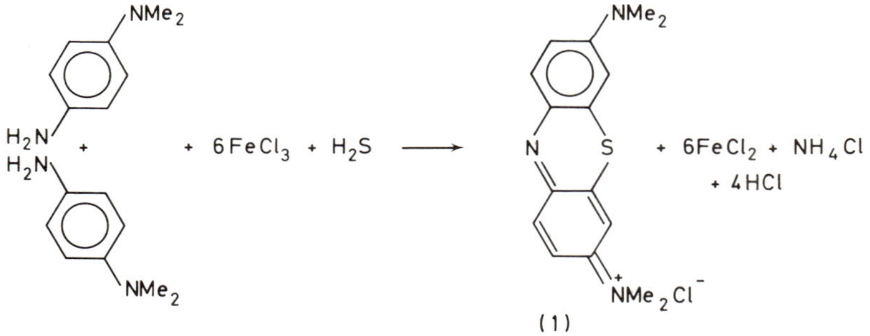

(1)

(b) *Range*—The method is applicable from p.p.b. to p.p.m. ranges.

(c) *Interference*—There is no interference from other normally occurring constituents. The method measures all soluble sulphides and hydrogen sulphide.

(d) *Sampling and determination*—A sampling train similar to the one used in the West–Gaeke procedure for SO_2 (Figure 4.2) can be used. The absorbing solution containing suspended cadmium hydroxide can be prepared by dissolving 4.3 g of $CdSO_4 \cdot 8H_2O$ in deionised water and adding 0.3 g NaOH and diluting to 1 litre. The absorption mixture must be stirred well to ensure homogenity before adding it to the impinger. Depending on the levels of H_2S in the air, samples are drawn at suitable rates for about one hour or longer. The absorption mixture is transferred to a volumetric flask, *p*-aminodimethylaniline–sulphuric acid solution and a drop of concentrated $FeCl_3$ solution are added and the mixture is made up to a known volume. After allowing 30 minutes for full colour development, the absorbance of methylene blue is measured at 670 nm. A reagent blank is also made under the same conditions. By referring to a standard calibration curve, the concentration of H_2S can easily be calculated.

(e) *Calibration*—The method can be calibrated using standard solutions of Na_2S made up to contain known amounts of H_2S. Standards containing 0, 1, 2, 3, 4 and 5 μg of H_2S are added to the adsorption mixture and the colour is developed as described above. The absorbances are measured at 670 nm, from which a calibration curve can be drawn.

(f) *Expression of results*—Since the concentration of H_2S generally encountered in ambient air is at the parts per billion* level, it is reported as shown below:

$$H_2S\ (\text{p.p.b.}) = \frac{\mu g\ (H_2S) \times 2 \times 719}{\text{volume of air in litres at 25 °C and 760 Torr}}$$

4.2.2.2 *Impregnated paper tape methods*

(a) *Principle*—The method depends on the formation of an insoluble coloured sulphide, when an air sample containing H_2S is drawn through paper tapes impregnated with solutions of metals. The optical density of the spot is proportional to the concentration of the metal sulphide precipitate formed, which in turn depends on the concentration of H_2S in the air sample[22-28]. Several salts have been used for impregnation such as lead acetate, silver nitrate, mercuric chloride, etc., with comparable effectiveness. The tape methods are ideally suited for field studies and have a working range of 0.001–50 p.p.m.

(b) *Interference*—Sulphur dioxide, NO_2, O_3, etc., tend to bleach the sulphide spot, thereby lowering the optical density. They may be eliminated by using appropriate scrubbing columns in the sampling train[23].

(c) *Preparation of impregnated filters*—Many kinds of filter have been used such as Whatman No. 41, No. 1 Chromatographic paper, Millipore membranes type AA, SS and SM, etc., to prepare impregnated papers. The general method of preparation requires immersing the filter material in a bath containing the metal solution (sometimes acidified) for about an hour and then drying in the dark either in a desiccator or with a current of warm air. The filters are stored in a desiccator in the dark.

(d) *Sampling and determination*—Samples are pulled through the filter at relatively slow rates for efficient collection. A pre-filter is always incorporated to remove airborne particulates. The density of the spot is measured using a

* p.p.b. in this review denotes parts per thousand million.

densitometer; by referring to a standard calibration curve, the amount of H_2S can be determined.

The method is calibrated using standard techniques of calibration such as the permeation tube method or gas dilution techniques.

4.2.2.3 Miscellaneous methods

There are several other methods of determining and monitoring H_2S based on a variety of physical properties. As an example, continuous monitoring of H_2S levels $>$ 0.002 p.p.m. has been achieved using a sulphide-sensitive $Ag|Ag_2S$ electrode[29]. A gas chromatographic technique has been proposed[30] and also many methods listed in Tables 4.1 and 4.2 are valid for monitoring H_2S. The details of these methods are beyond the scope of this article.

4.2.3 Nitrogen dioxide and nitric acid

Of the many oxides of nitrogen present in air (N_2O, N_2O_3, NO, NO_2), nitric oxide and nitrogen dioxide are the most important pollutants. They are produced by the combustion of coal, fuel oils, gasoline, natural gases, etc. in industrial operations and automobiles. They are important constituents in the formation of photochemical smog, and reside at concentrations of 20–50 p.p.m. Nitrogen dioxide is a strong eye irritant, while 150 p.p.m. may bring about severe irritation of the respiratory organs. The recommended methods for determining these gases are described below.

4.2.3.1 Griess–Saltzman method

(a) *Principle*—This method is based on the classic Griess reaction in which sulphanilic acid is diazotised by nitrous acid and coupled with α-naphthylamine to form a red azo dye. Saltzman[31-32] used *N*-(1-naphthyl)ethylenediamine as the coupling agent.

Air containing NO_2 is drawn through an absorbing solution which consists of 0.5% sulphanilic acid, 14% acetic acid and 0.002% *N*-(1-naphthyl)ethylenediamine dihydrochloride. A stable pink coloured dye is produced within 15 minutes and is measured spectrophotometrically at 550 nm. The method can readily be adapted for measuring NO by oxidising it to NO_2 in a bubbler containing potassium permanganate solution[33].

(b) *Range*—The method is sensitive from a few p.p.b. to *ca.* 5 p.p.m.

(c) *Interference*—A 10-fold ratio of SO_2 to NO_2 does not interfere. At a 30-fold ratio, SO_2 slowly bleaches the colour to a slight extent. The interference due to SO_2 can be eliminated by adding 1% acetone to the reagent, which retards the fading by forming a temporary addition product with SO_2. If strong oxidising or reducing agents are present the measurement of absorbance should be made as quickly as possible (in less than 2 h).

(d) *Sampling and determination*—The sampling train consists of a fritted absorber, a rotameter and a pump, in that order (Figure 4.4). It is absolutely essential to use a fritted bubbler and the pore size of the frit must be 60 μm or smaller. The sampling flow rate is also critical; it must not exceed 0.5 l min^{-1}. Since there is a substantial pressure drop because of the frit it is

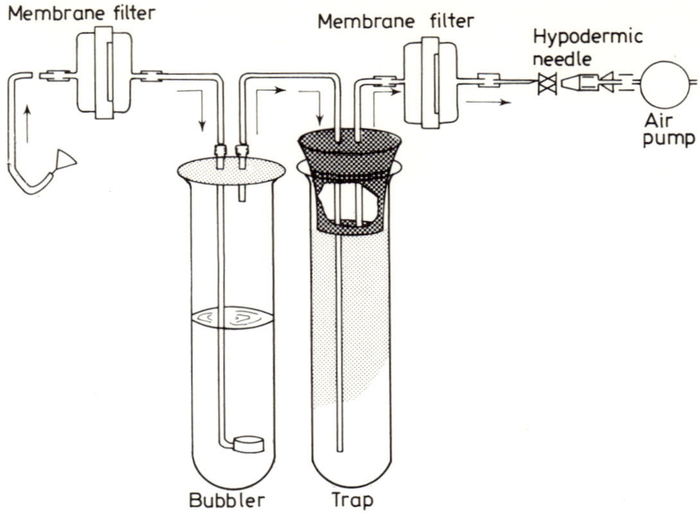

Figure 4.4 Sampling train for the determination of NO_2

important to calibrate the rotameter either by incorporating another rotameter upstream to the bubbler or by using a wet-test meter.

A suitable volume of the absorbing solution (sulphanilic acid–acetic acid–*N*-(1-naphthyl)ethylenediamine hydrochloride) is kept in the absorber and, depending upon the ambient levels of NO_2, air samples are pulled at a rate not exceeding 0.25 l min^{-1} for the desired period (generally 10–30 min). The colour develops during the process of collection itself. After collection, the colour development will be complete in 15 min. The absorbance of the colour developed is measured at 550 nm and the concentration of NO_2 is calculated by referring to a standard calibration curve.

(e) *Calibration*—The method can be conveniently calibrated by adding appropriate amounts of a standard $NaNO_2$ solution to the absorption mixture and measuring their respective absorbances.

At the present time there is some controversy about this method of calibration. Saltzman *et al.*[34] found in test air mixtures that 0.72 mole of $NaNO_2$ gives the same colour as 1 mole of NO_2, whereas Gill[35] found a nitrite equivalent of 0.58 and Stratmann and Buck[36] reported a value of 1. This controversy has not been resolved at the time of writing.

4.2.3.2 Jacobs–Hocheiser method[37]

This method has been accepted in the United States as the standard method for determination of nitrogen dioxide. In this method, NO_2 is collected by bubbling through an NaOH solution to form a stable solution of $NaNO_2$. The nitrite ion produced during sampling is determined colorimetrically by reacting the exposed absorbing solution with phosphoric acid, sulphanilamide and *N*-(1-naphthyl)ethylenediamine dihydrochloride—essentially the Saltzman method.

This method has some major drawbacks. Collection efficiencies of between 35 and 80% have been reported[38-42]. Also, during the absorption of NO_2 in NaOH, not all NO_2 forms nitrite ion, a portion of it forming NO_3^- according to the reaction:

$$2NO_2 + H_2O \rightarrow HNO_2 + HNO_3$$

The stoichiometry of this reaction has not been definitely determined. Thus, the method suffers from having a variable absorption efficiency and a variable, undefined stoichiometry.

Recently, Levaggi *et al.*[41] modified this method by using an entirely different absorbing solution consisting of a 1.5% solution of triethanolamine containing 0.3% butanol as the absorbing medium. The rest of the procedure is essentially the same as the Jacobs–Hochheiser method, and an absorption efficiency of >95%, with a conversion factor of 0.85 for the conversion of NO_2 (gas) to NO_2^- (ion), is reported.

4.2.3.3 The NO_2 *permeation method*

This sampling method is essentially the same as the one described for SO_2, i.e. permeation of NO_2 through a silicone membrane. The permeated NO_2 is collected in a highly efficient liquid absorber containing sodium hydroxide and thymol. The thymol acts as a reducing agent and almost quantitatively (93%) converts NO_2 gas to NO_2^-. The nitrite ion is determined by addition of *N*-(1-naphthyl)ethylenediamine hydrochloride solution and a mixture of sulphanilamide and phosphoric acid. The absorbance of the coloured product is determined at 540 nm. The method will detect as little as 5 p.p.b. of NO_2 collected over a 24 h sampling period, has a broad working range and is essentially unaffected by the presence of NO and other gaseous constituents of the air such as O_3, SO_2 and H_2S. Also, variations in environmental conditions have almost no effect on the measurement of NO_2^-. Since the absorbed species is quite stable, sampling times from a few hours to a week or more can be used.

4.2.4 Carbon monoxide

Carbon monoxide is a commonly encountered air pollutant and is also very widely distributed. The primary sources of CO are incomplete combustion of fuel gases and other organic substances, such as coal, fuel oils, gasoline, etc. Its toxicity is due to its ability to form a stable compound with haemoglobin in the blood. The affinity of CO for haemoglobin is about 200 times greater than that of oxygen. It is particularly dangerous in a closed environment because it cannot be detected by odour or any other visible physiological symptoms. The methods described below are commonly used for determining CO in ambient air.

4.2.4.1 Non-dispersive infrared spectroscopic method

(a) *Principle*—This method has been chosen in the U.S. as the reference method for the determination of CO. The method[7] is based on the absorption of i.r. radiation by CO. I.r. radiation, emitted from a suitable source, is split into parallel beams and directed through reference and sample cells. Both beams pass into matched cells, each containing a selective detector and CO. From the broad range of i.r. radiation passing through the sample and reference cells, CO in the sample cell absorbs only its characteristic frequencies to which the detector is sensitive. With a non-absorbing gas in the reference cell and with no CO in the sample cell, the signals from both detectors are balanced electronically. Any CO then introduced into the sample cell absorbs the radiation at its characteristic wavelengths, thus reducing the temperature and pressure in the detector cells and displacing a diaphragm. This displacement, proportional to the amount of CO in the sample cell, is detected electronically and amplified to provide an output signal. A schematic diagram of the non-dispersive i.r. analyser is given in Figure 4.5.

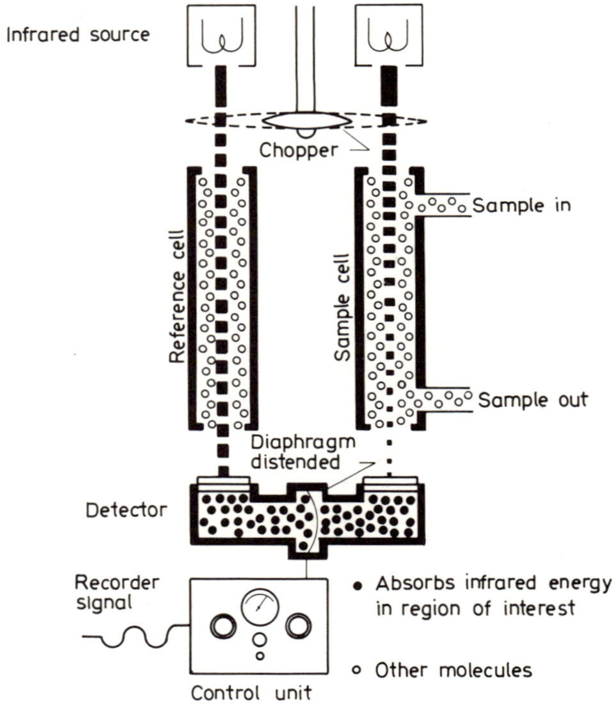

Figure 4.5 Non-dispersive infrared analyser for CO

(b) *Range*—The method has a sensitivity of 0.5 p.p.m. CO and a working range of 1–50 p.p.m., which is the range most commonly used for atmospheric sampling.

(c) *Interference*—The only serious interference is due to water vapour, which can be eliminated by passing the air sample through silica gel or some similar drying agent or by maintaining constant humidity in the sample and calibration gases.

(d) *Sampling and determination*—Initially the zero and the full-scale range of the instrument is set by passing through the sample cell only N_2 and a known concentration of CO in purified N_2, respectively. Air samples, dried by passing through a drying column, are drawn through the sample cell at a convenient rate using a suitable pump, flow control valve and flowmeter. The amount of CO is directly registered on the detector scale in p.p.m. or is determined by reference to a calibration curve.

(e) *Calibration*—The instrument can be calibrated by introducing gas mixtures of known CO concentration under the same conditions as the samples.

4.2.4.2 Miscellaneous methods

The classical manual method of determining CO is based on its reaction with iodine pentoxide at 135–150 °C to yield CO_2 and I_2 vapour. Methods based on both measurement of CO_2 and I_2 have been reported[44-47]. Liberated I_2 is titrated with $Na_2S_2O_3$ using starch as indicator; for higher sensitivity I_2 is oxidised to IO_3^- with Br_2, KI is added and the released I_2 is titrated with $Na_2S_2O_3$. By this method a six-fold improvement in sensitivity is achieved.

A gas chromatographic technique involving a flame ionisation detector has been reported[48] for measuring CO. In this method, CO is separated from other constituents by means of gas chromatography and is catalytically dehydrogenated to methane, which can be detected with high sensitivity in a flame ionisation detector.

Concentrations of CO at dangerously high levels can be determined rapidly in a semi-quantitative manner by the gas-detector tubes manufactured commercially by several firms (e.g. Drager in Germany, Unico and Mine Safety Appliance Co. in the U.S.A. and Kitagawa in Japan)[49]. One such tube is based on the reduction of ammonium molybdate to molybdenum blue by CO in the presence of a $PdCl_2$ catalyst.

4.2.5 Ozone and total oxidants

Ozone, photochemically formed at altitudes of 25 km, is essential for sustaining life on earth since it screens high-energy radiation from the sun and outer space. At the same time, it is an important pollutant in the lower atmosphere, produced in photochemical smogs by the reaction of O_2 and NO_2 with hydrocarbons. Even at concentrations as low as 1–2 p.p.m. it is a toxic and potent irritant, which can disturb the central nervous system, leading to bronchitis and headaches. The various methods used for monitoring ozone are described below.

4.2.5.1 Neutral buffered KI *method for total oxidants*

(a) *Principle*—This method measures the amount of O_3 and other oxidants such as NO_2, Cl_2, hydrogen peroxide and peroxyacyl nitrates. The method is based on absorption of these oxidants in a solution of KI buffered at a pH of 6.8. The oxidants react with KI, liberating equivalent amounts of I_2, the absorbance of which is measured[52] spectrophotometrically at 352 nm.

(b) *Range*—The method has a sensitivity of 0.01 p.p.m. and a range of 0.01–10 p.p.m.

(c) *Interference*—The method is subject to many interfering materials and also some ambiguities. Reducing dusts as well as gases such as SO_2, H_2S and thiols act as negative interferences[58].

Other oxidants such as NO_2, Cl_2, peroxyacyl nitrates and peroxides also react with KI to liberate I_2, but their response is not the same as equivalent concentrations of O_3. For example, NO_2 is known to give 10% of the response of an equivalent molar concentration of O_3, whereas peroxyacyl nitrate can give a response of approximately 50% of an equivalent amount of O_3[53,54].

Interference due to SO_2 may be eliminated by incorporating a chromic acid paper absorber in the sampling train upstream from the impinger.

(d) *Sampling and determination*—Assemble a sampling train consisting of a tube with chromic acid paper, a midget impinger, flowmeter and a pump. Place the absorbing solution (1.4% KH_2PO_4, 1.4% $Na_2HPO_4 \cdot 12H_2O$ and 1% KI adjusted to pH 6.8 ± 0.2 with NaOH) in the midget impinger and sample for up to 30 min at a rate of 0.5 to 3.0 l min^{-1}. The absorbing solution should be protected from sunlight.

The absorbing solution, after collection, is transferred to a colorimetric tube or cuvette and the absorbance is determined at 352 nm against distilled water as reference. The absorbance readings should be taken within 30–60 min after collection. Several other readings should be taken for extended periods of time because some organic oxidants react very slowly with KI. There may also be a gradual fading due to reductants in the air. Every few days, determine the blank correction by reading the absorbance of unexposed reagent.

(e) *Calibration*—The calibration curve can easily be obtained by adding known amounts of a standard I_2 solution to the absorbing solution and making up the volume to that used in the sample with the absorbing solution. The absorbances are measured at 352 nm, and are plotted against the exact calculated normalities of the corresponding I_2 solution. Beer's law is followed. Draw the straight line giving the best fit and from this graph determine the normality of the I_2 solution required to give exactly an absorbance of 1.0. This value multiplied by 1.224×10^{-5} gives the standardisation factor, M, defined as the number of μl of ozone required by 10 ml of the absorbing reagent to give an absorbance of exactly 1.0. For 1 cm cells, M should equal approximately 9.6.

(f) *Expression of results*—It is customary and convenient to express the total oxidant content in terms of O_3. It has been empirically determined that 1 mole of O_3 liberates 1 mole of I_2 by this procedure. The results for air samples can be calculated from the equation:

$$\text{Oxidant (expressed as } O_3 \text{ in p.p.m.)} = A \times M/V$$

where A = corrected absorbance; M = standardisation factor; and V = volume of air in litres per 10 ml of absorbing solution corrected to 25 °C and 760 Torr.

4.2.5.2 *Alkaline* KI *method for total oxidants*

This method is basically a modification of the method described above. The advantage of this method over the neutral iodide procedure is that a delay is permissible between sampling and completion of the analysis. In this method the oxidants are absorbed in an alkaline KI solution (1% KI in 1 M NaOH). A stable product is formed that can be stored for several days without loss. The iodine, equivalent to the amount of oxidant present, released by acidification of the absorbing solution is determined by measuring the absorbance at 352 nm. The interference due to SO_2 is eliminated by oxidising it to H_2SO_4 with H_2O_2. The procedures for sampling, determination, calibration and calculation are similar to the neutral KI method, the details of which can be found in Ref. 52.

4.2.5.3 *Chemiluminescence method for ozone*[50,51]

(a) *Principle*—The method is based on the chemiluminescence reaction of ozone with ethylene. Air mixtures containing ozone and ethylene are introduced simultaneously into a mixing zone, where the ozone and ethylene react to emit light, which is detected by a photomultiplier tube. The resulting photocurrent is amplified and is either read directly or displayed on a recorder. This method is the official reference method in the United States.

(b) *Range*—The method has a sensitivity of *ca.* 0.004 p.p.m. and has a broad working range of 0.005–5 p.p.m.

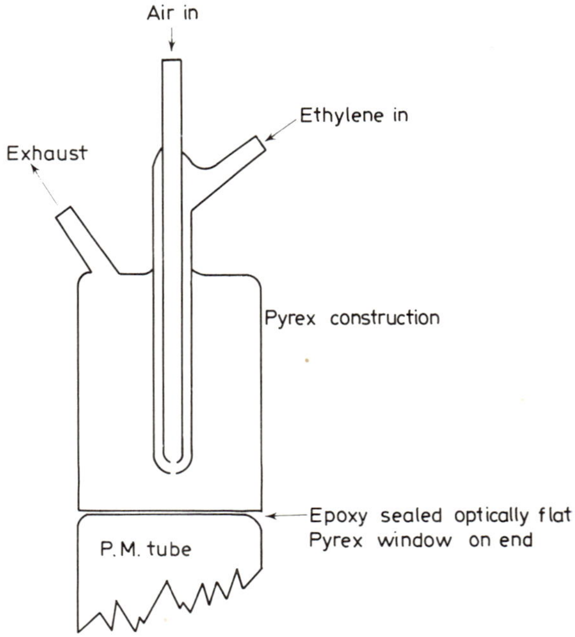

Figure 4.6 Ozone detector

(c) *Interference*—The method is specific for ozone with no interference from other oxidants or reductants found in air.

(d) *Sampling and determination*—A diagram of a typical ozone chemiluminescent detection system is shown in Figure 4.6. The ambient air is passed through a pre-filter and is then introduced into the detector cell. Its flow rate is controlled by a flowmeter to a suitable value between 0.1 and 1.5 l min^{-1}. Simultaneously, ethylene is introduced at a flow rate of between 1 and 50 ml min^{-1}. The intensity of the emitted chemiluminescence is proportional to the concentration of O_3 and is measured by the photomultiplier tube detection system and displayed on a recorder. By referring to a calibration curve or by calibrating the instrument, the concentration of ozone in air can be determined.

(e) *Calibration*—Atmospheres containing known amounts of ozone can readily be generated by passing ozone-free air into a quartz tube and irradiating it with a very stable low-pressure mercury lamp. Different ozone concentrations can be obtained by varying the level of irradiation[51]. The levels of ozone in the test atmospheres are determined by the neutral buffered KI method. These test atmospheres containing known amounts of O_3 are used to calibrate the chemiluminescence detector by the procedure described above.

4.2.5.4 Methods based on formation of aldehydes from ozonides[65-68]

Ozone reacts with olefins specifically to form ozonides, which decompose to yield aldehydes. The amount of aldehyde produced is directly proportional to the concentration of ozone. If the aldehyde produced can be determined accurately, then it is an indirect measure of ozone. Based on this principle, Bravo and Lodge[66] have developed a method in which they produce the ozonide of 4,4′-dimethoxystilbene, which cleaves to liberate anisaldehyde. Anisaldehyde is determined spectrophotometrically by the blue colour it produces when reacted with fluoroanthrene and trifluoroacetic acid.

Hauser and Bradley[68], in order to eliminate the use of poisonous trifluoroacetic acid in the above method, used instead 1,2-di-(4-pyridylethylene) in glacial acetic acid. The pyridine-4-aldehyde formed after ozonolysis was determined with the aldehyde reagent 3-methyl-2-benzothiazolinone.

Sachdev *et al.*[65] used the reaction of ozone with eugenol (4-allyl-2-methoxyphenol). Ozone preferentially attacks the olefinic bond in the allyl groups and releases formaldehyde. Formaldehyde is determined by the reverse West–Gaeke method[3] or by the chromotropic acid method[69]. It was found that the reverse West–Gaeke method is better suited in this application.

4.2.5.5 Miscellaneous methods

Several methods have been reported in which oxidants or O_3 are absorbed in neutral buffered KI solution and the liberated I_2 is determined by potentiometry, amperometry or coulometry[56-59].

Methods based on the absorption of u.v. or i.r. radiation have been proposed by Renzetti[60] and by Hanst *et al.*[61]. Bovee and Robinson[62] have reported a colorimetric method in which the air sample is passed at a rate of 2.8 l min^{-1} through a midget impinger containing 10 ml of 1% sodium diphenylamine sulphonate in 0.02% $HClO_4$. Ozone reacts with the absorption solution to form a colour, the absorbance of which is measured at 593 nm. According to Arnold and Mentzel[63], several oxidants react with the base *N*-tetramethyl-diaminodiphenylmethane to give different colours and this has been the basis of a method developed by Fuhrmann[64] for the continuous colorimetric determination of ozone. Hendricks and Larsen[70] have made a comparative study of seven different methods for the determination of O_3.

4.2.6 Fluorides

The most commonly encountered fluorine compounds in air are HF and SiF_4. The latter, in humid air, is slowly converted to HF. The most important sources of fluorides are industrial plants manufacturing phosphate fertilisers and aluminium. Iron foundries and brick and ceramics works also contribute considerably to pollution from fluorides, the toxicity of which are well known and documented. They are particularly notorious for the damage they cause to vegetation even at very low concentration. The procedure described below is the best available method for the microdetermination of fluorides.

4.2.6.1 The lanthanide–alizarin complexan–fluoride method[71]

(a) *Principle*—This method is based on the reaction of fluoride with lanthanum(III)–alizarin complexan to form a stable ternary complex, in a solution buffered to pH 4.2[72, 73]. The reaction is not subject to interference from other anions such as chloride, sulphate, phosphate or nitrate. Although several cations interfere, a method has been developed by West *et al.*[71] which successfully eliminates the interference in the process of sample collection.

(b) *Procedure*—The method involves aspirating the air sample containing fluoride through a pre-scrubber containing concentrated sulphuric acid maintained at approximately 70 °C, and then through an absorbing solution of distilled water or dilute alkali in a fritted glass scrubber. The pre-scrubber traps all of the cation, while fluoride complexes of metals such as Fe, Al, Zn and Cu dissociate, liberating HF which is volatilised under these conditions and is swept off to the next scrubber where it is absorbed, free from all interference. Lanthanum(III) and alizarin complexan solutions are added, after adjusting the pH to 4.2, and the absorbance of the resulting colour is measured at 618 nm. It has been determined by the authors that addition of a miscible organic solvent increases the sensitivity of the reaction. A 25% v/v mixture of acetone in the final solution was found to give the highest sensitivity.

(c) *Calibration*—This procedure can be readily calibrated using standard solutions of NaF, and developing and measuring colours as described above.

4.3 PARTICULATE INORGANIC POLLUTANTS

4.3.1 Dustfall and sootfall

The term dustfall may be defined as the amount of dust precipitated per unit area in a certain time interval. Dustfall measures the total amount of dust that settles spontaneously, together with particulates that are brought down by rain, snow and wind. Though the measurement and the definition of this parameter is rather crude, it is a fundamental measurement in the study of air pollution and considerable information can be gained about the particulates (such as the sinks for pollutants, nature and shapes of particulates, etc.). Generally, the particulates in this category are in the size range of 5–200 μm.

4.3.1.1 Gravimetric determination of dustfall

Dustfall measurements are usually made gravimetrically by collecting samples in a suitable container for a given period of time. There are no uniform specifications for the size and shape of these containers. The containers are often coated with vaseline or they may contain a liquid to prevent the possible loss of the collected dust due to wind action. In areas of heavy rainfall, precautions must be taken to prevent collected water from overflowing. A typical set-up for collecting dustfall is shown in Figure 4.7. The

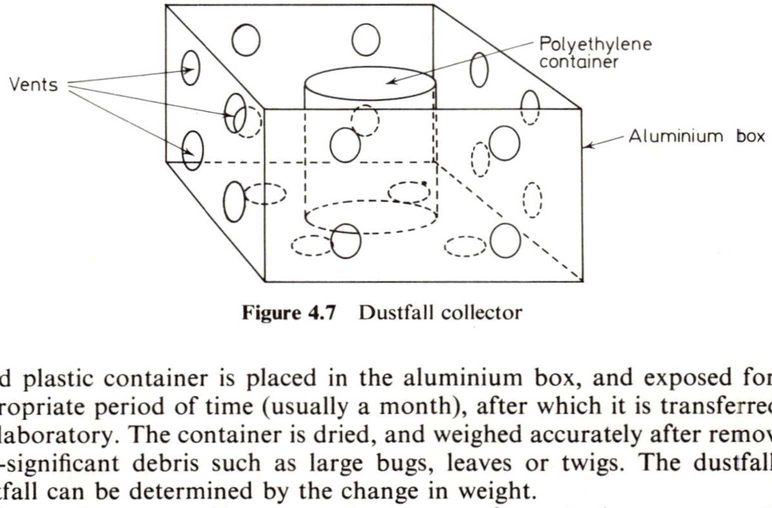

Figure 4.7 Dustfall collector

tared plastic container is placed in the aluminium box, and exposed for an appropriate period of time (usually a month), after which it is transferred to the laboratory. The container is dried, and weighed accurately after removing non-significant debris such as large bugs, leaves or twigs. The dustfall or sootfall can be determined by the change in weight.

The results are usually expressed as mg cm^{-2} month^{-1} or as ton mile^{-2} month^{-1}:

$$\text{mg cm}^{-2} = \frac{\text{g of dust} \times 100}{0.7855\,(\text{diameter of container in cm})^2}$$

or

$$\text{ton mile}^{-2} = \frac{\text{g of dust} \times 5660}{(\text{diameter of container in inches})^2}$$

4.3.1.2 Soluble and insoluble fractions of dustfall

For some studies it is advantageous to differentiate between the water-soluble and water-insoluble fractions of the dustfall. This is simply done by carefully washing down the inner surfaces of the dust container with hot distilled water, filtering through an ashless filter paper, evaporating the filtrate to complete dryness and weighing the residue to give the soluble fraction. The insoluble fraction can then be determined by difference.

4.3.1.3 Chemical and microscopic analysis of dustfall

Valuable information can be gained by chemical and microscopic analysis of the dust collected. The size, shape and even identity of many of the particles can be determined by simply observing a portion of the dust under a polarised light microscope. A small portion of the dust is suitably mounted on a glass slide by conventional techniques. Chemical analysis is also often desirable for gaining greater insight into the composition of the dustfall. The acidity or basicity of the water-soluble fraction can be determined readily with a pH meter. Sodium, potassium, calcium and other metals can be determined by flame photometry or atomic absorption spectroscopy. Metals of special interest can be determined by conventional methods such as spectrophotometry, polarography, emission and atomic absorption spectroscopy, etc. Details of these techniques are described in the following sections.

The amount of free carbon present in the dustfall is a good indicator of the soiling properties of the atmosphere and is also significant in physiological studies because of the well-known adsorptive capacity of carbon, which results in multiplication of the health hazards due to gaseous pollutants by concentration. The carbon content can be determined[74] by initially digesting the dust sample with nitric acid to destroy the organic matter and collecting the residue by filtering through a tared porcelain filter crucible. The crucible and the contents are dried and weighed. They are then ignited for 2 h at 700 °C, cooled and weighed. The loss in weight is reported as free carbon.

4.3.2 Airborne particulates

Airborne particulates or aerosols consist of either liquid or solid particles ranging in diameter from 0.01 μm or less, up to *ca.* 100 μm. The aerosols may be isolated by sedimentation, filtration, impingement, centrifugation, or by electrostatic or thermal precipitation. The evaluation of particulate burden is often based on the degree of soiling attributed to suspended dusts. Reduction in visibility, as measured photographically or by means of a transmissometer, gives a general measure of particulate loading. Such observations, though quite valuable in evaluating general air quality, provide very little specific information about the composition and nature of the pollutants. It is of greater importance to determine the identities of various pollutants and the amounts of the individual pollutants to evaluate the health effects and the toxic stress.

4.3.2.1 Dust stain index (or smoke stain index)

These indices measure the relative dust or soot burden in the atmosphere. A measured volume of air is drawn through a small spot on filter paper or on a membrane filter. The reflectance or the transmittance of the stain produced by the collected samples is then measured with a photometer.

Two different units are generally used for expressing the dust stain index, the 'Coefficient of Haze' (COH)[75] and the 'Reflectance Units of Dirt Shade' (RUDS)[76]. A COH unit has been defined as that quantity of particulate material which produces an optical density of 0.01 when measured by light transmission at 400 nm and when compared with the measurements on dust-free filter paper taken as 100%. The RUDS is the percentage reflectance obtained when 10 000 linear feet of air is drawn through the filter. The reflectance of the clean paper used as the reference standard is set to 100 on the reflectance meter.

Considerable difference of opinion exists regarding the relative merit of transmittance *vs.* reflectance measurements in evaluating dusts. It is of significance, however, that the report of the Working Party on Methods of Measuring Air Pollution and Survey Techniques of the Organisation for Economic Cooperation and Development has agreed that the estimation of smoke concentration from smoke stains can best be made by reflectometric measurements[77]. There were some attempts to correlate the weight of particulate collected to the darkness of the stain produced. A subgroup has studied this aspect and it is their conclusion that the darkness of the stain cannot be used to measure the mass of particulates collected.

4.3.2.2 Sampling of airborne particulates

(a) *Filtration*—This is the most commonly used method for sampling airborne particulates. Several kinds of sampler based on this principle are available commercially, suitable for holding filters of various sizes and shapes. Filters, made out of diverse materials such as filter paper, glass fibre, membranes of different kinds, asbestos, graphite and Teflon are commercially available. The appropriate filter material is chosen judiciously for specific tasks, taking into consideration factors such as the pore size, efficiency, pressure drop across the filter, contaminants in the filter, mechanical strength, reactivity of the filter material with the pollutant of interest, etc. As an example, glass fibre filters are widely used for sampling metallic particulates in air. However, these fillers have been found worthless for collecting sulphuric acid aerosols because of the alkaline sites present in the glass which neutralise the acid. Teflon and graphite filters are preferred for sampling H_2SO_4 aerosols.

Two kinds of sampler (see Figure 4.8) based on the filtration principle are commonly used, namely 'high-volume samplers' and the 'tape samplers'. As the name suggests, the high-volume samplers are equipped to collect large volumes of air through relatively large filters (8″ × 10″ sheets or 4″ diameter circles). The Environmental Protection Agency in the U.S.A. uses this kind of sampler in their NASN (National Air Sampling Network) stations.

For the routine analysis of airborne particulates, these samplers can run continuously for 24 h and they filter *ca.* 2000 m^3 of air during that time. In the tape samplers the particulates are collected on tapes having a width of 1″ or 1.5″. The samplers are collected on circular spots of 0.5″ or 1″ diameter.

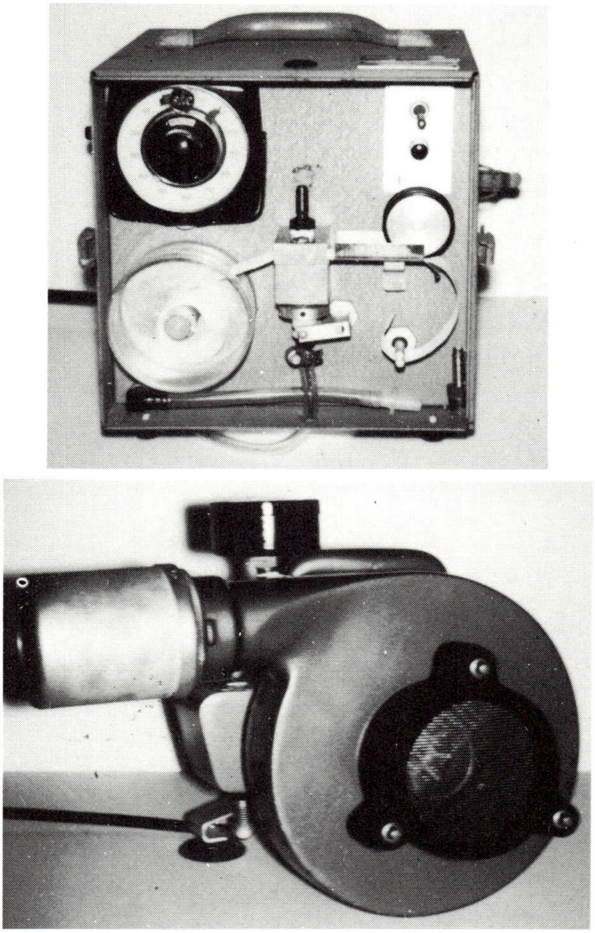

Figure 4.8 High-volume sampler and tape sampler

These samplers collect relatively low volumes of air of the order of 2–4 m^3 h^{-1}. Many commercially available tape samplers are automated in the sense that at a prefixed time the tape automatically advances to collect a new sample. They can also run continuously for days unattended. This type of sampler is especially suitable for measuring the dust stain index, microscopic studies and for ring oven techniques.

(b) *Electrostatic precipitation*[78, 79]—Samplers based on electrostatic precipitation consist of two spaced electrodes with a very high electrical potential across them such that a current flow is established. Many ions are liberated and maintained in the area between the electrodes. When the air containing the particulates is drawn between the electrodes, the particulates are charged because of collisions with the charged ions and are attracted to the collector electrodes. The efficiency of collection is dependent on the size, shape and type of electrodes, the electrode spacing, the velocity of the air stream and the potential across the electrodes, as well as the electrical resistivity of the particles. Electrostatic precipitators with an efficiency >99% for particles in the range of 0.01–10 μm are commercially available.

(c) *Impingers*—There are two types of impingers, the wet and the dry. In the first type the particles impinge on a surface submerged in a liquid and in the second the particles impinge on a surface exposed to the air. The impingers work on the principle that particles, when deflected against a body, tend to resist any change in direction because of their mass and inertia, and as a result collect on the body. The impingers are efficient only for particles >10 μm in diameter and are mainly used for microscopic work and qualitative identification. They are not suitable for quantitative work because of the poor collection efficiency below 5 μm.

4.3.3 Chemical analysis of airborne particulates

In the preceding section on gaseous inorganic pollutants, the analysis of important individual pollutants was discussed separately. However, this section, dealing with particulate pollutants, will be organised differently because of the large number of species involved and because many of the analytical techniques are common to most of the pollutants. Therefore, rather than discussing the pollutants individually, the analytical techniques employed for analysis will be discussed.

Table 4.3 Summary of the concentrations of the inorganic fraction of airborne particulates*

Inorganic fraction	*No. of stations*	*Concentration*/μg m^{-3}	
		Arithmetic average	*Maximum*
Nitrates	96	2.6	39.7
Sulphates	96	10.6	101.2
Ammonia	56	1.3	75.5
Antimony	35	0.001	0.160
Arsenic	133	0.02	—
Beryllium	100	<0.0005	0.010
Bismuth	35	<0.0005	0.064
Cadmium	35	0.002	0.420
Chromium	103	0.015	0.330
Cobalt	35	<0.0005	0.060
Copper	103	0.09	10.00
Iron	104	1.58	22.00
Lead	104	0.79	8.60
Manganese	103	0.10	9.98
Molybdenum	35	0.005	0.78
Nickel	103	0.034	0.460
Tin	85	0.02	0.50
Titanium	104	0.04	1.10
Vanadium	99	0.050	2.200
Zinc	99	0.67	58.00

* Air quality data from the U.S. National Air Sampling Network, 1964–65

The complexity of analysing particulates can be appreciated from Table 4.3, which lists typical concentrations of particles[80].

4.3.3.1 Atomic absorption spectroscopy

(a) *Principle*—When a solution containing metal ions is atomised either by aspiration into a flame or by injection into a heated zone, metal atoms in the ground state make up the bulk of the atomic population. These atoms absorb radiation at discrete wavelengths, which are characteristic for each element, and are thus raised to an excited state. The excited state atoms, in turn, emit radiation on returning to the ground state. The amount of energy thus absorbed and emitted is a function of the number of atoms present and hence may serve as a measure of the concentration of the respective elements in the sample.

(b) *Instrumental*—Basically atomic absorption spectrometers consist of a source, some means of atomisation, a monochromator, a detector and a mechanical chopper for modulating the source. The source is a hollow cathode lamp, the cathode of which is made of the element to be determined. Electrodeless discharge lamps are also available as sources for volatile elements such as As, Hg, Cd and Se. The sample to be measured is generally atomised in a flame using conventional atomisers and burners. Other methods for atomisation include cathode sputtering and electrically heating a graphite tube, a graphite rod or a tantalum boat in an atmosphere of argon or hydrogen[83, 84]. Radiation from the source traverses the atomic population representing the sample and enters the entrance slit of a monochromator, which separates the desired resonance line from the other lines that may be present. The net radiation intensity is measured by a conventional photomultiplier tube and amplifier. Usually the flame also emits radiation of the same wavelengths as the source. This background radiation is compensated by modulating the radiation from the source, and feeding the signal from the detector to a tuned a.c. amplifier, which rejects the d.c. signal contributed by emission from the flame. Modulation is accomplished by chopping the light mechanically and the amplifier is tuned to the chopping frequency. A schematic diagram of an atomic absorption spectrometer is given in Figure 4.9.

(c) *Sampling and determination*—Filtration is generally the method of

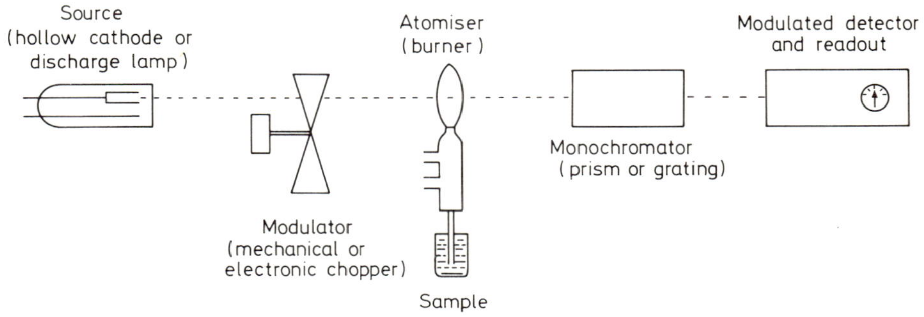

Figure 4.9 Schematic diagram of atomic absorption spectrometer

choice for sampling airborne particulates because of its simplicity and efficiency. The filtering material must be chosen with great care. Glass fibre filters, which are most commonly used, have particularly high levels of trace metal impurities[85] and are not well suited for critical work. Cellular and polystyrene membrane filters are recommended for sampling particulates, but for metals such as selenium and mercury, gas bubblers should be used. Mercury vapours are very conveniently collected by amalgamation with silver[86].

For most determinations the sample processing involves cutting out an appropriate portion of the filter and digesting it with a mixture of nitric and hydrochloric acids. The dissolved samples are filtered and the filtrate evaporated to dryness. The dried residue is then dissolved in a small volume of hydrochloric or nitric acid and made up to a known volume with deionised water[87, 88]. The solutions are aspirated into an oxy-acetylene flame, using an appropriate fuel-to-oxidant ratio. The absorbances are measured and referred to calibration curves to determine the concentration of the respective metals. Because of possible interference from the sample matrix it is preferable to carry out the determinations by the standard addition method. The details of sample processing and determination of metals using flameless atomic absorption spectrometry have been described in Refs. 89–91. These approaches are

Table 4.4 Atomic absorption spectroscopic methods

		Sensitivities		
		% *absorption**/p.p.m.		% *absorption*†/μg
Metal	*Wavelength*/Å	*Air–acetylene flame*	*Nitrous oxide–acetylene flame*	*Flameless atomisation*
Ag	3281	0.06		2×10^{-5}
Al	3092		1.00	1×10^{-4}
Ba	5536		0.40	5×10^{-5}
Be	2349		0.025	7×10^{-6}
Bi	2231	0.40		1×10^{-4}
Ca	4227	0.08	0.05	3×10^{-6}
Cd	2288	0.03		7×10^{-6}
Co	2407	0.15		9×10^{-4}
Cr	3579	0.10		4×10^{-5}
Cu	3247	0.09		2×10^{-5}
Fe	2483	0.12		2×10^{-4}
Hg	2536			8×10^{-4}
K	7665	0.04		3×10^{-6}
Mg	2852	0.01		2×10^{-7}
Mn	2795	0.06		2×10^{-5}
Na	5890	0.015		3×10^{-6}
Ni	2320	0.15		4×10^{-4}
Pb	2170	0.50		3×10^{-5}
Sn	2246	2.50		4×10^{-6}
V	3184	1.70		1×10^{-3}
Zn	2138	0.02		4×10^{-6}

* Obtained with Perkin-Elmer Model 403

† Sensitivities listed are by no means the maximum; sensitivities equal to or better than those listed can be obtained through proper control of parameters and impurities

still in the experimental stage and will not be discussed here. The sensitivities for different metals obtained with oxy-acetylene or nitrous oxide–acetylene flames and those with the flameless techniques are given in Table 4.4.

(d) *Interference*—Atomic absorption spectroscopy is free from the spectral interference encountered in flame photometry and the excitation interference common in emission spectroscopy. Chemical interference, such as that of phosphate on calcium, affects absorption and emission methods equally but may usually be overcome by the addition of a releasing agent such as EDTA, which serves to liberate the metal being determined from combination with an interfering species. A high-temperature flame such as N_2O–C_2H_2 can also be used to overcome some of the interference. It should be noted, however, that in the presence of a very high concentration of other salts or organic materials, the sensitivity may be reduced in some cases and enhanced in others. The decrease in absorption signal may be due to chemical interference such as the formation of refractory compounds or to changes in the physical properties of the sample solution such as viscosity and surface tension, which reduce the rate of atomisation. Enhancement of an absorption signal can take place by the reverse of the physical conditions mentioned above as well as by scattering of light by solid particles. The latter can be corrected to some extent by using the deuterium background corrector now available with most of the commercial atomic absorption instruments. Most physical interference factors can be overcome by an internal standardisation technique.

(e) *Advantages and disadvantages*—Atomic absorption spectroscopy is distinguished from most other analytical techniques by its sensitivity, speed, ease of operation, accuracy and relative freedom from physical and chemical interference.

Its principal disadvantage is that it cannot be applied to non-metals since their resonance lines lie in the vacuum-ultraviolet region. Also, simultaneous multi-element analyses are more difficult than in emission spectroscopy.

Until recently certain elements such as Al, Ba, Be, Si, Ti, V and W could not be determined because they form refractory compounds. This can now be done by atomising the sample in a nitrous oxide–acetylene flame. Other methods for improving the sensitivity include chelation and extraction of metals into organic solvents with subsequent aspiration of the organic phase[92], or increasing the effective optical path length by a system of mirrors or by using a long T-shaped quartz tube[93,94]. The use of flameless atomisation techniques has also increased the sensitivity enormously.

Atomic absorption is at present suitable for the determination of over 50 elements, 21 of which are listed in the Table 4.4. The precision, accuracy and reliability are equal to or better than other commonly used methods. Considerable progress has also been made in using atomic absorption for continuous monitoring[95].

4.3.3.2 Emission spectroscopy

Emission spectroscopy entails excitation of the sample to provide emission spectra of the metals present. This is usually accomplished by means of an arc or spark. A Lundegardh flame atomiser[96] has been used to detect some 34 elements, among which are some of those of greatest interest in air pollution monitoring systems. In addition to the excitation unit, the spectrometer includes a system for dispersing and isolating discrete portions of the spectrum (a prism or grating) and a receptor, such as photographic plate or a recording device for measuring the isolated radiant energy. The details of the instrumentation may affect its cost and convenience but generally the method remains one of the best approaches for the routine identification and estimation of metal in particulate samples. It is sensitive (microgram to nanogram range for most metals), qualitatively free from interference, rapid (once set up and applied routinely) and widely applicable. The equipment is relatively costly and considerable skill and experience are required of the spectroscopist.

Many important applications of emission spectroscopy are made in air pollution studies. The National Air Sampling Network applies emission and atomic absorption methods in studies of samples collected in various parts of the U.S.A. Cholak *et al*[97,98] have used spectrographic methods for broad studies of both airborne and settled dusts.

Cholak and Story[99] have applied emission methods for the determination of several metallic constituents of sootfall and ash. Clayton[100] has investigated spectrographically the material trapped by accordion-pleated filters and obtained data on 24 elements. Keenan and Byers[101] employed spectrographic methods for the estimation of 21 elements in samples collected with high-volume samplers. O'Neil[102] and Keenan and Holz[103] have described a spectrographic technique for determining beryllium in air.

4.3.3.3 Flame photometry

Flame photometry is now being more widely used than in the past because better methods of excitation have made more applications possible. Originally, flame photometry found use only in determining alkali or alkaline earth metals. With better methods of excitation it is now possible to use flame photometry in other determinations[104,105]. Flame photometry is sensitive and quite reliable, particularly when applied to air pollution problems where matrix effects are usually not serious. It is relatively inexpensive, rapid and simple to apply.

The flame, although a low-energy source compared with an arc or spark, is very stable and is capable of exciting intense emission of radiant energy from elements with low excitation potentials. The method has been applied for the analysis of several water-soluble atmospheric pollutants such as sodium, potassium, calcium, magnesium, strontium, manganese and iron.

Flame photometry may find wider application to air pollution work in the coming years, as improved instrumentation and hotter flames for excitation are becoming available. In addition, newly designed burners for introducing powdered samples may facilitate the direct study of dusts.

4.3.3.4 Electrochemical methods

(a) *Polarography*—This technique was first introduced by Heyrovsky[106], in 1922. Application of this method to a wide variety of analytical problems has made it an important tool in modern analytical chemistry. Polarography is fairly simple, rapid and reliable and has sensitivities comparable with those of the other methods. The method has the unique advantage of analysing both inorganic and organic species. Several metals can be determined simultaneously using polarography under favourable conditions. Polarographic instruments—manual, semi-automatic and completely automated—are readily available commercially.

The method is based on the application of a known voltage from an outside source across a sample cell consisting of a small polarisable electrode (generally a dropping mercury electrode) and a large non-polarisable electrode (generally a saturated calomel electrode). The dropping mercury electrode is the indicator electrode, serving to measure the concentration of the electroactive materials in the sample solution in which it is immersed, and the saturated calomel electrode is the reference electrode. When the voltage applied to the dropping mercury electrode exceeds the reduction potential of a cation, the cation is discharged on the dropping mercury electrode. Since the solution is quiescent, the discharge is controlled by the concentration of cation in the solution. When the diffusion current is plotted against the applied e.m.f., a polarogram is obtained. The decomposition potential, or preferably the half-wave potential, serves to identify the cation and the height of the diffusion current gives the concentration of each species. A relatively high concentration of a supporting electrolyte is necessary in the sample solution for suppressing migration and kinetic currents. The selection of different supporting electrolytes also provides a degree of flexibility for polarographic analyses because different supporting electrolytes give rise to different half-wave potentials for respective electroactive species. Sodium fluoride makes a good supporting electrolyte for air pollution investigations because the half-wave potentials of such important metals as copper, lead, cadmium, nickel, cobalt and manganese are well separated, with the exception of the nickel and zinc which have essentially the same half-wave potentials[107].

Polarography is sufficiently sensitive for air pollution studies. Except for detailed sample preparation, the method is quite simple and rapid. Nevertheless, polarography is used very little in air pollution studies. Dubois and Monkman[108] have used polarography to determine copper, lead, cadmium, nickel, cobalt, chromium, zinc and manganese simultaneously. Schneider and Breidenback[109] have evaluated the use of polarography for the determination of lead in air. A detailed survey of the literature reveals that the behaviour of some 80 elements have been studied polarographically, many of which can be directly applied to monitoring pollutants in the air. There have been several excellent reviews[111,112] on the subject and the interested reader is referred to them for more information.

The sensitivity and the resolution of conventional polarography has been vastly improved by scanning the potential at a rapid rate in the lifetime of one single mercury drop. This technique, in the form of cathode-ray polarography, has the effect of increasing the current flow and providing a smooth curve. By taking the derivative of the curve, the resolution can be greatly enhanced. Resolution of peaks of less than 50 mV separation is possible and detection limits are lowered to the 10^{-7} and 10^{-8} mol l^{-1} range.

The development of differential pulse polarography has served to enhance the sensitivity of determination of metal ions in solution. The pulsed signals are timed to coincide with the drop time of the dropping mercury electrode. The current flowing is measured immediately before the start and completion of the pulse. The change in current between the two times arises independently of the reduction of the electroactive species, and occurs on a time interval such that it is larger than the diffusion controlled current. The result of this approach is a peak for each species corresponding to the derivative of the standard form of the polarogram. This serves to enhance the resolution and sensitivity to levels approaching 10^{-9} mol l^{-1}. For details the reader is referred to Ref. 113.

(b) *Anodic stripping voltammetry*[114]—This technique can be applied to most metals and involves pre-concentration of the material by application of a large enough potential to the electrode system to plate out the metals on the electrode surface. The working electrode is usually a hanging mercury drop electrode (HMDE) or a graphite electrode with a thin mercury coating. Occasionally a platinum electrode may be used. When all the metals of interest have plated out from the solution, the potential is scanned in the reverse direction and the current flow is recorded. The magnitude of the current is proportional to the concentration of the species originally present in the solution. Recently[115], techniques similar to those employed in differential pulse polarography have been used in the anodic stripping method, giving better resolutions of the peak currents. The sensitivity of anodic stripping

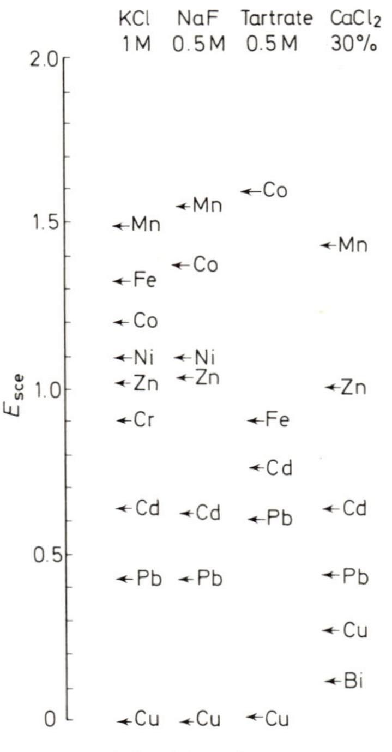

Figure 4.10 Selectivity of some important metals in polarography

Table 4.5 Detection limits (μg) for some common elements

Element	*Polarography**	*Anodic stripping**
Aluminium	0.01	
Arsenic	0.75	
Bismuth	0.01	0.02
Bromine	1.0	
Cadmium	0.05	0.00005
Chlorine	2.0	
Chromium	10.0	
Cobalt	0.01	0.002
Copper	0.02	0.00006
Erbium	0.01	
Gallium	0.01	
Germanium		1.0
Indium	0.01	
Iodine	1.0	
Iron	0.01	
Lead	0.01	0.00021
Molybdenum	10.0	
Nickel	0.01	
Niobium	4.0	
Selenium	5.0	
Silver	0.5	
Sulphur	10.0	
Tellurium	1.0	
Tin	0.05	
Tungsten	10.0	
Vanadium	0.01	
Zinc	0.01	0.00001
Zirconium	0.01	

* Detection limits (μ) for a 10 ml solution

voltammetry is remarkably high, permitting the analysis of metal solutions as dilute 10^{-12} mol l^{-1}.

Table 4.5 and Figure 4.10 clearly show the high sensitivity and selectivity obtained by these two electrochemical techniques for some metals important in air pollution studies.

4.3.3.5 Ring oven methods

The ring oven technique was introduced in 1954 by Weisz[116]. The ring oven is a simple, inexpensive apparatus which permits the separation, concentration and analysis of inorganic and organic substances at microgram to nanogram levels. The method offers great promise for the study of airborne particulates because of its sensitivity and reliability. It is very convenient for field studies and for 'on the spot' analysis.

(a) *Equipment*—The equipment needed includes a ring oven, an electric hair dryer or oven suitable for drying filter papers, several micropipettes and capillary pipettes. The ring oven consists of a heated ring, ordinarily with an

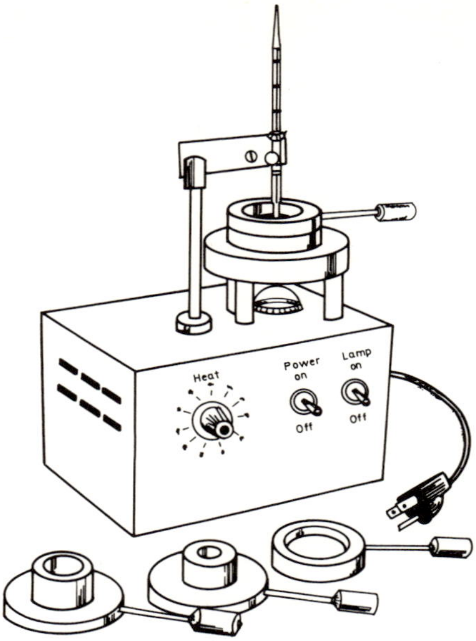

Figure 4.11 Ring oven

inner diameter of 22 or 33 mm, which serves as a means of fixing solute materials in a sharply defined circle. Although a suitable ring oven can easily be constructed in any good laboratory workshop, commercial instruments are readily available. One of these is shown in Figure 4.11. The heating areas of the ring oven are generally made of aluminium, but various other materials such as copper, steel or glass, can also be used.

(b) *Principle*—The ring oven method is a special refinement of spot test procedures that are carried out on filter paper. The sample (a solution or dust spot) is introduced at the exact centre of a filter paper, placed on the heated surface of the ring oven. The species of interest are washed to the ring zone using suitable solvents added to the centre of the paper. As the solution approaches the heated ring zone, the carrier solvent evaporates, depositing solute in a sharply defined ring. In this process, a 30- to 50-fold concentration of the sample is achieved. The filter paper is removed from the ring oven and cut into as many sectors as there are determinations to be run. The analysis is completed by adding appropriate specific reagents to the respective sectors to develop the definitive test colours, which are compared against a series of standards made in a similar fashion. The essential steps of a ring oven analysis are depicted in Figure 4.12. If the specific reagents are not available, the interferents are marked with suitable masking agents and then the colour of the desired species is developed.

(c) *Sampling and determination*—The ring oven technique can, because of its inherent sensitivity, readily be adapted for analysing airborne particulates. Samples are conveniently collected using a tape sampler. For most of the determinations only 0.5–2 m^3 of air sample is needed, which can be collected

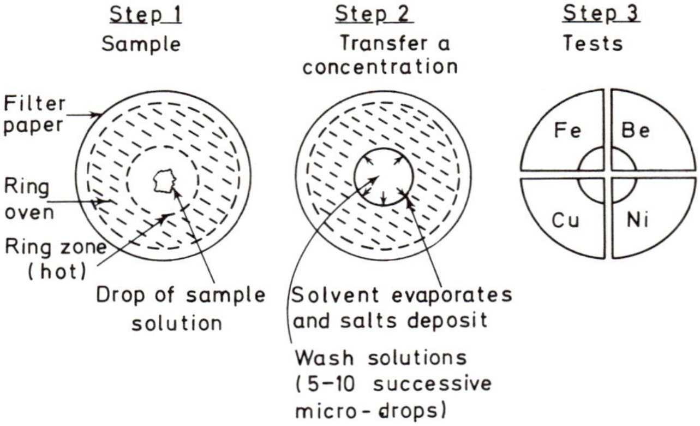

Figure 4.12 Steps for ring oven analysis

in 1–2 h with most of the commercially available tape samplers. The collection, concentration and the determination can all be done on the filter paper tape itself. The section of the tape containing the sample spot is washed with appropriate solvents and finally the ring is developed on the tape itself by applying the appropriate reagents. The concentration of the desired metal is then usually estimated by comparing the colour intensity of the ring with those of the standards prepared in the same way.

Particulate samples collected by the high-volume samplers can also be analysed as follows. A small disc of suitable size is punched out of the large filter and fixed to the centre of a filter placed on the ring oven. The disc is washed with appropriate solvents to dissolve the metals of interest, deposited on the ring zone and analysed as described above.

Only quantitative or high-quality filter paper should be used for ring oven studies. All papers studied were found to contain certain traces of selenium[117]. Except for studies of selenium itself, a selenium impurity in paper is not significant. Iron impurities, on the other hand, are troublesome and papers with low iron content should be used as much as possible. For those cases where contaminated paper must be avoided, the paper can be pretreated to remove the unwanted impurities[118]. West and Shendrikar[127] have developed a novel method of masking the iron impurities by precipitating iron as the basic acetate in the centre of the filter paper.

(d) *Air pollution studies with the ring oven technique*—During the past few years, several ring oven methods have been developed to determine a variety of pollutants in airborne particulates. The methods are specially suited for field studies or *in situ* analysis because the equipment is portable, power requirements are minimal, requires small samples and the analytical procedures are versatile and sensitive.

Methods for the individual determination of antimony[119], copper[120], beryllium[121, 122], zinc[123, 124], selenium[125], cadmium[126], lead[127, 128], vanadium[129, 130], manganese[131], sulphur[132], phosphates[133], sulphuric acid and total acidity[134] for application in air pollution studies have been worked out by West and co-workers. A scheme for the simultaneous analysis of eight different metals

Table 4.6 Ring oven methods

Element	*Reagent and colour*	*Optimum range*/mg	*Limit of detection*/μg	*Ref.*
Al	Morin Reaction (Fluorescence)	0.03–0.5	0.01	120
Sb	Phosphomolybdic acid (Blue)	0.1–1.0	0.08	122
Be	Morin–fluorescence	0.01–0.2	0.01	125
Co	1-Nitroso-2-naphthol (Red-brown)	0.08–2.0	0.05	121
Cu	Dithio-oxamide (Green-black)	0.1–1.0	0.08	
Cd	Ferrous dipyridyl (Iodide-pink)	0.08–2.0	0.05	127
Ni	Dimethylglyoxime (Pink)	0.1–2.0	0.08	125
Mn	Benzidine (Blue)	—	—	132
H_2SO_4	Fluorescence (Pink)	—	—	135
Pb	Dithizone	0.5–4.0	0.1	128
Zn	Dithizone (Red-violet)	0.1–1.5	0.05	125

in airborne particulates has been reported[124]. To adapt the ring oven technique to water analysis and to handle large volumes of reagents rapidly, without the fear of flooding the ring zone, an impervious barrier embedded in a ring made out of some high melting polymeric material has recently been developed[135]. Table 4.6 summarises applications of ring oven methods to the analysis of airborne particles.

4.3.3.6 Miscellaneous methods

(a) *X-Ray fluorescence spectroscopy*—In the past few years, considerable advances have been made in the use of X-ray methods for the characterisation and analysis of airborne particulates. Of the different X-ray techniques such as diffraction, absorption, fluorescence, etc., the fluorescence technique has been widely applied for the determination of inorganic airborne particulates.

X-Ray emission or fluorescent analysis is a non-destructive method, which requires little or no sample preparation. It is sufficiently sensitive to enable trace element determinations to be carried out on a milligram or so of sample. The technique is especially attractive because it has the inherent capability of identifying and determining a multitude of elements in a single sample with a scan rate as high as one element per minute. The method is also ideally suited for automation.

The analytical technique of X-ray emission spectroscopy depends on the ability to excite and accurately measure characteristic K and L X-rays emanating from the specimen. Prior to 1966, energy separation was usually achieved by using wavelength dispersive spectrometers. Bowman *et al.*[136] in 1966 first utilised semiconductor detectors [Si(Li) and Ge(Li)] for energy dispersive X-ray spectroscopy. Progress in electronic design over the past few years has significantly improved the energy resolution and performance of these detectors, thereby increasing the sensitivity of the methods to nanogram levels[137-141].

X-Ray fluorescence spectroscopy is not without defect. In application the ideal sample should be as thin as possible so as to minimise matrix effects. Suitable standards, containing known amounts of the desired metals, simulating the actual airborne particulates in size, shape and distribution, are essential for calibration. Until now, reliable dust standards were not readily available. Recently, Dharmarajan and West[142] have devised a simple and convenient method to generate standard dust samples in the laboratory which closely simulates the naturally occurring metal particulates. They have proved the feasibility and superiority of using these dust samples as standards for calibration[143]. Some special limitations apply to the collection of the sample because glass fibre and vinyl membrane filters cannot be used on account of the high concentrations of contaminating elements likely to be present. The principal deterrent to the use of X-ray fluorescence spectroscopy is the cost and complexity of the equipment required. Like neutron activation analysis, it is not a tool that can be casually incorporated into a programme. The high cost can be justified only on the basis of an assured flow of thousands of samples. Only in such circumstances can specially qualified personnel and the costly equipment be justified.

(b) *Neutron activation analysis*—Neutron activation analysis is a powerful, versatile tool for the study of airborne particulate samples. The great cost of the facility required for activation of samples limits the use of the technique. Likewise, the cost of counting equipment and the highly specialised training required dictates that the method be used only in special cases. Detailed discussion of the methods involved is beyond the scope of this review.

For general and detailed information on current techniques and procedures, the compendium *Air Pollution* by Stern is recommended[144].

References

1. West, P. W. and Gaeke, G. C. (1956). *Anal. Chem.*, **28,** 1916
2. Nauman, R. V., West, P. W., Tron, F. and Gaeke, G. C. (1960). *Anal. Chem.*, **32,** 1307
3. Lyles, G. R., Dowling, F. B. and Blanchard, V. J. (1965). *J. Air Pollution Control Assoc.*, **15,** 106
4. Scaringelli, F. P., Saltzman, B. E. and Frey, S. A. (1967). *Anal. Chem.*, **39,** 1709
5. Pate, J. B., Ammons, B. E., Swanson, G. A. and Lodge, J. P., Jr. (1965). *Anal. Chem.*, **37,** 942
6. Zurlo, N. and Griffini, A. M. (1962). *Med. Lavoro*, **53,** 330
7. Environmental Protection Agency. (1971). *U.S. Federal Registry*, **36,** 8189, April 30
8. O'Keeffe, A. E. and Ortman, G. C. (1966). *Anal. Chem.*, **38,** 760
9. Scaringelli, F. P., Frey, S. A. and Saltzman, B. E. (1967). *Amer. Ind. Hyg. Assoc. J.*, **28,** 260
10. Scaringelli, F. P., O'Keeffe, A. E., Rosenberg, E. and Bell, J. P. (1970). *Anal Chem.*, **42,** 871
11. (1970). *Annual Book of ASTM Standards*, Part 23, *Water; Atmospheric Analysis*, ASTM Designation-D-1355-60, 333 (Philadelphia: American Society For Testing & Materials)
12. Griever, P. M. and Cook, W. A. (1960). *A.M.A. Arch. Ind. Health*, **21,** 233

13. Nader, J. S. and Dolphin, J. L. (1959). *J. Air Pollution Control Assoc.*, **8,** 336
14. McKee, H. C. and Rollwitz, W. L. (1959). *J. Air Pollution Control Assoc.*, **8,** 339
15. Reiszer, K. D. and West, P. W. (1973). *Environ. Sci. Technol.*, **7,** 526
16. Hollowell, C. D. and McLaughlin, R. D. (1973). *Anal. Chem.*, **45,** 63A
17. Hollowell, C. D., Gee, G. Y. and McLaughlin, R. D. (1973). *Anal. Chem.*, **45,** 63A
18. Environmental Instrumentation Group, Lawrence Berkeley Laboratory. (1972). *Instrumentation for Environmental Monitoring: Air*, Vol. 1 (Berkeley, California: LBL)
19. Jacobs, M. B., Braverman, M. M. and Hochheiser, S. (1957). *Anal. Chem.*, **29,** 1349
20. Budd, M. S. and Bewick, H. A. (1952). *Anal. Chem.*, **24,** 1536
21. Jacobs, M. B. (1965). *J. Air Pollution Control Assoc.*, **15,** 314
22. Natusch, D. F. S., Sewell, J. R. and Tanner, R. L. (1974). *Anal. Chem.*, **46,** 410
23. Okita, T., Lodge, J. P., Jr. and Axelrod, H. D. (1971). *Environ Sci. Technol.*, **5,** 532
24. Lodge, J. P., Jr., Pate, J. B. and Huitt, H. A. (1963). *Amer. Ind. Hyg. Assoc. J.*, **24,** 380
25. Natusch, D. F. S. (1970). *Clean Air*, **4,** 69
26. Smith, A. F., Jenkins, D. A. and Cunningworth, D. E. (1961). *J. Appl. Chem.*, **11,** 317
27. Sanderson, H. P., Thomas, R. and Katz, M. (1966). *J. Air Pollution Control Assoc.*, **16,** 328
28. Wohlers, H. C. and Feldstein, M. (1966). *J. Air Pollution Control Assoc.*, **16,** 19
29. Strange, J. P. (1957). *Anal. Chem.*, **29,** 1878
30. Stevens, R. K., Mulik, J. D., O'Keeffe, A. E. and Krost, K. J. (1971). *Anal. Chem.*, **43,** 827
31. Saltzman, B. E. (1954). *Anal. Chem.*, **26,** 1949
32. Saltzman, B. E. (1960). *Anal. Chem.*, **32,** 135
33. Thomas, M. D., MacLeod, J. A., Robbins, R. C., Goettelman, R. C., Eldridge, R. W. and Rogers, L. H. (1956). *Anal. Chem.*, **28,** 1810
34. Saltzman, B. E. and Wartburg, A. F. (1965). *Anal. Chem.*, **37,** 1261
35. Gill, W. E. (1960). *Amer. Ind. Hyg. Assoc. J.*, **21,** 87
36. Stratmann, H. and Buck, M. (1966). *Int. J. Air Water Pollution*, **10,** 313
37. Jacobs, M. B. and Hocchheiser, S. (1958). *Anal. Chem.*, **30,** 426
38. Morgan, G. B., Golden, C. and Tabor, E. C. (1967). *J. Air Pollution Control Assoc.*, **17,** 300
39. Purdue, L. J., Dudley, J. E., Clements, J. B. and Thompson, R. J. (1972). *Environ. Sci. Technol.*, **6,** 153
40. Buck, M. and Stratmann, H. (1967). *Staub*, **27,** 11
41. Levaggi, D. A., Siu, D. and Feldstein, M. (1971). *J. Air Pollution Control Assoc.*, **23,** 30
42. Heuss, J. M., Nebel, G. J. and Colucci, J. M. (1971). *J. Air Pollution Control Assoc.*, **21,** 535
43. Reiszner, K. D. and West, P. W. (1974). Unpublished studies
44. Teague, M. C. (1920). *Ind. Eng. Chem.*, **12,** 964
45. Grant, G. A., Katz, M. and Haines, R. L. (1951). *Can. J. Technol.*, **29,** 43
46. Adams, E. G. and Simmons, N. T. (1951). *J. Appl. Chem.*, **1,** Suppl. 1, S20
47. Kainz, G. and Horwatitsch, H. (1960). *Z. Anal. Chem.*, **176,** 175
48. Porter, K. and Volman, D. H. (1962). *Anal. Chem.*, **34,** 748
49. Grosskopf, K. (1963). *Chem. Ztg.*, **87,** 38
50. Hodgeson, J. A., Martin, B. E. and Baumgardner, R. E. (1971). *Progress in Analytical Chemistry*, Vol. V (New York: Plenum)
51. Hodgeson, J. A., Stevens, R. K. and Martin, B. E. (1971). *Analysis Instrumentation Symposium* (Houston: Instrument Society of America)
52. Byers, D. H. and Saltzman, B. E. (1958). *J. Amer. Ind. Hyg. Assoc.*, **19,** 251
53. Saltzman, B. E. and Gilbert, M. (1959). *Anal. Chem.*, **31,** 1914
54. Altshuller, A. P., Schwab, C. M. and Bare, M. (1959). *Anal. Chem.*, **31,** 1987
55. Saltzman, B. E. and Wartburg, A. F. (1965). *Anal. Chem.*, **37,** 779
56. Galster, H. (1962). *Z. Anal. Chem.*, **186,** 359
57. Paneth, F. A. and Gluckhauf, E. (1941). *Nature*, **147,** 61
58. Schulze, F. (1966). *Anal. Chem.*, **38,** 748
59. Ehmert, A. (1949). *Z. Naturforsch.*, **4b,** 321
60. Renzetti, N. A. (1957). *Anal. Chem.*, **29,** 869
61. Hanst, P. L., Stephens, E. R., Scott, W. E. and Doerr, R. L. (1961). *Anal. Chem.*, **33,** 1113
62. Bovee, H. H. and Robinson, R. J. (1961). *Anal. Chem.*, **33,** 1115

63. Arnold, C. and Mentzel, C. (1902). *Ber.*, **35**, 1324
64. Fuhrmann, H. (1965). *Staub*, **25**, 266
65. Sachdev, S. L., Lodge, J. P., Jr. and West, P. W. (1971). *Anal. Chim. Acta*, **58**, 141
66. Bravo, H. A. and Lodge, J. P., Jr. (1964). *Anal. Chem.*, **36**, 761
67. Bufalini, J. J. (1968). *Environ. Sci. Technol.*, **2**, 703
68. Hauser, T. A. and Bradley, D. W. (1966). *Anal. Chem.*, **38**, 1529
69. Altshuller, A. P., Miller, D. L. and Sleva, S. F. (1961). *Anal. Chem.*, **33**, 621
70. Hendricks, R. H. and Larsen, L. B. (1966). *Amer. Ind. Hyg. Assoc.*, **27**, 81
71. West, P. W., Lyles, G. R. and Miller, J. L. (1970). *Environ. Sci. Technol.*, **4**, 487
72. Leonard, M A. and West, T. S. (1960). *J. Chem. Soc.*, 4477
73. Greenhalgh, R. and Riley, J. P. (1961). *Anal. Chim. Acta*, **25**, 179
74. McCarthy, R. and Moore, C. E. (1952). *Anal. Chem.*, **24**, 411
75. Hemeon, W. L., Haines, G. F., Jr. and Ide, H. M. (1953). *J. Air Pollution Control Assoc.*, **3**, 22
76. Gruber, C. W. and Alpaugh, E. L. (1954). *J. Air Pollution Control Assoc.*, **4**, 143
77. Working Party. (1965). *Methods of Measuring Air Pollution*, Report No. 17913 (Paris: Organisation for Economic Cooperation and Development)
78. Barnes, E. C. (1952). *Proc. U.S. Tech. Conf. Air Pollution*, 547 (New York: McGraw-Hill)
79. White, H. J. (1963). *Industrial Electrostatic Precipitation* (Reading, Mass.: Addison-Wesley)
80. Subcommittee on Environmental Improvement. (1969). *Cleaning Our Environment, The Chemical Basis for Action* (Washington: American Chemical Society)
81. Robinson, J. W. (1966). *Atomic Absorption Spectroscopy* (New York: Marcel Dekker)
82. Elwell, W. T. and Gidley, J. A. F. (1961). *Atomic Absorption Spectrophotometry* (Oxford: Pergamon Press)
83. Russel, B. J. and Walsh, A. (1959). *Spectrochim. Acta*, **15**, 883
84. Gatehouse, B. M. and Walsh, A. (1960). *Spectrochim. Acta*, **16**, 602
85. Hwang, J. Y. (1972). *Anal. Chem.*, **44**, 20A
86. Sachdev, S. L. (1974). *Determination of Mercury in Air.* Presented at the American Chemical Society, Analytical Chemistry Symposium, Los Angeles
87. Thompson, J. R., Morgan, G. B. and Purdue, L. J. (1969). *Analysis of Selected Elements in Atmospheric Particulate Matters by Atomic Absorption.* Presented at the Instrument Society of America Symposium, New Orleans
88. (1971). *Annual Book of ASTM Standards.* Part 32, E-2P1, 1052
89. Kirkbright, G. F. (1971). *Analyst*, **96**, 609
90. Amos, M. D. (1970). *Amer. Lab.*, August, 33
91. West, T. S. and Williams, X. K. (1969). *Anal. Chim. Acta*, **45**, 27
92. Sachdev, S. L. and West, P. W. (1970). *Environ. Sci. Technol.*, **4**, 751
93. Robinson, J. W. (1962). *Anal. Chim. Acta*, **27**, 465
94. Sachdev, S. L. Robinson, J. W. and West, P. W. (1967). *Anal. Chim. Acta*, **38**, 499
95. Loftin, H. P., Christian, C. M. and Robinson, J. W. (1970). *Spectrosc. Lett.*, **3**, 111
96. Cholak, J. and Hubbard, D. M. (1944). *Ind. Eng. Chem., Anal. Ed.*, **16**, 728
97. Cholak, J., Schafer, L. J. and Hoffer, R. F. (1950). *Arch. Ind. Hyg. Occupational. Med.*, **2**, 443
98. Cholak, J., Schafer, L. J. and Hoffer, R. F. (1952). *Arch. Ind. Hyg. Occupational Med.*, **6**, 314
99. Cholak, J. and Story, R. V. (1941). *J. Opt. Soc. Amer.*, **31**, 730
100. Clayton, G. D. (1953). *Determination of Atmospheric Contaminants* (New York: Amer. Gas Assoc.)
101. Keenan, R. G. and Byers, D. H. (1952). *Arch. Ind. Hyg. Occupational Med.*, **6**, 226
102. O'Neil, R. L. (1962). *Anal. Chem.*, **34**, 781
103. Keenan, R. G. and Holtz, J. L. (1964). *Amer. Ind. Hyg. Assoc. J.*, **25**, 254
104. Dean, J. A. (1960). *Flame Photometry* (New York: McGraw-Hill)
105. Herrmann, R. and Alkamade, C. T. J. (1952). *Flame Photometry* (Translated by P. T. Gilbert, Jr.) (New York: Wiley-Interscience)
106. Heyrovsky, J. (1922). *Chem. Listy*, **16**, 256
107. West, P. W., Dean J. and Breda, E. J. (1947). *Coll. Czech. Chem. Commun.*, **13**, 1
108. Dubois, L. and Monkman, J. L. (1964). *Amer. Ind. Hyg. Assoc. J.*, **25**, 485

109. Schneider, M. C. and Breidenbach, A. W. (1962). *Evaluation of a Polarographic Procedure for Lead in Air* (Cincinnati: Robert A. Taft Sanitary Engineering Center)
110. Flato, J. B. (1972). *Anal. Chem.*, **44,** 75A
111. Ellis, W. D. (1973). *J. Chem. Ed.*, **50,** A131
112. Maienthal, J. and Taylor, J. K. (1968). *Advances in Chemistry Series*, No. 73, 172 (R. F. Gould, editor) (Washington: American Chemical Society)
113. Heyrovsky, J. and Kuta, J. (1966). *Principles of Polarography* (New York: Academic Press)
114. Kemula, W. (1967). *Pure Appl. Chem.*, **15,** 283
115. Siegerman, H. and Dom. G. (1972). *Amer. Lab.*, **4,** 59
116. Weisz, H. (1954). *Mikrochim. Acta*, 140
117. West, P. W. and Cimmermann, Ch. (1964). *Anal. Chem.*, **36,** 2013
118. West, P. W., Llacer, A. J. and Cimmerman, Ch. (1962). *Mikrochim. Acta*, **6,** 1165
119. West, P. W. and Llacer, A. J. (1962). *Anal. Chem.*, **34,** 555
120. West, P. W. and Pitombo, R. M. (1967). *Anal. Chim. Acta*, **37,** 374
121. West, P. W. and Jungreis, E. (1969). *Anal. Chim. Acta*, **45,** 188
122. West, P. W. and Mohilner, P. R. (1962). *Anal. Chem.*, **34,** 558
123. West, P. W. and Thabet, S. K. (1967). *Anal. Chim. Acta*, **37,** 246
124. Dharmarajan, V. (1972). *Ph.D. Dissertation*, Louisiana State University
125. West, P. W. and Cimmerman, Ch. (1964). *Anal. Chem.*, **36,** 2013
126. Dharmarajan, V. and West, P. W. (1971). *Anal. Chim. Acta*, **57,** 469
127. West, P. W. and Shendrikar, A. D. (1971). *Anal. Chim. Acta*, **61,** 43
128 Jungreis, E. and West, P. W. (1969). *Israel J. Chem.*, **7,** 413
129. Jungreis, E. and West, P. W. (1969). *Anal. Chim. Acta*, **44,** 440
130. Sachdev, S. L. (1966). *Ph.D. Dissertation*, Louisiana State University
131. McDaniel, M. and West, P. W. (1974). *Anal. Chim. Acta*, **70,** 482
132. Rowland, R. and West, P. W. (1972). *Anal. Chim. Acta*, **61,** 152
133. West, P. W. and Sachdev, S. L. (1968). *Atmos. Environ.*, **2,** 331
134. West, P. W., Shendrikar, A. D. and Herrara, N. (1974). *Anal. Chim. Acta*, **69,** 43
135. West, F. K., Dharmarajan, V. and West, P. W. (1973). *Anal. Chim. Acta*, **66,** 482
136. Bowman, H. R., Hyde, E. K., Thompson, S. G. and Jared, R. C. (1966). *Science*, **151,** 562
137. Goulding, F. S., Walton, J. T. and Malone, D. F. (1969). *Nucl. Instrum. Methods*, **71,** 273
138. Goulding, F. S., Walton, J. T. and Pehl, R. H. (1970). *IEEE Trans. Nucl. Sci.*, **17,** 218
139. Landis, D. A., Goulding, F. S., Pehl, R. H. and Walton, J. T. (1971). *IEEE Trans. Nucl. Sci.*, **18,** 155
140. Goulding, F. S., Jaklevic, J. M., Jarrett, B. V. and Landis, D. A. (1972). *Advan. X-Ray Anal.*, **15,** 470
141. Giaque, R. D., Goulding, F. S., Jaklevic, J. M. and Pehl, R. H. (1973). *Anal. Chem.*, **45,** 671
142. Dharmarajan, V. and West, P. W. (1974). *Anal. Chim. Acta*, **69,** 43
143. Dharmarajan, V., Thomas, R. L. and West, P. W. (1974). Unpublished studies
144. Stern, A. C. (1974). *Air Pollution*, 3rd ed. (New York: Academic Press)

5
Developments in Forensic Analysis

E. F. PEARSON
The Home Office, London

5.1 INTRODUCTION

In discussing forensic analysis, the term 'forensic science' will often be used. In this review, forensic science is taken to mean the science which lies behind expert scientific evidence presented in the law courts. In the United States this field may be referred to as 'police science' or 'criminalistics', in France it is referred to as 'criminalistique' and in Germany 'kriminalistik'. The agreement between the precise areas of work covered is not exact and there are differences in the way scientific evidence is presented in the different countries. For convenience, however, forensic science and criminalistics will be taken to have the same meaning.

The task of the forensic scientist is to assist the court in deciding on a person's involvement in a crime. When an item is submitted to him for examination by the police, the scientist must first decide what he is dealing with. The item may be a paint flake, a glass fragment, a fibre, a hair, a piece of wood, plastic or rubber—in fact it could be almost anything. Items which are easy to identify when encountered in large amounts are often difficult to identify when only nanogram quantities are available. Secondly, the scientist must decide whether the item can be associated with the suspect. He must decide whether the paint flake or glass fragment removed from the clothing of a suspect is similar to items taken from the scene of the crime. He must decide, perhaps, whether a bloodstain found at the scene of a crime has characteristics similar to those possessed by the accused. Finally, and by no means least, he must decide for himself the value of the evidence he has discovered and be in a position to communicate his opinion, either orally or as a written statement, to the court examining the case.

Forensic science now embraces many forms of fundamental analysis and the techniques of the inorganic and organic chemist, toxicologist, biochemist, biologist, geologist and metallurgist are all needed. It is not possible to cover all of these disciplines adequately in a single review and accordingly this review is primarily concerned with recent developments in forensic chemistry.

5.2 THE WORK OF THE FORENSIC ANALYST

The work of the forensic scientist ranges from simple visual comparative examinations to fundamental analysis of trace material. In recent years there have been a number of attempts[1-4] to rationalise the service given by forensic science laboratories. Typically[3], investigators attempt to establish some sort of criteria by which, given a particular criminal justice organisation and crime pattern, an optimal forensic science service can be set up. To date, most of the studies have only achieved a simple documentation of the work load and type of a given laboratory[4]. As yet there is no evidence that investigators have been able to present their material in a form which ensures that the statistical information collected makes an effective contribution to the formulation of policy. No real attempt has been made to identify users' needs or to establish how the services could best be managed and co-ordinated from the results of the surveys. If nothing else, however, the studies are able to reveal trends.

There has been a great increase in some types of case in many countries. In some areas of work, demand is creating a need for the introduction of automated methods. The automated analysis of urine[5, 6] and tablets[7] for drugs has been reported. Work is well advanced on the automated analysis of blood and urine for ethyl alcohol[8] and on the automated determination of Rhesus groups[9]. Cases involving drugs, explosives, flammables and firearms are examples of types of case which are becoming increasingly important, while the need for the examination of paint chips, glass fragments, hair, fibres, plastics, rubber and dyes as well as blood, semen, pollen and wood continues to have a strong influence on the analytical techniques used in forensic science laboratories[10-12]. Ideally, the techniques used in a particular case should have those attributes desirable in all analytical work: they should be relatively inexpensive, rapid in operation, have application to a wide range of problems, be non-destructive and not need highly specialised staff. However, there has been such a demand for accurate analytical work on progressively smaller amounts of material that modern forensic science laboratories now possess highly sophisticated analytical equipment.

5.3 TECHNIQUES USED IN INORGANIC ANALYSIS

5.3.1 Atomic absorption spectroscopy

The availability of sensitive analytical methods has been a major factor in stimulating the interest of the forensic practitioner to the possibility of characterising samples through trace element composition. Atomic absorption spectroscopy is now at the forefront of these methods and offers relatively inexpensive equipment which readily yields quantitative information in concentration ranges which previously have had to be measured by neutron activation analysis or mass spectrometry. At first, the forensic scientist was interested in the use of atomic absorption spectroscopy to determine single elements such as thallium, lead and mercury. In the investigation of crimes involving counterfeit coins or theft of metals, the possibility of the analysis of simple alloys also attracted attention. More recently, the forensic scientist has begun to investigate the wider potential of atomic absorption equipment associated with the advent of hot hollow cathode lamps possessing a multi-element capability; electrodeless discharge lamps[13]; the possibility of using several lamps at the same time in one analysis and then analysing the absorption in spectrographic equipment designed to cope with emission; the development of non-flame atomisers[14, 15]; and the possibility of using atomic absorption equipment for measuring atomic fluorescence[16].

One of the difficulties sometimes encountered in the use of atomic absorption techniques is the need for some form of sample treatment prior to analysis[17]. Most methods for destroying organic matter fall into two classes: dry ashing and wet digestion. The relative merits of these oxidation methods have been studied extensively[18, 19]. However, much conflict of opinion still remains as to which is to be preferred. The samples encountered in forensic science can be extremely small and ashing may introduce an unacceptable level of impurity. With the introduction of non-flame atomisers, such as the graphite furnace, the removal of organic matter *in situ* in the atomiser itself becomes possible through a stepwise increase of temperature prior to recording the absorption signal of the element under investigation at the optimum temperature. This latter technique could introduce significant sources of error through matrix effects and it may be that the conventional techniques are the most suitable.

It is often necessary to separate the element of interest from the sample matrix by solvent extraction. Curry *et al.*[20] used a micro-sampling boat made of tantalum to determine thallium in biological material. Using this technique, nanogram amounts of thallium in 50–100 μl of blood or urine can be determined with high accuracy. An increase in sensitivity over conventional atomic absorption spectroscopy by a factor of 25 was achieved as a result of using the tantalum boat. The extraction procedure described by Berman[21], involving chelation of thallium with sodium diethyldithiocarbamate and extraction into isobutyl methyl ketone (MIBK), was found to give quantitative recovery of thallium at optimum pH values. The same authors report that MIBK is the most suitable solvent to use in the determination by atomic absorption spectrophotometry of other heavy metals such as lead and cadmium. Recently, thallium has been determined in urine and plasma by Delves cup atomic absorption[22].

Ammonium pyrrolidinedithiocarbamate (APDC) is a useful chelating agent for extracting a large number of elements. A review by Christian[17] contains details of optimum pH values at which some twelve elements have been extracted into MIBK with APDC.

Many laboratories use wet ashing methods for the determination of whole-blood lead. Delves[23] placed the blood in a cup together with 100-volume hydrogen peroxide and, after ashing the mixture, the cup was introduced directly into an air–acetylene flame. Rose and Willden[24] found a considerable improvement in the reproducibility of this method using aqua regia instead of hydrogen peroxide. Other workers have also used the Delves cup technique for blood lead analysis[25, 26].

Many elements of interest, because of their high toxicity (arsenic, selenium, antimony, cadmium and phosphorus), have been difficult to determine by conventional atomic absorption spectroscopy. However, recent developments with electrodeless discharge lamps[13, 27], which provide very high intensity sources, have improved the detection limits for these elements to a point where their determination has become possible in the samples encountered in forensic science.

The increased sensitivity obtainable with non-flame atomisers has extended the application of atomic absorption and the possibility of using the technique for the analysis of gunshot residues is being explored. The neutron activation methods originally developed by Ruch *et al.*[28] are expensive and require several days or even weeks to complete. In addition, lead does not give an isotope which is convenient for the measurement of radioactivity and hence cannot readily be determined by neutron activation analysis. Renshaw *et al.*[29] recently described the use of flameless atomic absorption to analyse lead, antimony and barium in gunshot residues. Using this technique the whole analysis may be completed in under one hour. In the experiments described, lead was complexed with APDC, barium with thenoyltrifluoroacetone (TTA) in MIBK and antimony(V) removed by adding ceric sulphate. Absolute detection limits for lead, antimony and barium are given as 1×10^{-10}, 1×10^{-9} and 1×10^{-10} g, respectively. In the case of barium it was necessary to line the surface of the carbon furnace with a tantalum liner[30]. Green and Sauve[31] have also reported the analysis of gunshot residues using atomic absorption spectrophotometry and Brunelle *et al.*[32] have reported a preliminary study of the elemental composition of pistol bullets.

Measurement of the trace elements in hair has been used in forensic practice both as a means of the characterisation of hair and as a means of detecting poisoning. Most of the work has been carried out using neutron activation analysis, but the increased sensitivity of modern atomic absorption techniques is now directing attention away from neutron activation analysis. Harrison *et al.*[33] used atomic absorption to determine copper, iron, magnesium and zinc in hair samples after digestion in nitric and perchloric acids. Renshaw *et al.*[34, 35] used an atomic absorption spectrophotometer equipped with a furnace atomiser to determine lead and copper concentrations along the length of single hairs. The work demonstrated that concentration gradients exist along the length of hairs. In the case of lead and copper the concentration gradient generally increased from the root to the distal end of the hair. Other metals such as zinc, cadmium and iron may be monitored in a similar manner. The distribution of elements along the length of a hair will probably depend on the mechanism by which the element enters and is removed from the hair.

The possibility of establishing either identity or non-identity in the comparison of two hair samples depends on the precision of the analysis, the trace element concentration variation among specimens from different individuals, the trace element concentration variation among specimens of hair from the same individual and the possibility of variation in trace element concentration along the length of a single hair. Clearly the statistical interpretation of results presents considerable difficulties, especially since there may be a variation in trace elements in the hair with time and because the correlations in the concentrations of trace elements over a single head of hair may be different from the correlations between the same elements in the population as a whole.

At the moment it is necessary carefully to cut hairs individually into sections prior to analysis and demands on manpower are, consequently, quite high. It is not sufficient to attempt to align a large number of hairs and cut them all into sections in one operation. Because of the large number of hair sections that it is necessary to analyse to obtain meaningful measures of the variations referred to, atomic absorption will not become highly attractive for the characterisation of hair until multi-element techniques become generally available. Multi-element hollow cathode discharge lamps are available, but with these the detection limits are not as good as single element lamps and the elements in the commercially available multi-element lamps may not necessarily be the ones of interest. Multi-channel spectrometer systems using equipment designed for use with emission may provide the answer.

Other applications of atomic absorption spectrophotometry of possible interest to forensic scientists include the determination of trace elements in fingernails[36, 37], paint[38, 39], fuels and lubricants[40, 41], pharmaceutical products[42], fibres[43], polymers[44] and paper[45].

5.3.2 Atomic emission spectrography

Emission spectrography is probably the most frequently used technique for the qualitative and semi-quantitative analysis of a large number of elements in many types of sample encountered in forensic science. The speed and simplicity of analysis together with the relatively low cost of the equipment are among features which help to maintain interest in the technique[46], despite the appearance of methods having much higher sensitivity. In many applications the major disadvantage of emission spectrography has not been lack of sensitivity so much as lack of reproducibility. Photographic plates are normally used for recording spectra and semi-quantitative analyses result from a visual examination of the plates. Plates are sometimes examined using a microdensitometer. The wide variety of samples which the spectrograph is required to analyse in forensic practice has generally prohibited the use of electronic means of detection.

If sufficient sample is available, the normal practice is to burn a few milligrams of sample between carbon or copper electrodes. Recently, attention has been focused on the analysis of very small samples. Evans and Waller[47] attached samples with a minimal mass of *ca.* 25 μg to pointed graphite electrodes using a nitrocellulose based adhesive. Harvey[48] used cupped electrodes and diluted the sample with powdered graphite or ammonium chloride. He also used a spherical mirror and a compound lens as a condensing system to increase the light-gathering power of the spectrograph. Christopher[49] has recently described a purely qualitative technique which combines the use of pointed electrodes with simple condensing optics. Christopher describes the use of carbon fibres to bridge the gap between the electrodes and uses an argon sheath to reduce cyanogen band emission. The method described by Harvey[48] is probably the more suitable for semi-quantitative analysis.

The application of a liquid layer on a solid electrode surface during sparking is said to produce improved analytical results when compared with data obtained under identical conditions without the liquid layer[50]. Examples are presented which show both improved sensitivity and precision and decreased interferences. A solute in the solution layer was used as an internal reference instead of the conventional matrix element in the sample.

The possible applications of the emission spectrograph have been extended through the use of high-energy laser beams to vaporise the sample. A monochromatic laser beam is optically guided and focused on to the surface of the sample placed 1–2 mm below the electrodes. Vaporisation of the sample occurs at the focal point and the vapour plume rises causing the electrodes, which already have an applied potential across them, to discharge. The parameters of the discharge may be altered to obtain arc-like or spark-like conditions. Usually less than 1 μg of sample is used and in absolute amounts the detection limits are of the order of 10^{-10}–10^{-11} g. The technique is extremely simple in operation and can be made fairly reproducible because of the degree of control over the conditions. Quantitative analysis of samples is still difficult, however, because of the problem in standardisation and difficulties of incorporation of an internal standard in the material of interest. Conventional methods of quantitation cannot be used without limiting the method.

Neuninger[51] described the application of laser beam emission spectroscopy to samples encountered in forensic chemistry. He used equipment designed by Moenke and Moenke[52] and describes the qualitative analysis of metals, glass fragments, paint and valuable objects. Manura and Saferstein[53] have presented results which give an indication of the value of the technique of laser beam emission spectroscopy in the analysis of paint chips. They carried out a semi-quantitative analysis of 112 paint chips of various colours and taken from wrecked, late-model automobiles. Of the 112 paints analysed, 11 pairs of paints which were physically identical could be distinguished by the emission spectrograph, 4 pairs of paints which were physically identical were also identical by the emission spectrograph and 3 pairs of paints which were not physically identical were identical by the emission spectrograph. The authors concluded that the technique was a useful supplement to microscopic examination.

Initial attempts at quantitative analysis using laser beam emission have yielded promising results[54,55]. A suitable method entails measuring optical densities of a line from each of the elements detected and then summing the optical densities of all of the lines and expressing the contribution of each element as a percentage of the total. Butterworth[55] was able to carry out a quantitative analysis of a glass sample by mixing *ca.* 1 mg of glass with graphite. Eleven elements were determined and the relative standard deviation varied between 3 and 11% with the exception of copper (31%). These figures are sufficiently good to enable even the more modern window glasses to be differentiated. Quantitative analysis is simplified in the case of glass analysis because the concentration of silicon is constant to a good approximation, enabling it to be used as an internal standard.

Butterworth[55] also analysed 10 samples each of red, white and green paints produced by different manufacturers to conform to a standard colour. The results showed that it was possible to distinguish all 10 of the green paints from each other, whilst the red and white paints could be divided into 9 and 4 groups, respectively. The discrimination compared favourably with that achieved by other methods[56].

Emission spectroscopy using a *Q*-switched ruby laser has been applied to the study of rare earth elements[57]. Scandium, yttrium, europium and ytterbium in a sodium chloride matrix could be detected down to p.p.m. concentrations in one laser shot.

The potential advantages of inductively coupled plasma–optical emission analytical spectroscopy for simultaneous multi-element determinations have recently been compared with those of atomic absorption and fluorescence spectroscopy[58]. In one set of operational conditions, detection limits for 16 elements were in the ng ml^{-1} to fractional ng ml^{-1} range for 100 μl samples. The technique looks very promising and its usefulness in forensic science will no doubt be soon explored fully.

5.3.3 Spark source mass spectrometry

The technique of spark source mass spectrometry has been in use for many years but, surprisingly, it is only comparatively recently that it has been employed for the accurate and precise determination of trace element concentrations. The technique is well suited to the analysis of a wide range of materials and can be used to determine simultaneously many elements at concentrations of a few p.p.m. The interpretation of spectra is simpler than in most other spectral techniques and the amount of sample handling required is small. In the analysis of some materials, techniques such as X-ray fluorescence and atomic absorption spectroscopy are attractive providing sufficient material is available. Spark source mass spectrometry is well suited to the analysis of milligram amounts of material, particularly glass and metal fragments.

Chastagner[59] and Swift[60] have developed techniques for the analysis of very small samples, but German and Scaplehorn[61] were the first to demonstrate the application of spark source mass spectrometry to routine analyses in forensic science. These workers were able to determine 15 elements in glass fragments weighing between 1.5 and 2.5 mg. With care, using photoplates for detection, a precision of between 5 and 10% relative standard deviation was reported to be possible. In order to achieve this level of precision it was found to be important to control the spark parameters closely. The position of the electrodes was found to be particularly important, and in the experiments described it was adjusted by means of a template to standardise the distance from the first slit and a low-power microscope to standardise the position of the electrodes with respect to each other. Changes in sensitivity with spark gap width were investigated by Magee and Harrison[62, 63]. For metallic matrices, gap effects were said to be of the order of 5–10% for most elements. In graphite matrices, changes of sensitivity of up to a factor of 3 were recorded. A correlation between gap effects and ionisation potential was found and the ratio of ionisation potentials between the matrix and a particular measured species appeared to exert a definite influence on the behaviour of sensitivity changes with gap width.

Using the method described by German and Scaplehorn[61], it has been shown possible[64] to divide a group of glasses that were indistinguishable on the basis of refractive index and density alone. The glasses chosen for analysis had values of refractive index and density in the range of modern mass production window glasses where changes in trace elements might be expected to be smallest. The same method can be used in the analysis of headlamp and container glasses.

Because of the possibility of rapid, automated analyses, attention has recently been directed from photoplate detection to electrically recorded spectra processed by an on-line computer[65]. At the moment, electrical detection of spectra does not have the precision necessary to discriminate adequately where the range of population values of a parameter covers a narrow concentration range. In these cases photoplate detection is still preferred. Line densities are measured with a microdensitometer. Where the range of concentrations is fairly broad, such as is the case with the analysis of metal fragments or container glass fragments, then the speed of analysis obtained when using electrical recording of spectra may be preferred.

Materials such as hair, synthetic fibres, tobacco, blood, serum and fingernails have also been analysed using spark source mass spectrometry[66, 67]. The amounts of material used were larger than those normally encountered in forensic science laboratories. Blood was reported to show an even elemental composition over normal human subjects. Other material to be analysed using the technique include human kidneys and dried plant leaves[68] and tobacco ashes[69]. Thallium has been measured in human tissue with a sensitivity of 10^{-12} g using an isotope dilution method[70]. Impurities in the reagents used for wet ashing and memory effects in the mass spectrometer caused interference.

5.3.4 Scanning electron microscopy

The use of the scanning electron microscope (SEM) enables the forensic scientist to examine specimens at much higher magnifications than those possible with optical microscopy and without the difficulties of specimen preparation associated with conventional electron microscopy. A number of workers have investigated the potential of the SEM in forensic investigations and claim the technique has distinct advantages[71-77]. The main feature of the SEM in forensic practice is the great depth of focus which is some 300 times better than optical systems at similar magnifications. The standard SEM picture is almost stereoscopic. The SEM can be equipped with an X-ray spectrometer which may be of the energy dispersive or wavelength dispersive type and which enable both a qualitative and quantitative analysis to be made. The potential of other effects produced by the primary electron beam in the SEM such as ultraviolet, visible and infrared luminescence are beginning to receive attention, although relatively little has been reported as yet. The SEM has not yet reached the stage of being a major analytical tool in forensic science, but its use is increasing rapidly for the non-destructive analysis of small samples and as an aid to the interpretation of optical microscopic observations. Comparison by split-field SEM has been reported by van Essen and Morgan[78]. Using this technique it is possible to match two specimens side by side in the SEM in a single examination rather than viewing them separately and taking pictures.

Taylor[75,76] has used the technique to distinguish between a number of acrylics which are difficult to differentiate by other means. Since commercial Orlon staple may be composed of a number of different fibres (of similar type), the composition of the staple could form a basis for the characterisation of a garment. Taylor has also done some work on natural fibres and has tentatively suggested[76] that wool, cashmere and mohair can be differentiated by measuring the scale thickness. Wool scales were found to have a mean thickness of 1 μm, cashmere scales a thickness of 0.5 μm and mohair scales a thickness of 0.25 μm. Measurements of the standard deviation were being made. The SEM has also been used to examine fabrics fractured by the passage of bullets from firearms and fabrics cut by sharp objects[79].

The technique has been used to examine specimens of lacquered human hair. It has been possible to distinguish polymer based lacquers from those based on shellac, but it has not been possible to differentiate different brands of lacquer[76]. Verhoeven[80] examined hairs with the SEM and concluded that there was a definite improvement over optical means in the identification of structure and surface morphology.

The SEM seems particularly suited to the analysis of paint flakes[74,76,81].

The technique can, by raster scanning, present a visual display of the general layering and texture in a paint flake and display the distribution of each element in turn. Both qualitative and quantitative analyses can be carried out using a probe facility. The major disadvantage of the technique in the analysis of paint is the level of detection (*ca.* 0.1% by weight). Nevertheless, in terms of absolute sensitivity the detection level can represent 10^{-12} g when small areas are examined. Large single-layered paint flakes are probably more effectively analysed by other means such as emission spectroscopy and the SEM is not particularly well suited to the analysis of paint smears.

Smale[81] has recently reported a study of 400 paints as a result of about 200 comparisons made in the course of routine investigation of crime. Out of these paints, 4 pairs from unrelated sources were found that were the same in colour, texture and composition. The report does not include details of how many of the paints it was possible to distinguish by using colour and texture alone, so it is not possible to determine in an objective way the value of the SEM from these experiments. Size of sample is a major difficulty in trying to be objective about discriminating between paint fragments on the basis of a visual colour match. When colour differences become very small it is necessary, in the absence of viable instrumentation, to specify colour appearance using actual colour chips such as those found in a colour atlas. If the size of sample decreases to below *ca.* 1 mm^3 there is an increased possibility of an incorrect visual colour comparison being made between samples which are close to one 'just noticeable difference' apart in colour. Such small differences could easily be produced by normal weathering of a sample. Presumably the real value of the SEM in the analysis of paint flakes will depend on the sample size, the instrument being more valuable in the analysis of smaller fragments for which visual comparison becomes more difficult.

Many types of evidence have been subject to examination using the SEM, but as yet relatively few have received concentrated attention. Davies[73] has used the SEM to distinguish between duck feathers and feathers from a gallinaceous bird, such as the domestic chicken. Williams[74] has reported the analysis of ruby glass believed to have come from a Field Marshall's baton. The glass contained an unusually high amount of antimony and good agreement was obtained between the queried sample and the control. The detection levels of the technique are not such, however, that modern mass production window glasses are likely to be differentiated but it may be useful for distinguishing basically different types of glass.

The technique has been used for examining the anatomical features of small fragments of timber[76]; in the characterisation of soil through a study of diatoms present[77, 83]; and for examining the distribution and identification of coccoliths in the whiting component of putty and in other everyday materials such as paper, carpet backing, flour and animal feedstuffs[84]. The value of the occurrence and distribution of coccoliths has still to be determined. Other uses include comparison of characteristic markings which can be found on cartridge cases, including firing pin impressions and imprints made by the breechblock, extractor and ejector of a weapon[72]. Rodgers and Jacob[85] have reported the examination of counterfeit coins. One of the most useful applications of the SEM is the examination of metal fractures[74, 76, 87] and the technique has been used to match fine striations such as those left by machine tools and hacksaws on metal fragments. Bullet wounds[86] have also been investigated and it has been found possible to demonstrate the presence of lead around the point of entry into the flesh.

5.3.5 Neutron activation analysis

The application of the technique of neutron activation analysis to forensic science has received much attention. The advantages of the technique are said to include rapid, non-destructive analysis of trace amounts of material. However, despite extensive research the technique has not found widespread application in routine forensic work and although the technique has much to offer, its use in crime detection has yet to be established[88].

By far the most popular method of activation used in forensic activation analysis has been exposure to low-energy (thermal) neutrons in a nuclear reactor. Much of the early work was carried out on the trace elements detectable in hair[89-91]. However, recent studies[92] have shown that early optimism about the usefulness of the technique was premature. There have always been difficulties in the interpretation of results because of the variation of trace elements over a single head of hair not just at any one moment in time but also over a period of time, depending on factors such as diet, environment and washing habits. In many cases only one or two hairs are transferred from a criminal in the course of a crime and each hair must then be treated as an individual piece of evidence. The realisation that the concentration of many elements can vary considerably along a single hair has meant that the application of neutron activation to the analysis of trace elements in hair has had to be reassessed. The next step would seem to be to see if the distribution pattern of an element along the length of a hair together with the overall concentration level can help in the characterisation of the hair. However, even the best sodium iodine crystals are not sufficiently sensitive for it to be possible to analyse fractions of a single hair for several elements simultaneously.

Coleman and Wood[93] showed that glass fragments could be analysed non-destructively using neutron activation. The concentrations of 18 elements could be determined from a short 1 min irradiation followed by a 4–5 day irradiation in a neutron flux of *ca.* 4×10^{12} neutrons cm^{-2} s^{-1}. Coleman and Weston[95] used this technique in casework, but the large quantity of ^{24}Na and the complexity of the induced γ-spectra produced problems[93]. Later work was directed towards radiochemical separation of the irradiated material[94,96]. Goode *et al.*[97] developed a fully automated technique for the radiochemical separation of irradiated glass fragments which was based on sequential solvent extraction of 25 elements as 6 groups of neutral chelates. The technique was designed to deal with fragments weighing a few milligrams. Coleman and Goode[98] used the statistical technique developed by Parker[99] to show that with rarer specimens of glass it is possible to distinguish between batches of glass produced in any given factory. They suggest a batch of glass is equivalent to several tons of production.

Harrison and Gilroy[100] devised a method for the determination of lead, antimony and barium in gunshot residues by a suitable colour test. The colour test is relatively insensitive and much effort has been put into the application of neutron activation analysis to the measurement of antimony and barium in gunshot residues. Ruch *et al.*[28] were first to apply the technique. In the most important study that has so far been carried out on gunshot residues using neutron activation analysis[101], barium and antimony were measured following activation and radiochemical separation. The levels were determined in handlifts from persons who had not fired a weapon (260 handlifts) and also on the hands of those persons who had fired a wide range of handguns (353 handlifts). It was concluded that generally the barium and antimony levels in blanks were of the order of 0.1 μg and 0.01 μg, respectively. The levels were about an order of magnitude higher in handlifts taken after a single firing. It was not possible to establish a correlation between the amount of residue and the calibre of the weapon. Although work is still being reported[102] in which the neutron activation of gunshot residues is followed by radiochemical separation, attention is now being directed towards doing this type of analysis by atomic absorption spectrophotometry.

Neutron activation analysis is particularly effective when applied to biological material because the matrix does not interfere by becoming active. The technique is useful, therefore, for detecting possible exposure to some of the elements of interest to the forensic toxicologist[103], particularly arsenic. Liebscher and Smith[104] have established base level concentrations. Pearson and Pounds[105] showed that where arsenic is administered over a period of time, arsenic grows into hair from the root and its presence in sectional lengths is an indication of the uptake of arsenic by the body over a period of time.

The trace element content of bullet lead and jacket material[106,107], paint[108], paper[109] and fibres[110] has recently been studied using neutron activation analysis.

5.4 STATISTICAL INTERPRETATION OF ANALYTICAL DATA

The amounts of sample encountered in forensic practice are often quite small. It is not unusual to have to deal with paint and glass fragments weighing a few milligrams or less, materials such as hair or fibre often in the microgram range and with nanogram quantities of drug extracts. When an elemental analysis is needed, qualitative or semi-quantitative methods are often sufficient where elements differ markedly from one sample to another for a given type of material. However, in order to distinguish between samples of different origin which contain fairly similar amounts of the same elements, the precise quantitative measurement of concentrations down to and sometimes below 1 p.p.m. is necessary if the analysis is to be meaningful. This can present a challenging statistical problem as well as a challenging analytical problem. There needs to be a variation in the parameter being measured (the concentration of an individual element perhaps) in the population of samples under consideration (all the glass in window panes, for example) in order to distinguish between samples. Heterogeneity in individual samples, correlation between the measured parameters as well as analytical errors can, however, make the interpretation of numerical data difficult.

Parker[99, 112] and Parker and Holford[113] were first to present useful techniques to establish similarity or difference between samples. Parker suggested the use of a discrepancy index C, defined as

$$C = \sum_{i=1}^{N} D_i^2 \tag{5.1}$$

where

$$D_i = \frac{X_i - Y_i}{(\theta_i^2 + \phi_i^2)^{\frac{1}{2}}} \tag{5.2}$$

X_i and Y_i are the mean measured values of some parameter obtained for sample and control and θ_i and ϕ_i are the standard deviations of the mean of sets of N_1 and N_2 readings taken on the sample and control, respectively. If both θ_i and ϕ_i are accurately known and if the sample and control originate from the same source, C is distributed as χ^2 with N degrees of freedom; N is the total number of attributes measured. Parker and Holford[113] have suggested mathematical techniques to use when two or more parameters are correlated. The techniques for use with uncorrelated parameters are simple to use but in practice those for use with correlated parameters are so complex and require so much information (i.e. many samples must be analysed) that it is better to search for parameters which are uncorrelated. The techniques developed by Parker are now finding application in forensic science[96, 114].

The Bayesian approach to the identification of evidence has received some attention[115, 116], but has not found wide application probably because of the requirement for some *a priori* assessment of probability. A method described by Newton and Spurrell[117, 118] has been applied to the statistical analysis of the concentrations of trace elements in fibres[110]. Lawson and Framan[119] use the χ^2 test to present data in the form of a contingency table. Unfortunately the idea of contingency leads to the anomaly that two different indices of disparity are available, depending on the view taken about which of the two samples being compared should be regarded as the expected value and which as the observed value. In addition, the idea of replacing an intercomparison amongst a small number of subjects by a large number of comparisons between pairs of subjects is contrary to good statistical practice for the reason that these comparisons are not independent of one another. The approach of Lawson and Framan, therefore, seems to be a retreat on the work described by Parker[99, 112].

A completely different approach to the statistical problems encountered in comparisons arising in forensic science has recently been gaining in popularity. The probability that two randomly selected samples will be discriminated when a single attribute determination is made on each of them is a good measure of the effectiveness of that attribute in forensic science. This probability is therefore termed the 'discriminating power' of the attribute. Jones[120] showed how the discriminating power of individual blood grouping systems or of a series of uncorrelated systems could be calculated from the relevant population distributions. This concept has now been extended to correlated and continuous variables[121, 122] and seems appropriate for a wider application.

5.5 TECHNIQUES USED IN ORGANIC ANALYSIS

5.5.1 Organic mass spectrometry

Spectroscopic methods, particularly i.r. and u.v. absorption spectroscopy, have been used extensively in the identification of organic compounds. In addition, chromatographic methods have provided inexpensive and simple means of identifying and quantifying mixtures. Recently, however, a need has arisen for the isolation and identification of drugs in pharmaceuticals, body fluids and post mortem material. The advantages of mass spectrometry in this area are that it has greater sensitivity than the conventional methods, it can easily be coupled to a gas chromatograph, which can be used to isolate the compound of interest, and the output from the mass spectrometer can be fed into a computer and quickly compared with reference spectra for identification purposes. If necessary the whole procedure can be automated and a computer–gas chromatography–mass spectrometry system can be used for rapid screening[123, 124]. At the moment, major drawbacks are the relatively high cost of the equipment and the existence of relatively few reference mass spectra, in contrast to the large number of i.r. and u.v. spectra available. Even the most extensive compilations of mass spectral data contain less than 20 000 spectra[125].

Mass spectrometry can either be used to confirm the presence of a particular drug or to identify an unknown. There are many types of mass spectrometer and the particular use to which it is to be put must be very carefully considered before purchase. Typically, a low-resolution instrument used for the characterisation of drugs has a resolution of *ca.* 500 while a high-resolution instrument has a resolution of 10 000. The latter are generally used to determine the mass of an ion to several places of decimals in order to assign an unequivocal molecular formula to it. The use of mass spectrometry is not yet widespread in forensic science laboratories, but the next decade could well see dramatic changes. Those laboratories that have purchased equipment seem to have gone for low-resolution mass spectrometers with a view to automation of routine analyses. When it is necessary to identify a completely unknown compound it may be possible to hire time on a high-resolution instrument. In many cases, however, the analysis will simply involve confirmation of an identification based on the size, shape, colour and markings of a manufactured product such as a tablet or capsule. This type of analysis is straightforward and microgram amounts of material can easily be detected.

Mass spectrometry has been used for the detection of drugs and their metabolites without the need for preliminary sample preparation[126-129]. New sources of ions for mass spectrometry such as chemical ionisation, field ionisation and field desorption ionisation are of significance primarily because they extend the range of mass spectrometry to less volatile materials[130]. The application of chemical ionisation mass spectrometry looks particularly promising. This technique has now been utilised as a rapid and sensitive means of identifying heroin and its common diluents[131]. The procedure requires no sample preparation or prior chromatographic treatment and its sensitivity is said to permit a rapid identification of microgram quantities of illicit heroin preparations. Jain and Green[127] used direct multiple mass fragmentography to analyse amphetamines, barbiturates, phenothiazines, alkaloids and volatile poisons. The limits of detection for most drugs are said to be within the nanogram range.

It is frequently necessary to purify extracts before identification can be attempted and the purification stage usually involves the use of either thin layer[132], paper or gas chromatography. Most of the literature has centred around linked gas chromatography–mass spectrometry (GC–MS) systems[133-139]. One of the difficulties encountered in GC–MS work is that gas chromatography is carried out at relatively high pressure while mass spectrometry is carried out at very low pressures. The usual ways of getting round the pressure barrier are either to use a stream splitter, which allows only a fraction of the gas chromatograph effluent into the mass spectrometer, or an enrichment device which preferentially removes the carrier gas. Such enrichment devices include the Watson–Biemann diffusion separator[140], the Ryhage jet separator[139] and the membrane separator[141]. Finkle *et al.*[142, 143] have examined the applications of computerised GC–MS to forensic toxicology.

The applications of mass spectrometry to the identification of materials of interest to the forensic scientist have included amines[144-146], phenothiazines[147], benzodiazepines[148], cannabinoids[149], barbiturates[150, 151], alkaloids[152] and anti-epileptic agents[153].

5.5.2 Infrared spectroscopy

No attempt is made here to review the extensive literature on i.r. spectroscopy since this article is concerned only with recent developments. Infrared spectroscopy is, of course, a valuable aid to the analyst for the identification of a wide variety of materials. However, when a forensic scientist has to identify a completely unknown material the work involved in searching through a large number of reference spectra can be quite considerable. In these cases, identification requires the availability of a comprehensive collection of reference spectra and means for retrieval[154]. In order to speed up retrieval, systems based on film feature cards using microfilm storage[155], edge-punched cards[156] or body-punched cards have been used. More recently, libraries of spectroscopic data have been stored on magnetic tape and collections searched by computer[157]. Computer retrieval of spectra is very effective, but the cost of the system can often limit its applicability. Feature cards are a most useful means of spectrum retrieval for the small to average size laboratory, although at the moment there are some difficulties in manufacture. Using a feature card system suitable for a collection of up to 10 000 i.r. curves, correct identification within 1–2 min is achieved in 95% of searches[155].

Since the introduction of attenuated total reflection[158] and its modification, multiple attenuated total reflection, i.r. reflectance techniques have found application for obtaining spectra of substances difficult or impossible to obtain by transmission. Attenuated total reflection has been used in the examination of solid surfaces such as paper coatings and paints[159, 160], fibres and fibre blends[161, 162] and in the identification of tablets and capsules containing barbiturates[163] and sulphonamides[164]. In most cases encountered in forensic science there is insufficient material to obtain a good reflectance spectrum. It is more usual to obtain spectra from a halide disc prepared by grinding the sample with KCl or KBr. For small samples weighing more than a few micrograms the lead disc[165] or the micropellet[166] techniques can be used successfully. Care must be taken since grinding is known to produce variations in halide disc i.r. spectra[167].

Considerable effort has been expended in combining i.r. spectroscopy with gas chromatography[168]. Following chromatographic separation the isolated component can be transferred to KBr for microdisc preparation[169] or the chromatographic effluent can be examined on-line in the vapour phase[170]. Using the latter technique it is said to be possible to identify components in unresolved peaks from the chromatograph, although a rapid scan usually has to be traded for an unacceptable loss of resolution. On-line identification following a gas chromatographic separation does not yet appear to be as sensitive as the microdisc technique. Recent advances in i.r. Fourier transform spectroscopy[171] are said to have gone a long way to solving the problems of on-line identification of gas chromatographic effluents since spectra can be produced with speeds of *ca.* 1/300 of conventional spectra[172]. Computer signal averaging can be used to increase the signal-to-noise ratio and a high scale expansion is said to enable the analysis of nanogram amounts of material to be carried out.

Infrared identification following the pyrolysis of small amounts of paint resins and other polymeric materials has been reported[173]. It seems unlikely, however, that the conditions of pyrolysis described are sufficiently reproducible for general use.

It is not possible in this article to present a comprehensive review of even the most recent applications of i.r. spectroscopy. A few applications include the identification of fibres[174], paints[300], lubricating greases[175], oil[176] and fragrance and flavour compounds[177].

5.5.3 Ultraviolet spectroscopy

In the past few years there has been a consolidation of data in the field of u.v. absorption spectroscopy[178-180] and the extensive literature on the subject is only briefly referred to here. Spectra are now available to the forensic scientist on a wide range of compounds of interest, including large numbers of drugs, and the putrefaction products of human viscera. Because of the large amount of data available and the simplicity of the technique, u.v. absorption spectrophotometry remains one of the principal methods for the identification of basic compounds and drugs in forensic science.

Stevens and Evans[178] report that there are no differences in the wavelength of the maxima in the u.v. absorption spectra of bases between 400 and 200 nm if the spectra are plotted using either 0.5 N or 0.1 N sulphuric acid as the solvent medium. The stronger acid (0.5 N) was preferred for the isolation of bases from visceral extracts to avoid the possibility of neutralisation. $E_{1\,\text{cm}}^{1\%}$ values of maxima were determined in 0.1 N sulphuric acid solution.

Smalldon[181] has described two capillary microcell accessories for the routine examination of small samples in solution by visible and u.v. spectroscopy. The microcells can be used to obtain absorption spectra from solutions containing 10^{-5}–10^{-7} g of absorbing solute compared with the range 10^{-3}–10^{-5} g necessary with the usual 1 cm path-length cell. Some possible applications of the technique to forensic science are discussed by Smalldon and a spectrum obtained from the dyestuffs from a 3 mm stroke of black ballpoint ink is shown. By cutting a second window in the microcell at right angles to the one used to obtain the u.v. absorption spectra, a fluorescence spectrum can be obtained.

Reflectance spectra have not been widely used in forensic science but the use of an integrating sphere which serves as a reflectance attachment to a visible and u.v. spectrophotometer has been described. The spectra presented are from 1 × 2 mm paint chips, which is not unrealistic for those encountered in forensic science, but the spectra of two red paints, for example, show little difference in the u.v. region although there is a definite difference in the visible spectrum. The authors do not say whether it was possible to distinguish the two red paints visually[182]. Kevern[183] has also described the adaptation of a spectrophotometer to record the reflectance spectra of small areas. However, a colorimeter which records tristimulus values is more likely to be satisfactory in the long run for characterising small fragments. A fibre optic colorimeter described by Ishak has recently been adapted to record tristimulus values for paint flakes of *ca.* 1 mm^2 in area[184].

5.5.4 Fluorescence spectroscopy

In order to deal with small quantities of material, techniques which have a high intrinsic sensitivity are much sought after in forensic science. Fluorescence spectroscopy has interested the forensic scientist since early workers in the field[185-187] showed that it was possible to detect small quantities of oils using fluorescence. Matthews[188] showed that fluorescence could be used in the examination of pitch, bitumen, mineral oil and grease. Lack of commercially available equipment has impeded the development of applications of the technique until recently, but the use of the technique in forensic science has increased fairly rapidly in the past few years and the rate of increase shows no sign of slowing down. It could not be said that the developments have been as dramatic as those in the field of organic mass spectrometry, but the technique does appear to have a future in the well equipped forensic science laboratory.

Fluorescence is conventionally scanned as a variable of either excitation or emission wavelengths. In the emission spectrum the excitation wavelength is held constant and the variation of emission intensity with wavelength is recorded. The excitation spectrum is obtained by varying the excitation wavelength and recording the variation of emission intensity at a fixed wavelength. A number of workers have applied fluorimetric procedures to the determination of amphetamine and related compounds[189-191]. The method seems well suited to automated procedures. Hayes[6] has reported a completely automated fluorimetric determination of amphetamine in urine. The dihydrolutidine derivative produced by the reaction of amphetamine with formaldehyde and acetylacetone is excited at 410 nm and the resulting fluorescence intensity measured at 476 nm. The working limit of detection of amphetamine in fresh urine is 0.5 mg l^{-1}. Other workers[5, 192] have applied spectrophotofluorimetric methods to the determination of morphine in urine. Blackmore *et al.*[5] described an automated method for the determination of morphine in urine following conversion of morphine to pseudomorphine at pH 9.1 with aqueous potassium ferricyanide.

Lloyd[193-195] has recently described the use of a commercially available spectrofluorimeter fitted with a synchronous scan mode intended for absorption measurements to obtain synchronous excitation of fluorescence emission. In this technique the excitation monochromator and the monochromator analysing the fluorescence spectrum are scanned synchronously with a fixed, predetermined wavelength interval between them. Highly reproducible and characteristic spectra are obtained. Lloyd's work shows that the spectra obtained are particularly useful in the examination of traces of mineral oils and related materials. Waste oil is a widely distributed material and may well be transferred at scenes of crime. A synchronously excited emission spectrum may be contributed to by many fluorescent compounds and the spectrum, therefore, serves as a fingerprint in which the presence and concentrations of the compounds may be compared in juxtaposition. Such a comparison is less readily made between families of spectra excited at various fixed wavelengths. With standard commercial equipment spectra can be obtained with *ca.* 10 μg of oil. When microtechniques are applied these quantities are reduced to *ca.* 0.1 μg, under which circumstances some of the compounds responsible for a significant part of the fluorescence are present in quantities as small as a picogram[193].

The fluorescence and phosphorescence characteristics of some antihistamines[196] and of amphetamine and barbiturate preparations[197] have been reported. The spectrofluorimetric analysis of ballpoint ink in forensic science has been discussed by Kelly[198]. In the study of thin films, internal reflection techniques can be employed in fluorescence spectrometry and are said to achieve gains in sensitivity of thousands over conventional fluorescence[199].

In the search for methods capable of analysing microscopic specimens the technique of microfluorimetry has been developed. This technique combines the well established techniques of fluorescence microscopy and fluorescence spectroscopy. The development of microspectrofluorimeters has been reported in the literature[200, 201] and it is of interest that it is relatively easy to convert a fluorimeter to measure the phosphorescence of a substance[202]. Whitehouse[203] claims that microfluorimetry can provide an objective measurement of the colour of fluorescent materials and Williams[12] states the apparatus is used frequently for fibre identification in which it often happens that two fibres of apparently the same colour have different fluorescent spectra. Bowd *et al.*[204-206] have reported the use of microspectrofluorimetric techniques for the determination of LSD and related ergolines and of cannabinols. The cannabinols could be detected in concentrations down to 10 ng ml^{-1}.

5.5.5 Radioimmunoassay

The use of the technique of radioimmunoassay is increasing rapidly. The principle of analysis used is that of isotope dilution, the extracted drug being used to compete with a given amount of the same drug with a radioactive component for a limited amount of anti-drug antibody. After reaction, uncombined drug molecules are isolated by one of a number of methods such as the addition of charcoal; the addition of a second antibody (the double antibody method), which removes the drug–antibody molecule; or by initially fixing the antibody to the drug to the walls of a plastic tube and decanting the free uncombined drug from the tube. Addition of charcoal seems to be the most popular method. The amount of drug present is calculated from the amount of radioactivity removed from solution. Counting procedure depends on the isotope being used. To date the isotopes of interest have been ^{3}H, ^{14}C and ^{125}I.

The main advantage of radioimmunoassay over spectroscopic and chromatographic methods is its high sensitivity, although developments in organic mass spectrometry may produce even higher sensitivities than those achieved by radioimmunoassay. The most important interferences are other drugs or metabolites of the parent drug. Cross reactions can often be predicted, but sometimes totally unrelated molecules may react. The technique is attractive for measuring the concentration of digoxin in blood[207] and LSD in blood[208-210] in pg ml^{-1} amounts.

Other immunoassay techniques that have been used include spin immunoassay[211], haemagglutination inhibition[212, 213] and enzyme inhibition[214]. In a large-scale study of routine screening methods for heroin, some 50 000 urine samples were analysed by thin-layer chromatography, spectrofluorimetry and with an automated morphine analyser. Additional studies included the efficacy of hydrolysis, gas chromatography and haemagglutination inhibition immunoassay[213]. The data suggested that where morphine excretion is the single parameter of interest, immunological techniques requiring little or no instrumentation may surpass the instrumental techniques in sensitivity, specificity and economy.

5.5.6 Chromatographic techniques

Chromatographic techniques are probably used in more cases in forensic science than any other technique. Gas, paper and thin layer chromatography are now the work-horse techniques of most laboratories involved in forensic work. In these techniques the emphasis is now on standardisation of methods in order that data banks can be built up of the chromatographic characteristics of large numbers of compounds. The newer chromatographic techniques such as pyrolysis gas chromatography and high-pressure liquid chromatography have rapidly been applied to the development of new methods in forensic science.

5.5.6.1 Gas chromatography

A gas chromatograph must by now be standard equipment in most forensic science laboratories, however remote. Unfortunately, the wide use of the technique has brought about a situation in which the analyst is confronted with a bewildering number of phases to choose from. In spite of this only a small number of phases are in common use. The most popular non-polar phases are still SE-30 and Apiezon L, while Carbowax 20M and CDMS are most frequently used when a polar phase is required[215]. In an attempt to assess the usefulness of eight of the most popular columns, the retention indices and discriminating power of some 62 drugs have been determined on these columns[215]. The combined discriminating power and correlation coefficient for pairs of phases were also calculated. The most popular columns used together, SE-30 and Carbowax 20M/KOH, are shown to be not as good as the combination SE-30 and DEGS/KOH. The usefulness of derivative formation is examined in the same paper. A theoretical study of the 'informing power' of a chromatographic method and its use as a quality criterion has been made by Massart and Smits[216]. They use a quantity P_{inf}, introduced into analytical chemistry by Kaiser[217], and relate this to the resolution of a chromatographic system.

The characterisation of hydrocarbons using gas chromatography is well known. In cases in which arson is suspected, debris from the scene of the fire is extracted and examined in order to determine whether or not an incendiary material is present. The incendiary material may include aviation spirit, motor vehicle petrol, light oil, heavy oil or diesel oil. Capillary columns are often used for the identification of possible incendiaries; however, a number of new techniques have appeared which may prove of value. These include gel-permeation chromatography[218], differential gas chromatography[219, 220] and a method utilising a high pressure, high temperature liquid sampling valve[221].

The use of gas chromatography has been extensively investigated for the analysis of alcohol in blood and urine[222-224]. There has been some interest in standards of practice of blood alcohol analyses[12, 225], the stability of alcohol in stored blood[226, 227] and in automated methods for the determination of alcohol in blood and urine[8]. It will be interesting to see whether automated methods for the determination of blood alcohol levels eventually centre around gas chromatographic methods or around the use of the enzymic method[228, 229]. The instruments available for determining breath alcohol have been reviewed by Lovell[230].

The use of the electron capture detector[231] as a selective detector is well known, as is the enhanced sensitivity of the detection of a drug when derivatives containing halogen such as perfluoracyl are prepared. Derivative formation is not widely used in forensic science and usually other methods are preferred.

5.5.6.2 Paper and thin layer chromatography

Paper and thin layer chromatographic systems are used widely in forensic science and recently emphasis has been placed on building up data collections[232-236]. Important considerations are obviously speed, sensitivity, reproducibility, the differentiation of related molecules and the existence of correlations between systems when more than one system is used to assist identification.

A very informative comparison of the results obtained from several laboratories has been carried out by Dhont and co-workers[237, 238]. In a series of combined experiments the standard deviation for the determination of inter-laboratory thin layer chromatography R_f values for single- and multi-component systems were found to be 0.12 and 0.06, respectively. Only a small number of determinations were carried out in each laboratory and presumably, therefore, the values of standard deviation obtained are on the high side. However, when using reference standards, calculation of corrected R_f values showed that the standard deviations reduced to 0.02 and 0.03 for the same systems. These results could lead to improvements which could have a dramatic effect on the usefulness of data banks.

A recent paper by Moffat and Smalldon[239] discusses the optimum use of paper and thin layer methods for the identification of basic drugs. The R_f values of 100 drugs were determined in eight systems. The drugs were said to represent 98% of drugs encountered in forensic science laboratories. The discriminating power[240] of each system was found and the correlations between the systems determined at a range of possible values of error. The use of multiple R_f values for identification purposes when paper and thin layer methods are used has also been examined by Connor[241]. Independence of R_f values was said to be enhanced by using chromatographic systems which were qualitatively different from each other. Thin layer systems are quite widely used in forensic science laboratories for the separation of dyes from both natural and synthetic fibres, inks and illicit drug preparations. Unfortunately, little seems to have been published recently in these fields.

5.5.6.3 Pyrolysis gas chromatography

Forensic scientists are mainly interested in the use of pyrolysis gas chromatography as an analytical technique and not with its application to kinetic studies of thermal degradation although, of course, the two subjects cannot be completely separated. There are a number of reviews covering the applications of pyrolysis gas chromatography[242-245]. The requirements of a good analytical pyrolyser are the production of degradation products which are as nearly unique to the sample as possible, reproducible and capable of successful separation and elution in the gas chromatograph. Sometimes these requirements conflict. Basically the pyrolyser should be capable of heating the sample very rapidly so that an insignificant proportion of the sample decomposes before the 'pyrolysis temperature' is reached[246]. Alternatively, the sample must be heated at a readily reproducible rate. For applications where comparison with pyrograms obtained from other units is not necessary, a precise temperature in the region 500–800 °C in an environment constant with respect to gas flow and volume is all that is required[82]. One of the problems, however, is that although the technique is being used successfully and reproducibly both as a fingerprint technique and for quantitative analysis in individual laboratories, the diversity of pyrolysis and chromatographic conditions used is leading to inter-laboratory disagreement. This prevents the accumulation of data banks and requires each individual laboratory to build up its store of data[111]. Recently, a system has been described that will permit inter-laboratory comparison of pyrograms[247]. After investigation of various pyrolysers and column packings, a Curie point pyrolyser and a solid-phase packing, Porapak Q, were chosen as being most suitable. The use of a combined Curie point pyrolyser–quadrupole mass filter–1024 channel signal averaging system has been used for obtaining fingerprint pyrograms[248].

In forensic science work, pyrolysis gas chromatography has been mainly used in the identification of solid materials such as paint, plastics, hair and natural and synthetic fibres. The most suitable types of pyrolyser for this have been the hot filament, Curie point and furnace[249]. Although there is still some discussion, the Curie point pyrolyser[250, 251] seems to be emerging as the most useful unit[247]. Other forms of heating have included electric discharge[252] and laser light[253]. Laser beam heating has some advantages in terms of short heating time, the small area of specimen required and minimal contamination problems, but the major disadvantage of this method is the breaking down of the material into very small molecules and the consequent loss of the ability of the technique to provide fingerprint pyrograms.

Pyrolysis gas chromatography is useful as an aid to the identification of polymers in paint since it is insensitive to pigments and only requires a small sample size (5–10 µg). Several methods for the characterisation of paint have been described[247, 254, 255]. A wide variety of textile fibres are encountered in routine forensic science, including cotton, rayon, silk, wool, polyamines, polyesters, acrylics, polyurethanes, cellulose acetates, poly(vinyl alcohols), poly(vinyl chloride), polyethylene, polypropylene, fibre blends and copolymers. At the present time, comparative morphology is carried out in all cases and this may be followed by solvent tests[256]. Instrumental methods are only of value if they help identify fibres difficult to distinguish by traditional methods. At the moment the most frequently encountered fibre types are probably wool and acrylic. To be useful, instrumental techniques must distinguish between wool from different sources and between the different acrylic and modacrylic fibres available. Janiak and Damerau[257] have published pyrograms of natural and synthetic fibres while Bortniak *et al.*[258] have concentrated on attempting to distinguish between acrylic and modacrylic fibres. These workers were able to separate 27 acrylics and 14 modacrylic fibres into groups of 12 and 3, respectively.

Other materials from which pyrograms have been produced include rubber[259-261], plastics[262, 263], bitumens[264], alkaloids[265] and bacteria[266]. When using pyrolysis gas chromatography to obtain fingerprint pyrograms it must be remembered that microgram scale samples are required to minimise secondary reactions; larger samples limit the possibility of reproducing the results.

5.5.6.4 High-pressure liquid chromatography

In 1941, Martin and Synge[267] predicted that the highest performance in liquid chromatography should be obtained by using very small particles and a high pressure difference across the length of the column. It has only been comparatively recently, however, that suitable equipment and experimental conditions have been developed to the point where high-pressure liquid chromatography rivals gas chromatography in its speed, sensitivity and power of resolution[268, 269]. Undoubtedly as equipment becomes more widely available applications of the technique will expand rapidly. A variety of detectors is available and versatile equipment with u.v. and fluorescence detectors seem most popular for forensic applications of the technique. Knox[270] has reviewed the equipment available for high-pressure liquid chromatography and Michaelis *et al.*[271] have discussed the principles underlying the technique and reviewed its applications. The use of high-pressure liquid chromatography as a standard practice in the forensic science laboratory with emphasis on theoretical factors has been examined by Cashman and Thornton[272].

Jane and Wheals[273] found that in terms of speed and specificity high pressure liquid chromatography compared favourably with other chromatographic methods for the determination of LSD in illicit preparations. A spectrofluorimetric detector was found more suitable than u.v. detection because of the need for a preliminary column clean-up with the latter. Using a microcell, quantities of LSD as low as 10 pg in a 5 μl injection could be detected for aqueous solutions. In the separation of LSD from other ergot alkaloids the retention time of LSD on the Corasil C18 column (inlet pressure 750 psig) was 4.4 min. The technique seems particularly well suited to studies of drugs and their metabolites and has opened up new possibilities for the analysis of polar and thermally unstable drugs. The use of the technique for the analysis of barbiturates, diphenylhydantoin and their hydroxylated metabolites has been reported[274]. Column inlet pressures ranged from 700 to 1600 psig and the eluting peaks were detected with a u.v. detector. Cashman and Thornton[275] have described a method for the separation and quantitation of heroin, monoacetylmorphine HCl and morphine. The retention time of heroin on a Porasil T column at 300–500 psig was 7 min, while those of monoacetylmorphine and morphine were 8.5 and 13 min, respectively. The system described was unable to detect injected amounts of less than 25 μl of a mixture of all three containing 1 mg each per ml in 1 : 1 chloroform–methanol. However, with improved equipment the authors thought that the injected sample size could be less than 20 ng. The separation of analgesics[276, 277], benzodiazepines[278] and alkaloids and other compounds[279] has been reported.

High-pressure liquid chromatography seems well suited to the qualitative and quantitative analysis of hydrocarbon mixtures[280]. The use of the technique in the separation of polynuclear hydrocarbon mixtures in used engine oils has been described[281]. Detection limits of the procedure using both u.v. and fluorescence detectors have been determined and the effects of temperature and solvent composition on the separations studied. The response with the u.v. detector compared favourably with that obtained when detection is made with a flame ionisation detector following gas chromatographic separation.

An interface which accepts the effluent from a liquid chromatograph column, isolates the solute from the carrier solvent and introduces the sample into the ion source of a mass spectrometer has been described[282].

5.6 FUTURE DEVELOPMENTS

The trend in recent years has been towards progressively larger instrumentation assisted by computer control. Undoubtedly, the next few years will see the introduction of automated methods of analysis in routine applications such as the analysis of blood and urine for alcohol and drugs. It is possible that new techniques such as the ion microprobe and ESCA, with their high sensitivities and elemental discrimination[283-286], will prove helpful in solving specific problems. For example, it may prove possible to scan for a number of elements across the diameter of a single hair using the ion microprobe. This could show how elements are distributed in hair and influence the treatment of hair prior to analysis as well as provide assistance in the interpretation of results in cases of poisoning. Because of the high expense of these techniques, however, it seems unlikely that they will come into use in routine work. Instead techniques such as multi-element atomic absorption spectroscopy are more likely to interest the forensic analyst.

The organic chemist is likely to find that the high-pressure liquid chromatograph and the organic mass spectrometer become essential instrumentation, to be used more and more as routine equipment as well as to provide assistance in difficult analyses. Nevertheless, the more traditional chromatographic methods together with i.r., u.v. and visible spectroscopy will continue to form a basis of routine organic analysis for some time to come.

Textbooks

Because forensic science covers such a broad area of science it is not possible to present anything like a comprehensive list of textbooks. Just a few of the more recent textbooks which the forensic analyst might find useful to have on the library shelf are given in Table 5.1.

Table 5.1 Some textbooks dealing with subjects of interest to the forensic analyst

Title	*Author(s)*	*Ref.*
Atomic Absorption Spectroscopy—A Practical Guide (1970)	Reynolds and Aldous	287
Advances in Activation Analysis (1969)	Lenihan and Thompson	288
Neutron Activation Analysis (1972)	Soete, Gijbels and Hoste	289
The Forensic Examination of Paints and Pigments (1968)	Crown	290
Laser Microspectral Analysis (1973)	Moenke and Moenke	52
Fibres, Films, Plastics and Rubbers—Handbook of Common Polymers (1971)	Roff and Scott	291
Automation in Analytical Chemistry (1966)	Skeggs (editor)	293
Synthetic Dyes (1971)	Gurr	294
Computer-Based Management Information Systems (1970)	Krauss	295
Radioimmunoassay Methods (1970)	Kirkham and Hunter	296
Automated Enzyme Assays (1970)	Roodyn	297
Liquid Scintillation Counting (1972)	Crook, Johnson and Scales (editors)	298
Practical Hints on Infra-Red Spectrometry from a Forensic Analyst (1971)	de Faubert Maunder	299
Infra-Red Spectroscopy—Its Use in the Coatings Industry (1969)	Chicago Society for Paint Technology	300

Table 5.1 *continued*

Title	*Author(s)*	*Ref.*
Analytical Profiles of Drug Substances (1972)	Florey (editor)	301
Infra-Red Analysis of Polymers, Resins and Additives. An Atlas (1971)	Hummel	302
The Isolation and Identification of Drugs (1969)	Clarke (editor)	303
Drug Analysis by Chromatography and Microscopy (translation) (1973)	Stahl (editor)	236
Identification Techniques in Gas Chromatography (1970)	Leathard and Shurlock	305
Gas Chromatography (translation) (1970)	Szepesy	306
Manual of Hydrocarbon Analysis (1968)	ASTM	307
Advances in Forensic and Clinical Toxicology (1972)	Curry	308
Crime and Information Theory (1970)	Willmer	309
Infra Red and Ultra Violet Spectra of Some Compounds of Pharmaceutical Interest (1972)	Association of Official Analytical Chemists	292
Trace Analysis by Mass Spectrometry (1972)	Ahearn	304
Flame Emission and Atomic Absorption Spectrometry, Vol. 2 (1972)	Dean and Rains	310
Analytical Profiles of Drug Substances (1972)	Florey	311

References

1. Joseph, A. (1968). *Crime Laboratories—3 Study Reports* (LEAA: City University of New York)
2. Parker, B. and Peterson, J. (1970). *Physical Evidence Utilization in the Administration of Criminal Justice* (LEAA NI-0032: University of California)
3. Krendel, E. S., Drummer, R. M. and Freifelder, L. R. (1971). *Management Planning for Forensic Science Laboratories* (LEAA NI71-070-6: University of Philadelphia)
4. Benson, W. R., Stacy, J. E., Jr. and Nicol, J. D. (1972). *J. Forensic Sci.*, **16,** 25
5. Blackmore, D. J., Curry, A. S., Hayes, T. S. and Rutter, E. R. (1971). *Clin. Chem.*, **17,** 896
6. Hayes, T. S. (1973). *Clin. Chem.*, **19,** 390
7. Patterson, D. A. and Gomm, P. J. (1972). *Proc. Int. Meeting, Edinburgh, Automated Colour Reactions, Int. Microform J. Legal Med.*, Vol. 8, No. 3, 8, B-7
8. Solon, J., Watkins, J. and Mikkelson, L. (1972). *J. Forensic Sci.*, **17,** 447
9. Culliford, B. J. (1971). *The Examination and Typing of Bloodstains in a Crime Laboratory*, 85 (LEAA: Washington)
10. Curry, A. S. (1972). *Nature*, **235,** 369
11. Kirk, P. L. (1970). *J. Forensic Sci.*, **10,** 97
12. Williams, R. L. (1973). *Anal. Chem.*, **45,** 1076A
13. Dagnall, R. M. and Sharp, B. L. (1972). *MTP International Review of Science, Physical Chemistry Series One*, Vol. 12, *Analytical Chemistry Part 1*, 67 (T. S. West, editor) (London: Butterworths)
14. Kirkbright, G. F. (1971). *Analyst*, **96,** 609
15. Kerber, J. D., Russo, A. J., Peterson, G. E. and Ediger, R. D. (1973). *Atom Absorp Newsl.*, **12,** 106
16. Barnett, W. T. and Kahn, H. L. (1972). *Anal. Chem.*, **44,** 935
17. Christian, G. D. (1969). *Anal. Chem.*, **40,** 24A
18. Gorsuch, T. T. (1959). *Analyst*, **84,** 135

19. Gorsuch, T. T. (1962). *Analyst*, **87,** 112
20. Curry, A. S., Read, J. F. and Knott, A. R. (1969). *Analyst*, **94,** 744
21. Berman, E. (1967). *Atom. Absorp. Newsl.*, **6,** 58
22. Shklolnik, G. M. and Bevill, R. F. (1973). *Atom. Absorp. Newsl.*, **12,** 112
23. Delves, H. T. (1970). *Analyst*, **95,** 431
24. Rose, G. A. and Willden, E. G. (1973). *Analyst*, **98,** 243
25. Hicks, J. M., Gutierrez, A. N. and Worthy, B. E. (1973). *Clin. Chem.*, **19,** 322
26. Fernandez, F. J. and Kahn, H. L. (1973). *Atom. Absorp. Newsl.*, **12,** 70
27. Fernandez, F. J. (1973). *Atom. Absorp. Newsl.*, **12,** 93
28. Ruch, R. R., Guinn, V. P. and Pinker, R. H. (1964). *Nucl. Sci. Eng.*, **20,** 381
29. Renshaw, G. D., Pounds, C. A. and Pearson, E. F. (1973). *Atom. Absorp. Newsl.*, **12,** 55
30. Renshaw, G. D. (1973). *Atom. Absorp. Newsl.*, **12,** 158
31. Green, A. L. and Sauve, J. P. (1972). *Atom. Absorp. Newsl.*, **11,** 93
32. Brunelle, R. L., Hoffman, C. M. and Snow, K. B. (1970). *J. Assoc. Offic. Anal. Chem.*, **53,** 470
33. Harrison, W. W., Yurachek, J. P. and Benson, C. A. (1969). *Clin. Chim. Acta*, **23,** 83
34. Renshaw, G. D., Pounds, C. A. and Pearson, E. F. (1972). *Nature*, **238,** 162
35. Renshaw, G. D., Pounds, C. A. and Pearson, E. F. (1973). *J. Forensic Sci.*, **18,** 143
36. Harrison, W. W. and Tyree, B. B. (1971). *Clin. Chim. Acta*, **31,** 63
37. Barnett, W. B. and Kahn, H. L. (1972). *Clin. Chem.*, **18,** 923
38. Henn, E. L. (1973). *Atom. Absorp. Newsl.*, **12,** 109
39. Searle, B., Chan, W., Jensen, C. and Davidow, B. (1969). *Atom. Absorp. Newsl.*, **8,** 126
40. Mostyn, R. A. and Cunningham, A. F. (1967). *J. Inst. Petroleum*, **53,** 101
41. Kashiki, M., Yamazoe, S. and Oshima, S. (1971). *Anal. Chim. Acta*, **53,** 95
42. Smith, R. V. (1973). *Amer. Lab.*, **5,** 27
43. Price, J. P. (1972). *Atom. Absorp. Newsl.*, **11,** 1
44. Oliver, M. (1971). *Atom. Absorp. Newsl.*, **10,** 12
45. Kerber, J. D., Koch, A. and Peterson, G. E. (1973). *Atom. Absorp. Newsl.*, **12,** 104
46. Dyfverman, A. M. and Borjesson, K. (1972). *Proc. Int. Meeting, Edinburgh, Analysis of Paint Flakes by Emission Spectroscopy, Int. Microform J. Legal Med.*, Vol. 8, No. 2, 6, C-3
47. Evans, J. A. and Waller, W. N. (1966). *J. Forensic Sci. Soc.*, **6,** 189
48. Harvey, C. E. (1968). *J. Forensic Sci.*, **13,** 269
49. Christopher, A. J. (1972). *Microchem. J.*, **17,** 470
50. Barnes, R. M. and Malmstadt, H. V. (1974). *Anal. Chem.*, **46,** 66
51. Neuninger, H. (1970). *Jena Review*, **4,** 235
52. Moenke, H. and Moenke, L. (1973). *Laser Microspectrochemical Analysis, English Translation* (London: Hilger)
53. Manura, J. J. and Saferstein, R. (1973). *J. Assoc. Offic. Anal. Chem.*, **56,** 1227
54. Webb, M. S. and Webb, R. J. (1971). *Anal. Chim. Acta*, **55,** 67
55. Butterworth, A. (1972). *Laser Arc Emission Spectroscopy, Proc. Int. Meeting, Edinburgh, Int. Microform J. Legal Med.*, Vol. 8, No. 2, 5, A-8
56. May, R. W. and Porter, J. (1972). *J. Forensic Sci. Soc.*, **15,** 137
57. Ishizuka, T. (1973). *Anal. Chem.*, **45,** 538
58. Nixon, D. E., Fassel, V. A. and Knisely, R. N. (1974). *Anal. Chem.*, **46,** 210
59. Chastagner, P. (1969). *Anal. Chem.*, **41,** 797
60. Swift, P. (1968). *Proc. Soc. Anal. Chem.*, 142
61. German, B. and Scaplehorn, A. W. (1972). *J. Forensic Sci. Soc.*, **12,** 367
62. Magee, C. W. and Harrison, W. W. (1973). *Anal. Chem.*, **45,** 852
63. Magee, C. W. and Harrison, W. W. (1973). *Anal. Chem.*, **45,** 220
64. Dabbs, M. D. G., German, B., Pearson, E. F. and Scaplehorn, A. W. (1973). *J. Forensic Sci. Soc.*, **13,** 281
65. Brown, R., Powers, P. and Wolstenholme, W. A. (1971). *Anal. Chem.*, **43,** 1079
66. Harrison, W. W., Clemena, G. G. and Magee, C. W. (1971). *J. Assoc. Offic. Anal. Chem.*, **54,** 929
67. Harrison, W. W., Clemena, G. G. and Yurachek, J. P. (1969). *Anal. Chem.*, **41,** 1666
68. Evans, C. A. and Morrison, G. H. (1968). *Anal. Chem.*, **40,** 869

69. Jones, R. M., Kuhn, W. F. and Varsel, C. (1968). *Anal. Chem.*, **40,** 10
70. Zink, P. (1972). *Proc. Int. Meeting, Edinburgh, The Normal Concentrations of Thallium in Human Tissues, Int. Microform J. Legal Med.*, Vol. 8, No. 3, 7, D-12
71. Bradford, L. W. and Devaney, J. (1970). *J. Forensic Sci.*, **15,** 110
72. Korda, E. J., MacDonnell, H. L. and Williams, J. P. (1970). *Police Sci.*, **61,** 453
73. Davies, A. (1970). *J. Forensic Sci. Soc.*, **10,** 165
74. Williams, R. L. (1971). *Proc. Symp. Chicago, Evaluation of the SEM*, 539 (IITRI: Chicago)
75. Taylor, M. E. (1971). *Proc. Symp. Chicago, Applications of the SEM*, 545 (IITRI: Chicago)
76. Taylor, M. E. (1973). *J. Forensic Sci. Soc.*, **13,** 269
77. Vasan, V. L., Stewart, W. D., Jr. and Wagner, J. B., Jr. (1973). *J. Assoc. Offic. Anal. Chem.*, **56,** 1206
78. van Essen, C. and Morgan, J. E. (1973). *Proc. Symp. Chicago, A Split Field Comparison SEM*, 159 (IITRI: Chicago)
79. Paplauskas, L. (1973). *J. Police Sci. Admin.*, **1,** 362
80. Verhoeven, L. E. (1972). *J. Crim. Law Criminol. Police Sci.*, **63,** 125
81. Smale, D. (1973). *J. Forensic Sci. Soc.*, **13,** 5
82. Brauer, G. M. (1970). *Techniques and Methods of Polymer Evaluation*, Vol. 2, 41 (New York: Marcel Dekker)
83. Ross, R. and Sims, P. (1971). *Scanning Electron Microscopy Systematic Evolutionary Applications*, Vol. 4, 155 (London: Academic Press)
84. Weston, N. T. and Downs, N. S. (1972). *Proc. Int. Meeting, Edinburgh, The Forensic Significance of Chalk, Int. Microform J. Legal Med.*, Vol. 8
85. Rodgers, P. G. and Jacob, J. R. G. (1971). *J. Forensic Sci.*, **16,** 92
86. Boehm, E. (1971). *Proc. Symp. Chicago, Application of the SEM to Forensic Medicine*, 554 (IITRI: Chicago)
87. Thornton, J. I. and Mitosinka, G. T. (1971). *J. Forensic Sci. Soc.*, **11,** 197
88. Smith, H. (1973). *J. Forensic Sci. Soc.*, **13,** 137
89. Perkons, A. K. and Jervis, R. E. (1966). *J. Forensic Sci.*, **11,** 50
90. Coleman, R. F., Cripps, F. H. and Stimson, A. (1965). *Interim Report on the Determination of Trace Elements in Single Hairs by Neutron Activation*, 0-78/65 (Aldermaston: AWRE)
91. Coleman, R. F., Cripps, F. H., Stimson, A. and Scott, H. D. (1966). *The Determination of Trace Elements in Human Hair by Neutron Activation*, 0-86/66 (Aldermaston: AWRE)
92. Cornelis, R. (1972). *Med. Sci. Law*, **12,** 188
93. Coleman, R. F. and Wood, G. A. (1968). *The Value of Trace Analysis in the Comparison of Glass Fragments*, 0-3/68 (Aldermaston: AWRE)
94. Goode, G. C., Wood, G. A. and Coleman, R. F. (1969). *The Multi-element Analysis of Glass Fragments by Neutron Activation*, 0-54/69 (Aldermaston: AWRE)
95. Coleman, R. F. and Weston, N. T. (1968). *J. Forensic Sci. Soc.*, **8,** 32
96. Goode, G. C., Wood, G. A., Brooke, N. M. and Coleman, R. F. (1971). *Multi-element Analysis of Glass Fragments by Neutron Activation*, 0-24/71 (Aldermaston: AWRE)
97. Goode, G. C., Baker, C. W. and Brooke, N. M. (1969). *Analyst*, **94,** 728
98. Coleman, R. F. and Goode, G. C. (1973). *J. Radioanal. Chem.*, **15,** 367
99. Parker, J. B. (1966). *J. Forensic Sci. Soc.*, **6,** 33
100. Harrison, H. C. and Gilroy, R. (1959). *J. Forensic Sci.*, **4,** 184
101. Schlesinger, H. L., Lukens, H. R., Guinn, V. P. and Hackleman, R. P. (1970). *Special Report on Gunshot Residues Measured by NAA*, GA-9829 (San Diego: Gulf Atomic)
102. Rudzitis, E., Kopina, M. and Wahlgen, M. (1973). *J. Forensic Sci.*, **18,** 93
103. Smith, H. (1973), *J. Radioanal. Chem.*, **15,** 71
104. Liebscher, K. and Smith, H. (1968). *Arch. Environ. Health*, **17,** 881
105. Pearson, E. F. and Pounds, C. A. (1971). *J. Forensic Sci. Soc.*, **11,** 229
106. Guy, R. D. and Pate, B. D. (1973). *J. Forensic Sci.*, **18,** 87
107. Lukens, H. R. (1970). *Forensic Neutron Activation Analysis of Bullet Lead*, GA-10141 (San Diego: Gulf Atomic)
108. Schlesinger, H. L. (1970). *Forensic Neutron Activation Analysis of Paint*, GA-10142 (San Diego: Gulf Atomic)

109. Lukens, H. R. (1970). *Forensic Neutron Activation Analysis of Paper*, GA-10113 (San Diego: Gulf Atomic)
110. Bush, D., Butterworth, A., Pounds, C. A. and Pearson, E. F. (1973). *J. Radioanal. Chem.*, **15,** 245
111. Schaler, R. C. and Prichard, W. W. (1972). *J. Assoc. Offic. Anal. Chem.*, **55,** 832
112. Parker, J. B. (1967). *J. Forensic Sci. Soc.*, **7,** 134
113. Parker, J. B. and Holford, A. (1968). *Appl. Statistics*, **17,** 237
114. Dabbs, M. D. G. and Pearson, E. F. (1972). *Proc. Int. Meeting, Edinburgh, The Statistical Treatment of Results from the Measurement of Refractive Index and Density, Int. Microform J. Legal Med.*, Vol. 8, No. 2, 6, A-5
115. Kingston, C. R. (1965). *J. Amer. Statistical Assoc.*, **60,** 1028
116. Kingston, C. R. (1966). *J. Crim. Law Criminol Police Sci.*, **57,** 93
117. Newton, R. G. and Spurrell, D. J. (1967). *Appl. Statistics*, **16,** 51
118. Newton, R. G. and Spurrell, D. J. (1967). *Appl. Statistics*, **16,** 165
119. Lawson, D. D. and Framan, E. P. (1973). *J. Forensic Sci.*, **18,** 110
120. Jones, D. A. (1972). *J. Forensic Sci. Soc.*, **12,** 355
121. Smalldon, K. W. and Brown, C. (1973). *J. Forensic Sci. Soc.*, **13,** 307
122. Smalldon, K. W. and Moffat, A. C. (1973). *J. Forensic Sci. Soc.*, **13,** 291
123. Baczynskyj, L., Duchamp, D. J., Zieserl, J. F., Jr. and Axen, U. (1973). *Anal. Chem.*, **45,** 479
124. Holland, J. F., Sweeley, C. C., Thrush, R. E., Teets, R. E. and Bieber, M. A. (1973). *Anal. Chem.*, **45,** 308
125. *Eight Peak Index of Mass Spectra*, (1970). (Aldermaston: Mass Spectrometry Data Centre)
126. De, P. K. and Umberger, C. J. (1968). *Developments in Applied Spectroscopy*, Vol. 7A, 267 (E. L. Grove, editor) (New York: Plenum Press)
127. Jain, N. C. and Green, D. E. 1972). *Proc. Int. Meeting, Edinburgh, Fully Automated Detection of Drugs and Their Metabolites, Int. Microform J. Legal Med.*, Vol. 8
128. Milne, G. W. A., Fales, H. M. and Axenrod, T. (1971). *Anal. Chem.*, **43,** 1815
129. Zlatkis, A., Bertsch, W., Lichtenstein, H. A., Tishbe, A., Shunbo, F., Liebich, H. M., Coscia, A. M. and Flischer, N. (1973). *Anal. Chem.*, **45,** 375A
130. Chait, E. M. (1972). *Anal. Chem.*, **44,** 77A
131. Chao, J. M., Saferstein, R. O. and Manura, J. (1974). *Anal. Chem.*, **46,** 296
132. Rix, M. J., Webster, B. R. and Wright, J. C. (1969). *Chem. Ind.* (*London*), 452
133. Ryhage, R., Wilkstrom, S. and Waller, G. (1965). *Anal. Chem.*, **37,** 435
134. Hammar, C. G. (1971). *Acta Pharm. Sueccica*, **8,** 129
135. Hammar, C. G. and Hesling, R. (1971). *Anal. Chem.*, **43,** 298
136. van Cauenberghe, K., Vandewalle, M. and Versele, M. (1968). *J. Gas Chromatogr.*, **6,** 72
137. Blumer, M. (1968). *Anal. Chem.*, **40,** 1590
138. Junk, G. A. (1972). *Int. J. Mass Spectrom. Ion Phys.*, **8,** 1
139. Ryhage, R. (1964). *Anal. Chem.*, **36,** 759
140. Watson, J. T. and Biemann, K. (1965). *Anal. Chem.*, **37,** 844
141. Black, D. R., Flath, R. A. and Teranishi, R. (1969). *J. Chromatogr. Sci.*, **7,** 284
142. Finkle, B. S., Taylor, D. M. and Bonelli, E. J. (1972). *Proc. Int. Meeting, Edinburgh, Applications of Computerised GC–MS in Forensic Tox., Int. Microfilm J. Legal Med.*, Vol. 8
143. Finkle, B. S., Taylor, D. M. and Bonnelli, E. J. (1972). *J. Chromatogr. Sci.*, **10,** 312
144. Bellman, S. W., Turczan, J. W. and Kram, T. C. (1970). *J. Forensic Sci.*, **15,** 261
145. Blomquist, M. Bonnichsen, R., Fri, C. G., Marde, Y. and Ryhage, R. (1971). *Z. Rechtsmedizin*, **69,** 52
146. Bonnichsen, R., Maekly, A. C., Marde, Y., Ryhage, R. and Shubert, B. (1970). *Z. Rechtsmedizin*, **67,** 19
147. Gilbert, J. N. T. and Millard, B. J. (1969). *Org. Mass Spectrom.*, **2,** 17
148. Schwartz, M. A., Vane, F. M. and Postma, E. (1968). *Chem. Pharmacol.*, **17,** 965
149. Claussen, R. T. and Korte, F. (1966). *Tetrahedron Suppl.*, **7,** 88
150. Bohn, G. and Rucher, G. (1969). *Arch. Toxicol.*, **25,** 95
151. Gilbert, J. N. T., Millard, B. J. and Powell, J. W. (1970). *J. Pharm. Pharmacol.*, **22,** 897

152. Wheeler, D. M. S., Kinstle, T. H. and Rinehart, K. L. (1967). *J. Amer. Chem. Soc.*, **89**, 4494
153. Locock, R. A. and Coutts, R. T. (1970). *Org. Mass Spectrom.*, **3**, 735
154. (1900). *The Sadtler Standard Spectra* (Philadelphia: Sadtler)
155. Curry, A. S., Read, J. F. and Brown, C. (1969). *J. Pharm. Pharmacol.*, **21**, 224
156. McArdle, C. and Skew, E. A. (1965). *Symposium on Identification of Drugs and Poisons* (London: Pharmaceutical Press)
157. Erley, D. S. (1968). *Anal Chem.*, **40**, 894
158. Fahrenfort, J. (1961). *Spectrochim. Acta*, **17**, 698
159. Szymanski, H. A. (1967). *Progress in Infra-Red Spectroscopy*, Vol. 3, 39 (New York: Plenum Press)
160. Denton, S. (1965). *J. Forensic Sci. Soc.*, **5**, 112
161. Carlson, D. S. Suprunchuk, T. and Wiles, D. M. (1970). *Can. Text. J.*, **73**, 78
162. Basch, A. and Tepper, E. (1973). *Appl. Spectrosc.*, **27**, 268
163. Atkinson, J. E. (1971). *J. Pharm. Pharmacol.*, **23**, 781
164. Edwards, D. (1971). *J. Pharm. Pharmacol.*, **23**, 956
165 Cleverly, B. (1967). *Med. Sci. Law*, **7**, 148
166. Fox, R. H. and Schuetzman, H. I. (1968). *J. Forensic Sci.*, **13**, 397
167. Clarke, E. C. G. (1969). *Isolation and Identification of Drugs*, 117 (London: Pharmaceutical Press)
168. Welti, D. (1970). *Infra-Red Vapor Spectra* (Philadelphia: Sadtler)
169. Curry, A. S., Read, J. F., Brown, C. and Jenkins, R. W. (1968). *J. Chromatogr.*, **38**, 200
170. Penzias, G. J. (1973). *Anal. Chem.*, **45**, 890
171. Silk, C. (1973). *Lab. Equipment Digest*, 82
172. Rudzitis, E. (1971). *J. Crim. Law Criminol. Police Sci.*, **62**, 298
173. Smalldon, K. W. (1969). *J. Forensic Sci. Soc.*, **9**, 135
174. Rouen, R. A. (1970). *J. Forensic Sci.*, **15**, 410
175. Marino, J. H., Root, J. C. and Thomas, K. L. (1971). *Infra-Red Spectrophotometric Analysis of Additives Used in Lubricating Grease* (Philadelphia: NLGI)
176. Mattson, J. S. (1971). *Anal. Chem.*, **43**, 1872
177. Morris, W. W. (1973). *J. Assoc. Offic. Anal. Chem.*, **56**, 1037
178. Stevens, H. M. and Evans, P. D. (1973). *Acta Pharmacol. Toxicol.*, **32**, 525
179. Stevens, H. M. (1973). *J. Forensic Sci. Soc.*, **13**, 119
180. Daglish, C. (1969). *Isolation and Identification of Drugs*, 84 (E. G. C. Clarke, editor) (London: Pharmaceutical Press)
181. Smalldon, K. W. (1971). *Med. Sci. Law*, **11**, 193
182. Paul, F. W., Dougherty, P. M., Bradford, L. W. and Parker, B. (1971) *J. Forensic Sci.*, **16**, 241
183. Kevern, R. M. (1973). *J. Oil Colour Chem. Assoc.*, **56**, 318
184. Ishak, I. G. H. (1971). *J. Oil Colour Chem. Assoc.*, **54**, 129
185. Bayes-Cope, A. D. (1955). *Analyst*, **80**, 839
186. Rajeswaran, A. and Kirk, P. L. (1962). *Microchem. J.*, **6**, 21
187. Parker, C. A. and Barnes, W. J. (1960). *Analyst*, **85**, 3
188. Mathews, P. J. (1970). *J. Appl. Chem.*, **20**, 87
189. Nix, C. R. and Hume, A. S. (1970). *J. Forensic Sci.*, **15**, 595
190. Stewart, J. T. and Lotti, D. M. (1971). *J. Pharm. Sci.*, **60**, 461
191. Sawicki, E. and Carnes, R. A. (1968). *Anal. Chim. Acta*, **41**, 178
192. Takemori, A. E. (1968). *Biochem. Pharmacol.*, **17**, 1627
193. Lloyd, J. B. F. (1971). *J. Forensic Sci. Soc.*, **11**, 83
194. Lloyd, J. B F. (1971). *J. Forensic Sci. Soc.*, **11**, 153
195. Lloyd, J. B. F. (1971). *J. Forensic Sci. Soc.*, **11**, 235
196. Wilson, D. L., Wirz, D. R. and Schenk, G. H. (1973). *Anal. Chem.*, **45**, 1447
197. Miles, C. I. and Schenk, G. H. (1973). *Anal. Chem.*, **45**, 130
198. Kelly, J. H. (1973). *J. Police Sci. Admin.*, **1**, 175
199. Harrick, N. J. and Loeb, G. I. (1973). *Anal. Chem.*, **45**, 687
200. Parker, C. A. and Rees, W. T. (1962). *Analyst*, **87**, 83
201. Parker, C. A. (1969). *Analyst*, **94**, 161
202. Parker, C. A. and Hatchard, I. C. G. (1962). *Analyst*, **87**, 664
203. Whitehouse, M. J. (1972). *Proc. Int. Meeting, Edinburgh, Applications of Microfluorimetry, Int. Microfilm J. Legal Med.*, Vol. 8

204. Bowd, A., Hudson, J. B. and Turnbull, J. H. (1971). *J. Chem. Soc. B*, 1509
205. Bowd, A., Byrom, P., Hudson, J. B. and Turnbull, J. H. (1971). *Talanta*, **18**, 697
206. Bowd, A., Hudson, J. B. and Turnbull, J. H. (1973). *J. Chem. Soc. Perkin Trans. II*, 1312
207. Phillips, A. P. (1973). *Clin. Chim. Acta*, **44**, 333
208. Taunton-Rigby, A., Sher, S. E. and Kelley, P. R. (1973). *Science*, **181**, 165
209. Van Vunakis, H., Farrow, J. T., Gjika, H. B. and Levine, L. (1971). *Proc. Nat. Acad. Sci. U.S.A.*, **81**, 1483
210. Loeffler, L. J. and Pierce, J. V. (1973). *J. Pharm. Sci.*, **62**, 1817
211. Leute, R., Ullman, E. F., Goldstein, A. and Herzenberg, L. A. (1972). *Nature New Biol.*, **236**, 93
212. Adler, F. L. and Liuv, C. T. (1971). *J. Immunol.*, **106**, 1684
213. Sohn, D., Simon, J., Hanna, M. A., Ghali, G. V., Tolba, R. A. and Melkonian, V. (1973). *Anal. Chem.*, **45**, 1498
214. Rubenstein, K. E., Schneider, R. S. and Ullman, E. F. (1972). *Biochem. Biophys. Res. Commun.*, **47**, 846
215. Moffat, A. C., Stead, A. H. and Smalldon, K. W. (1974). *J. Chromatogr.*, **90**, 19
216. Massart, D. L. and Smits, R. (1974). *Anal. Chem.*, **46**, 283
217. Kaiser, H. (1970). *Anal. Chem.*, **42**, 24A
218. Albaugh, E. W. and Talerico, P. C. (1972). *J. Chromatogr.*, **74**, 233
219. Folmer, O. F., Jr. (1972). *Anal. Chim. Acta*, **60**, 37
220. Stuckey, C. L. (1972). *Anal. Chim. Acta*, **60**, 47
221. Stavinoha, L. L. and Newman, F. M. (1972). *J. Chromatogr. Sci.*, **10**, 583
222. Glendening, B. L. and Harvey, R. A. (1969). *J. Forensic Sci.*, **14**, 136
223. Reed, D. and Cravey, R. H. (1971). *J. Forensic Sci. Soc.*, **11**, 263
224. Curry, A. S., Walker, G. W. and Simpson, G. S. (1966). *Analyst*, **91**, 742
225. Bradford, L. W., Biasotti, A. A., Finkle, B. S., Harding, D. M. and Smith, W. C. (1971). *J. Forensic Sci. Soc.*, **11**, 127
226. Brown, G. A., Neylan, D., Reynolds, W. J. and Smalldon, K. W. (1973). *Anal. Chim. Acta*, **66**, 271
227. Smalldon, K. W. and Brown, G. A. (1973). *Anal. Chim. Acta*, **66**, 285
228. Goldberg, L. and Rydberg, U. (1965). *Proc. Technicon Symp. Automated Enzymatic Micro Determination of Ethanol in Blood and Urine* (London: Technicon)
229. Syed, R. A. (1967). *Proc. Technicon Symp. Enzymatic Determination of Ethanol by an Automated Fluorometric Technique* (New York: Technicon)
230. Lovell, W. S. (1972). *Science*, **178**, 264
231. Leathard, D. A. and Shurlock, B. C. (1970). *Identification Techniques in Gas Chromatography*, 165 (London: Wiley)
232. Curry, A. S. (1969). *Isolation and Identification of Drugs*, 43 (E. G. C. Clarke, editor) (London: Pharmaceutical Press)
233. Fox, R. H. (1969). *Isolation and Identification of Drugs*, 31 (E. G. C. Clarke, editor) (London: Pharmaceutical Press)
234. Sunshine, I. (1969). *Handbook of Analytical Toxicology* (Cleveland: Chemical Rubber Co.)
235. Macek, K. (1972). *Pharmaceutical Applications of Thin Layer and Paper Chromatography* (Amsterdam: Elsevier)
236. (1973). *Drug Analysis by Chromatography and Microscopy* (E. Stahl, editor) (Ann Arbor: Ann Arbor Science)
237. Dhont, J. H., Vinkenborg, C., Compaan, H., Ritter, F. J., Labadie, R. P., Verweij, A. and Zeeuw, R. A. (1970). *J. Chromatogr.*, **47**, 376
238. Dhont, J. H., Vinkenborg, C., Compaan, H., Ritter, F. J., Labadie, R. P., Verweij, A. and Zeeuw, R. A. (1972). *J. Chromatogr.*, **71**, 283
239. Moffat, A. C. and Smalldon, K. W. (1974). *J. Chromatogr.*, **90**, 9
240. Moffat, A. C., Smalldon, K. W. and Brown, C. (1974). *J. Chromatogr.*, **90**, 1
241. Connors, K. A. (1974). *Anal. Chem.*, **46**, 53
242. Jones, C. E. R. and Reynolds, G. E. J. (1969). *Reports Prog. Appl. Chem.*, **54**, 518
243. Perry, S. G. (1969). *Advan. Chromatogr.*, **7**, 221
244. McKenney, R. W. (1969). *Ancillary Techniques of Gas Chromatography*, 55 (L. S. Ettre and W. H. McFadden, editors) (New York: Wiley)
245. Sarner, S. F. (1972). *J. Chromatogr. Sci.*, **10**, 65

246. Levy, R. L. and Fanter, D. L. (1969). *Anal. Chem.*, **41,** 1465
247. May, R. W., Pearson, E. F., Porter, J. and Scothern, M. D. (1973). *Analyst*, **98,** 364
248. Meuzelaar, H. L. C. and Kistemaker, P. G. (1973). *Anal. Chem.*, **45,** 587
249. Coupe, N. B., Jones, C. E. R. and Perry, S. G. (1970). *J. Chromatogr.*, **47,** 291
250. Szymanski, A., Salinas, C. and Kwitowski, P. (1960). *Nature*, **188,** 403
251. Leathard, D. A. and Shurlock, B. C. (1970). *Identification Techniques in Gas Chromatography*, 114 (London: Wiley)
252. Sternberg, J. C. and Little, R. L. (1966). *Anal. Chem.*, **38,** 321
253. Folmer, O. F., Jr. and Azarraga, L. V. (1969). *J. Chromatogr. Sci.*, **7,** 665
254. Jain, N. C., Fontan, C. R. and Kirk, P. L. (1965). *J. Forensic Sci. Soc.*, **5,** 102
255. Heinonen, K. and Komsi, I. (1970). *Kemian Teollisus*, **27,** 631
256. (1970). *Identification of Textile Materials*, 6th ed., 54 (Manchester: Textile Institute)
257. Janiak, R. A. and Damerau, K. A. (1968). *J. Crim. Law Criminol. Police Sci.*, **59,** 434
258. Bortniak, J. P., Brown, S. E. and Sild, E. H. (1971). *J. Forensic Sci.*, **16,** 380
259. Foxton, A. A., Hillman, D. E. and Mears, P. R. (1969). *J. Inst. Rubb. Ind.*, 179
260. Krishen, A. (1972). *Anal. Chem.*, **44,** 494
261. Krishen, A. and Tucker, R. G. (1974). *Anal. Chem.*, **46,** 29
262. Nematollahi, J., Guess, W. and Autaun, J. (1970). *Microchem. J.*, **15,** 53
263. May, R. W., Pearson, E. F. and Scothern, M. D. (1974). *Pyrolysis Gas Chromatography* (London: Society for Analytical Chemistry)
264. Poxon, D. W. and Wright, R. G. (1971). *J. Chromatogr.*, **61,** 142
265. Kingston, C. R. and Kirk, P. L. (1965). *Bull. Narcot.*, **17,** 19
266. Meazelaa, H. L. C. and Veld, R. A. (1972). *J. Chromatogr. Sci.*, **10,** 213
267. Martin, A. J. P. and Synge, R. L. M. (1941). *Biochem. J.*, **35,** 1358
268. Kirkland, J. J. (1969). *J. Chromatogr. Sci.*, **7,** 7
269. Huber, J. F. K. (1969). *J. Chromatogr. Sci.*, **7,** 85
270. Knox, J. H. (1973). *Lab. Pract.*, **22,** 52
271. Michaelis, A. F., Cornish, D. W. and Vivilecchia, R. (1973). *J. Pharm. Sci.*, **62,** 1399
272. Cashman, P. J. and Thornton, J. I. (1971). *J. Forensic Sci. Soc.*, **11,** 115
273. Jane, I. and Wheals, B. B. (1973). *J. Chromatogr.*, **84,** 181
274. Anders, M. W. and Latorre, J. P. (1970). *Anal. Chem.*, **42,** 1430
275. Cashman, P. J. and Thornton, J. I. (1972). *J. Forensic Sci. Soc.*, **12,** 417
276. Stevenson, R. L. and Burtis, C. A. (1971). *J. Chromatogr.*, **61,** 253
277. Wolford, J. C., Dean, J. A. and Goldstein, G. (1971). *J. Chromatogr.*, **62,** 148
278. Scott, C. G. and Bommer, P. (1970). *J. Chromatogr. Sci.*, **8,** 446
279. Schmit, J. A. (1971). *Modern Practice of Liquid Chromatography*, 386 (J. J. Kirkland, editor) (New York: Wiley)
280. Snyder, L. R. (1969). *J. Chromatogr. Sci.*, **7,** 595
281. Vaughan, C. G., Wheals, B. B. and Whitehouse, M. J. (1973). *J. Chromatogr.*, **78,** 203
282. Lovins, R. E., Ellis, S. R., Tolbert, G. D. and McKinney, C. R. (1973). *Anal. Chem.*, **45,** 1553
283. Bankert, S. F., Bloom, S. D. and Sauter, G. D. (1973). *Anal. Chem.*, **45,** 692
284. Hercules, D. M., Cox, L. E., Onisick, S., Nickols, G. D. and Carver, J. C. (1973). *Anal. Chem.*, **45,** 1973
285. Evans, C. A. (1972). *Anal. Chem.*, **44,** 67A
286. Nordling, C. (1972). *Angew. Chem. Int. Ed. Engl.*, **11,** 83
287. Reynolds, R. J. and Aldous, K. (1970). *Atomic Absorption Spectroscopy—A Practical Guide* (London: Griffen)
288. Lenihan, J. M. A. and Thomson, S. J. (1969). *Advances in Activation Analysis* (New York: Academic Press)
289. De Soete, D., Gijbels, R. and Hoste, J. (1972). *Neutron Activation Analysis* (New York: Wiley)
290. Crown, D. A. (1968). *The Forensic Examination of Paints and Pigments* (Illinois: Thomas)
291. Roff, W. J. and Scott, J. R. (1971). *Fibres, Films, Plastics and Rubbers—Handbook of Common Polymers* (London: Butterworths)
292. (1972). *Infra Red and Ultra Violet Spectra of Some Compounds of Pharmaceutical Interest* (Washington: Association of Official Analytical Chemists)
293. Skeggs, L. T., Jr. (1966). *Automation in Analytical Chemistry* (New York: Mediad Inc.)
294. Gurr, E. (1971). *Synthetic Dyes* (New York: Academic Press)

295. Krauss, L. I. (1970). *Computer Based Management Information Systems* (New York: American Management Association)
296. Kirkham, K. E. and Hunter, W. M. (1971). *Radioimmunoassay Methods—European Workshop, Edinburgh* (Edinburgh: Churchill)
297. Roodyn, D. B. (1970). *Automated Enzyme Assays* (Amsterdam: North Holland)
298. Crook, M. A., Johnson, P. and Scales, B. (1972). *Liquid Scintillation Counting* (London: Heyden)
299. de Faubert Maunder, M. J. (1971). *Practical Hints on Infra-Red Spectrometry from a Forensic Analyst* (London: Hilger)
300. Chicago Society for Paint Technology. (1969). *Infra-Red Spectroscopy—Its Use in the Coatings Industry* (Philadelphia: Federation of Societies for Paint Technology)
301. Florey, K. (1972). *Analytical Profiles of Drug Substances* (New York: Academic Press)
302. Hummel, D. O. (1971). *Infrared Analysis of Polymers, Resins and Additives—An Atlas* (New York: Wiley)
303. Clarke, E. G. C. (1969). *The Isolation and Identification of Drugs* (London: Pharmaceutical Press)
304. Ahearn, A. J. (1972). *Trace Analysis by Mass Spectrometry* (New York: Academic Press)
305. Leathard, D. A. and Shurlock, B. C. (1970). *Identification Techniques in Gas Chromatography* (New York: Wiley)
306. Szepesy, L. (1970). *Gas Chromatography* (Translation) (London:Iliffe)
307. (1968). *Manual on Hydrocarbon Analysis*, 2nd ed. (Philadelphia: American Society for Testing Materials)
308. Curry, A. S. (1972). *Advances in Forensic and Clinical Toxicology* (Cleveland: Chemical Rubber Co.)
309. Willmer, M. A. P. (1970). *Crime and Information Theory* (Edinburgh: University Press)
310. Dean, J. A. and Rains, T. C. (1971). *Flame Emission and Atomic Absorption Spectrometry*, Vol. 2 (New York: Marcel Dekker)
311. Florey, K. (1972). *Analytical Profiles of Drug Substances* (New York: Academic Press)

6
Developments in Separation Methods using Extraction Chromatography

J. STARÝ
Czech Technical University, Prague

6.1 INTRODUCTION

Extraction chromatography (EC) is the separation technique in which an organic non-polar extractant is retained on an inert support (hydrophobised silica gel, paper, organic polymer) while a polar mobile phase is used as the eluting agent. Initially, this technique was called reversed phase chromatography or, alternatively, reversed phase partition chromatography to distinguish it from conventional partition chromatography, where the organic phase is the mobile phase.

The distribution of species in question between the fixed organic phase and the mobile aqueous phase is given by the distribution ratio D; the influence of inter-phase, which plays an important role in conventional partition (paper) chromatography, is avoided in EC. For this reason it is possible to predict easily the chromatographic behaviour of individual species from the D values determined in batch liquid–liquid distribution conditions.

EC was introduced by Carrit[1], who used dithizone in carbon tetrachloride solution supported by granular cellulose acetate for the isolation of heavy metals from aqueous solutions. The rapid development of this method sprang from the first papers of Siekierski and co-workers[2-4], who separated zirconium from niobium and individual rare earths on a column filled with hydrophobised silica gel supporting tri-n-butyl phosphate.

EC has several advantages over liquid–liquid distribution as well as other types of chromatographic method. Because of the dynamic arrangement of EC, it is possible to separate elements that are very similar, such as the rare earths or transplutonium elements; with dialkyl phosphoric acids EC separations are more selective than those using the most efficient ion-exchange procedures. The apparatus used is very simple, only a small amount of extractant is needed and one column can be used several times.

EC finds applications for solving several theoretical problems, e.g. for the determination of the D values of transplutonium elements when only a very limited amount is available, and for the determination of the composition and stability constants of metal complexes in both aqueous and organic phases.

The use of EC is limited chiefly by the choice of suitable extractants and experimental conditions. For practical reasons, extraction systems with slow kinetics are unsuitable. Also, the use of organic reagents and solvents miscible with water is not advantageous; in these cases the mobile phase has to be saturated with the extractant used. The use of highly polar solvents has to be avoided because of their small degree of sorption on the supporting material. There are also limitations regarding the pH range over which the separations can be carried out. Acidic extractants are transferred into the aqueous phase at high pH values because of their dissociation, whereas some basic extractants dissolve in acidic solutions (association with hydroxonium cation).

Although EC is a relatively new method, it has become very popular in recent years; many separations, previously unsuccessful, are now easily achieved by this method. Among reviews in this field[5-15], mention should be made of those by Katykhin[6, 13] and Cerrai *et al.*[10, 15]. A bibliography with 381 references covering the work on EC up to the beginning of 1971 has been compiled by Eschrich and Drent[16], and a bibliography on thin-layer chromatography including EC has recently been published by Brinkman *et al.*[17].

The present review deals with developments and trends in EC during the past four years.

6.2 THEORY

The theoretical treatment of EC depends on the type of technique applied. Column chromatography can be used for the separation of micro- as well as macro-amounts, whereas paper and thin-layer chromatography find their greatest application in the separation of trace amounts.

6.2.1 Column chromatography

The distribution ratio D in column chromatography can be expressed by the following equation

$$D = (V_e - V_f)V_{org}^{-1} \tag{6.1}$$

where V_e is the elution volume for the species investigated (volume of the eluent at the maximum of the elution peak). V_f is the free volume of the column and V_{org} is the volume of the stationary (organic) phase in the column.

The elution volume is expressed either in cm^3, drops or in number of free volumes of the column used. The elution volume of lanthanides or actinides is sometimes given as the gadolinium (californium) number. This number shows the position of the elution volume of the element in question with respect to the position of the elution volume of gadolinium (californium).

Equation (6.1) can be used directly for the determination of V_e for the species studied for the given column when the D value is known from liquid–liquid distribution studies. V_e generally increases with the increase of D value and V_{org}; the individual species are eluted from the column in the order of increasing D value.

When a solution of chelating agent, HA, is used as the stationary phase, V_e can be calculated directly from the value of the extraction constant, K_{ex}, of the metal chelate studied according to

$$V_e = K_{ex}\frac{[HA]^n_{org}}{[H^+]^n}V_{org} + V_f \tag{6.2}$$

where n is the charge of the metal ion in question. In this case, V_e increases with the increase of the K_{ex} value, V_{org} and the equilibrium concentration of the chelating agent used, whereas an increase of H^+ concentration leads to a decrease of V_e.

The equilibrium concentration of the organic reagent $[HA]_{org}$ after the volume V_e has passed through the column can be calculated from its initial concentration, c_{HA}, from the equation:

$$[HA]_{org} = c_{HA}\left[1 + \frac{V_e}{V_{org}}K_D(HA)^{-1}(1 + K_a[H^+]^{-1} + K_H^{-1}[H^+])\right]^{-1} \tag{6.3}$$

where

$$K_D(HA) = \frac{[HA]_{org}}{[HA]}$$

$$K_a = \frac{[H^+][A^-]}{[HA]}$$

$$K_H = \frac{[H^+][HA]}{[H_2A^+]}$$

From the above equation it is evident that for $[HA]_{org} > 0.99c_{HA}$ and $K_D(HA) > 100$, the pH value has to fulfill the following condition:

$$pK_H - \log K_D(HA) + 2 + \log\frac{V_e}{V_{org}} < pH$$

$$< pK_a + \log K_D(HA) - 2 - \log\frac{V_e}{V_{org}} \tag{6.4}$$

In the case when the chelating agent used does not form the H_2A^+ species, the separation can be carried out even in a very acidic medium.

The eluting agent can also contain a complexing agent, HB, which forms non-extractable complexes with the metal ion studied of the type MB_u; in this case the elution volume can be calculated from

$$V_e = K_{ex}\frac{[HA]^n_{org}}{[H^+]^n}V_{org}(1 + \sum_{u=1} \beta_u[B^-]^u)^{-1} + V_f \tag{6.5}$$

where β_u is the overall stability constant of the complex MB_u and $[B^-]$ is the equilibrium concentration of the anion of complexing agent, HB. Equations (6.2) and (6.5) have been verified by Šebesta[18-20], who also showed their applicability to the determination of K_{ex} and β_u values using EC data.

Recently, a new technique called displacement EC has been developed[21-24]. In this method the metal ion forming a chelate with lower value of $(1/n)\log K_{ex}$ is displaced from the column by metal ions that form chelates with higher $(1/n)\log K_{ex}$ values. Thus the following order of displacement has been observed in the case of metal diethyldithiocarbamates [values of

$(1/n) \log K_{ex}$ for carbon tetrachloride as organic solvent are given in brackets][25]: thallium(I) (−0.5), zinc (1.4), cadmium (2.9), lead (4.0), copper (7.0) and mercury (15.0).

6.2.2 Paper and thin-layer chromatography

The distribution ratio D of species investigated using paper and thin-layer chromatography is given by equation (6.6):

$$D = \frac{\sigma_m}{\sigma_s}\left(\frac{1}{R_F} - 1\right) = \frac{V_f}{V_{org}}\left(\frac{1}{R_F} - 1\right) \tag{6.6}$$

where σ_m and σ_s are the cross-sections of the mobile and stationary phases, respectively, and R_F is the retention factor. From equation (6.6) it follows that the higher the distribution ratio D, the lower is the R_F value. Verification of the above equation with D values, determined by batch liquid–liquid distribution studies, shows generally lower agreement than those using column chromatography. The reason lies probably in the fact that in several of the procedures described, the precipitation of solid extractant and/or metal complex formed is not fully avoided.

For the study of the structure and composition of the species formed, a parameter, R_M, is used which is a function of R_F:

$$R_M = \log\left(\frac{1}{R_F} - 1\right) \tag{6.7}$$

By comparing equations (6.6) and (6.7), it is evident that the R_M value is directly proportional to $\log D$. This value can be used to express the influence of pH, concentration of reagent, charge of metal ion, etc.

6.2.3 Separation factors

The separability of two metals, M and M′, is characterised by the separation factor, a, which is related to the distribution ratios, D and D', as follows:

$$a = \frac{D}{D'} = \frac{V_e - V_f}{V'_e - V_f} \tag{6.8}$$

where V_e and V'_e are the elution volumes of the metals M and M′, respectively.

From a practical point of view, it is important to know the difference ΔV_e between the elution volumes V_e and V'_e. By combining equations (6.1) and (6.8), we obtain

$$\Delta V_e = V_e - V'_e = (a - 1)D'V_{org} \tag{6.9}$$

From equation (6.9) it follows that for a constant value of a, ΔV_e increases with the increase of the distribution ratio D' of the first eluted metal M′ and with increase of the V_{org} value. The optimum conditions for the separation are reached for $D' \approx 0.5–1$ ($V_{org} \approx V_f$). If the D' value is very low, the difference ΔV_e is too small in comparison with the V_f value, whereas for high D' values the elution peaks become rather broad.

For the EC of metal chelates, the separation factor can be calculated from the corresponding extraction constants K_{ex} and K'_{ex} of both chelates according to

$$a = \frac{K_{ex}[HA]_{org}^{n-n'}}{K'_{ex}[H^+]^{n-n'}} \tag{6.10}$$

where n and n' are the charges of the metal ions M^{n+} and $M^{n'+}$, respectively. The separation factor is influenced by the presence of the complexing agent, HB. In this case

$$a_c = \frac{K_{ex}[HA]_{org}^{n-n'}(1 + \sum_{u=1} \beta'_u[B^-]^u)}{K'_{ex}[H^+]^{n-n'}(1 + \sum_{u=1} \beta_u[B^-]^u)} \tag{6.11}$$

where β_u and β'_u are the stability constants of the non-extractable complexes, MB_u and $M'B_u$, respectively.

Suitable complexing (masking) agents and optimum conditions for the separation can be found from known values of extraction[26] and stability constants[27]. For example, from the K_{ex} values of zinc ($\log K_{ex} = 2.30$) and cadmium dithizonates ($\log K_{ex} = 2.14$), it follows that $a(Zn/Cd) \approx 1.5$ with carbon tetrachloride as the organic solvent. However, the selectivity of the separation can be increased by using a suitable complexing agent, e.g. oxalic acid. The stability of zinc oxalate complexes ($\log \beta'_1 = 4.9$; $\log \beta'_2 = 7.6$) is greater than those of the cadmium complexes ($\log \beta_1 = 3.7$; $\log \beta_2 = 5.7$). Using 0.1 M oxalic acid–sodium oxalate (pH = 4.1), the value of the separation factor, $a_c(Cd/Zn) \approx 45$, calculated according to equation (6.11) has been verified experimentally[18].

For the separation of chemically similar elements such as lanthanides and transplutonium elements, the same form of the metal complex MB_u predominates in the aqueous phase. In this case, equation (6.11) simplifies, because $n = n'$, to

$$a_c = \frac{K_{ex}\beta'_u}{K'_{ex}\beta_u} \tag{6.12}$$

The K_{ex} values as well as the β_u values of metal complexes with most complexing agents increase with an increase of the atomic number of tervalent lanthanides and actinides. From this reason the higher the ratio K_{ex}/K'_{ex} the lower is β'_u/β_u and the separation factor is lower in the presence of complexing agents than in their absence.

6.2.4 Theoretical plates

The efficiency of the EC column for separation purposes can be judged from the number of theoretical plates, N, which is usually calculated from Glueckauf's equation[28]:

$$N = 8\frac{V_e^2}{W^2} = \frac{l}{\text{HETP}} \tag{6.13}$$

where W is the width of the elution peak at 1/e times the maximum solute concentration, l is the length of the column bed and HETP is the height equivalent to a theoretical plate. For the separation of chemically similar elements, usually 200 or more theoretical plates are required.

Recently, detailed studies concerning the dependence of the HETP on various factors have been published[29-32]. The HETP depends firstly on the column preparation and the type of supporting material used (see further). It was found that the HETP generally decreases with a decrease of grain size of the supporting material, the flow rate of the elution solution and the loading of the column with the extractant. It decreases with an increase of temperature or D value[10, 15, 29-32].

The extent to which particle size and flow rate may be decreased is limited by the necessity to use high pressures and by the time needed for the separation. The effect of the loading has been studied in detail by Horwitz and Bloomquist[30]. The HETP decreased from 0.61 mm for a loading with 1.2 mmol of di-(2-ethylhexyl) phosphoric acid per gram of Celite to 0.19 mm for a loading with 0.15 mmol of the above extractant per gram of Celite. Further diminution of the loading increased the HETP value because the surface of the supporting material was not fully covered by the extractant.

For practical reasons (the high elution volumes and large time needed for the separation), the D value has to be lower than 10 for columns having $V_{org} \approx V_f$. Although an increase of temperature decreases HETP, the separation factor decreases for certain systems at elevated temperatures. The formation of gas bubbles at higher temperatures also increases the HETP.

The HETP depends also on the type and the concentration of the extractant used. Lower HETP values were observed in columns with tri-n-butyl phosphate or tri-n-octylamine in comparison with columns supporting di-(2-ethylhexyl) phosphoric acid owing to the effect of the diffusion and kinetics of the extraction.

An increase of the concentration of the extractant leads to a decrease of the HETP. For this reason, undiluted extractants or nearly saturated solutions of extractants in suitable organic solvents are used most frequently.

The type and the concentration of eluting agent also have an influence on the HETP. It can be expected generally, that complexation in the aqueous phase will increase the HETP. For this reason, strong mineral acids are the best eluting agents.

6.3 TECHNIQUE

6.3.1 Supporting material

Support materials in EC are usually employed in a granular form and the process of the extraction and back-extraction takes place in a thin film of the stationary organic phase coated on the surface of the supporting material. In order to increase the capacity of the support, it is necessary to use it as finely granulated particles with a large specific volume.

The most important supporting materials and their trade names are listed in Table 6.1. Hydrophobised silica-base materials and fluoropolymers are the

Table 6.1 Supporting materials in EC

Class	*Trade name*
Kieselguhr	Celite, Chromosorb, Hyflo Supercel, KSK, KSS (hydrophobised with dichlorodimethylsilane)
Silica gel	SiO_2 gel (hydrophobised)
Polytetrafluoroethylene	Algoflon, Fluon, Fluoropak , Fluoroplast, Haloport, Teflon
Polytrifluorochloroethylene	Daiflon, Hostaflon C2, Kel-F, Plaskon, Teflex, Voltalef
Polyethylene	Alathene, Alkathene, Lupolene, Microthene, Mipor granules, Polyeth, Tenite
Polyisopropylene	Moplen, Profax
Poly(vinyl chloride)	Exon, Geon, Lucoflux, Opalon, Vinylite Q, Vipla
Copolymer of poly(vinyl chloride) and poly(vinyl acetate)	Corvic
Copolymer of styrene and divinylbenzene	Amberlist XAD-2, Porapak Q

best support materials; polythylene has a low density which makes column preparation difficult; the copolymer of poly(vinyl chloride) and poly(vinyl acetate) melts at temperatures higher than 60 °C and the copolymer of styrene and divinylbenzene swells in organic solvents. Cellulose, hydrophobised glass powder, porous Teflon, polyethylene and polystyrene or mixtures of various materials have been recommended for special purposes. Recently, polyurethane foam in the form of small cubes or cylinders has been successfully utilised in EC by Braun and co-workers[33-38].

6.3.2 Column preparation

The preparation of an EC column is a very critical step; an ideal column would allow elution without any applied overpressure.

The commonest way of preparation involves treatment of the support with the extractant before packing the column. The treatment usually consists of mixing the powdered supporting material with a known amount of extractant or with its solution in an organic solvent which can subsequently be removed by evaporation. The treated support is then slurried with an aqueous phase and added to the column in small portions, applying gentle pressure with a glass rod. An improved technique has been described by Horwitz and Bloomquist[30]: a suspension of treated support is placed in the column reservoir; at first, the material floats on the surface, but with stirring it becomes wetted with the aqueous phase and starts to fall into the column; the flow of liquid then compacts the bed. The application of pressure gives a more uniform bed compaction.

Filling with slurries of untreated support has also been used. After the column has been prepared, the extractant is percolated through the column and excess is removed by elution with an aqueous solution.

Packing with dry treated or untreated support has also been recommended. In this case the dry support is poured into the column, light tapping on the tube wall being applied in order to achieve tight, homogeneous filling. The solution of the extractant is then pipetted on to the top of the column. The organic phase is immediately covered with the elution solution and both solutions are sucked through the column under moderate pressure until the organic phase is sorbed on to the whole column support. When the extractant used is lighter than water, it is soaked into the column support and then covered with the elution solution[18-20].

Braun and Farag[39] recently recommended a vacuum technique of packing, which is based on the application of suction to one side of a closed column previously filled with the dry treated support. After evacuation of the column, water is allowed to fill it from the other side. Using this technique, a good homogeneous bed with improved flow characteristics has been obtained.

The types of column used in EC are practically the same as in ion-exchange chromatography. Apparatus for continuous EC separation has been described by Moskvin and Tsaritsyna[40, 41].

6.3.3 Paper and thin-layer preparation

Strips of paper (Whatman No. 1, S 2043a) are dipped into the solution of the extractant in a suitable organic solvent. The excess of solution is allowed to drain off and the solvent is evaporated[42].

Chromatoplates are usually prepared by dipping ordinary microscope slides or poly(vinyl chloride) sheets into a suspension of a solution of the extractant in a suitable organic solvent and supporting material. On leaving the plates for some minutes in air to allow the organic solvent to evaporate, a film of impregnated supporting material adheres to the slides[43].

6.4 ANALYTICAL APPLICATIONS

6.4.1 Elements and simple covalent compounds

The selective extraction of elements into non-polar solvents in the elementary state (S, Se, Br, I, At, rare gases) or as oxides (RuO_4, OsO_4) or halides ($GeCl_4$, SnI_4, $AsCl_3$, $AsBr_3$, AsI_3, $SbBr_3$, SbI_3, $SeBr_4$, SeI_4, $TeBr_4$, TeI_4, $HgCl_2$, $HgBr_2$ and HgI_2) can be used for the EC separation of these elements. Thus iodine (after the oxidation of iodide by sodium nitrite) can be extracted into benzene supported on microthene; the iodine can be eluted from the column with 0.1 M ascorbic acid. The method has been utilised for the isolation of iodine from synthetic food-stuffs[44].

Amberlyst XAD-2 loaded with benzene has been used for the isolation of arsenic(III) and germanium(IV) from 10–12 M HCl[45]. An alternative method[46] consisted of extracting arsenic(III) from a mixture of 4.8 M HBr and 3.8 M HCl into *o*-xylene using a Fluoroplast column.

6.4.2 Metal chelates

The optimal conditions for the EC separation of various metals in the form of metal chelates can be easily calculated from known values of extraction constants[26, 47]. The choice of suitable chelating agent for EC is limited by their solubility in aqueous phases and by the kinetics of extraction and back-extraction. Chelating agents forming extractable metal chelates are mostly solids (with the exception of dialkyl phosphoric acids) and, therefore, their solutions in suitable organic solvents are used as the stationary phase.

The distribution ratio, D, of metal chelates usually increases with an increase of pH value; the mixture of metals to be separated is thus extracted into the stationary phase at a high pH and the elution of individual metals is carried out using more acidic solutions [the order of the elution is given by the increasing $(1/n) \log K_{ex}$ values]. Sometimes masking agents forming stable water-soluble complexes may be used advantageously in the elution solution.

Values of dissociation and distribution constants of chelating agents have been listed[26] for the ionic strength $I = 0.1$ at 20–25 °C.

6.4.2.1 Dialkyldithiocarbamates

The use of dialkyldithiocarbamic acids in EC is limited because of their low stability against decomposition in acid media and high solubility in alkaline solutions. The metal chelates formed are substantially more stable, and displacement EC between various diethyldithiocarbamates has been utilised for the concentration of microimpurities of heavy metals and for the sub-stoichiometric determination of various metals[21-24]. More stable reagents of this group (e.g. pyrrolidinedithiocarbamic acid, hexamethylenedithiocarbamic acid) have not yet been used in EC.

6.4.2.2 Diarylthiocarbazonates

Solutions of diphenylthiocarbazone (dithizone, H_2Dz) have been used for the EC separation of various metals or radionuclides (see Table 6.2)[19, 48-52]. From pK_a (4.5) and $\log K_D(HA)$ values (4.5 for CCl_4 and 6.0 for $CHCl_3$), it follows that the elution of metals can be carried out for $V_e/V_{org} \approx 10$ at pH values below 6 and 7.5 for carbon tetrachloride and chloroform, respectively [see equation (6.4)]. Alkaline solutions completely elute dithizone from the column, whereas organic solvents elute the metal chelates which can then be determined spectrophotometrically[52].

Di-(β-naphthyl)thiocarbazone [$pK_a + \log K_D(HA) = 12.7$ for CCl_4] remains quantitatively on the column even at higher pH values.

Table 6.2 Separation of metals using diphenylthiocarbazone (dithizone, H_2Dz)

Metals separated	*Support*	*Stationary phase*	*Elution*	*Ref.*
Fe–Co	Styrene–divinylbenzene	H_2Dz (satd.) in CCl_4	Fe: 0.2 M CH_3CO_2H + 0.2 M CH_3CO_2Na Co: 4 M HCl	48
Fe–Zn	,,	,,	Fe: 0.2 M CH_3CO_2H + 0.2 M CH_3CO_2Na Zn: 0.1 M HCl	49
Au^{III}–Pt^{IV}	,,	,,	Pt: 0.1 M HCl Au: remains in the column	50
Hg–Zn, Cd, Ni	,,	0.3% H_2Dz in chlorobenzene	Zn, Cd, Ni: 0.1 M HCl Hg: remains in the column	51
Hg–Cu	Voltalef 300	10^{-3}% H_2Dz in $CHCl_3$	Cu: 0.05 M NaOH Hg: $CHCl_3$	52
Ag–Hg	Celite 545	5×10^{-3} M H_2Dz in CCl_4	Ag: 2% KSCN in 0.05 M H_2SO_4 Hg: 6% KI (pH = 4)	19
Cd–Ag	,,	,,	Cd: 0.1 M HNO_3 Ag: 6% KI (pH = 4)	
Zn–Cd	,,	,,	Zn: 0.1 M $H_2C_2O_4$ + $Na_2C_2O_4$ (pH = 4.1) Cd: 0.1 M $H_2C_2O_4$	
Pb–Bi–Po	,,	,,	Pb: 0.1 M HCl Bi: 0.3 M HCl Po: 1.5 M HCl	

6.4.2.3 *Chelates with hydroxamic acids*

N-Benzoyl-*N*-phenylhydroxylamine (HBPHA) is the most important chelating agent of this group because of its stability towards heat, light and mineral acids. Using chloroform as the solvent for HBPHA, the elution of individual metals can be carried out at pH < 7 [p$K_a = 8.15$; log K_D(HA) $= 2.3$]. A survey of such EC separations is given in Table 6.3[20,53,54].

Recently, the use of neotridecanohydroxamic acid (HNDHA) has been recommended for the separation of transuranium elements and for the determination of ^{237}Np and ^{239}Pu in urine[55].

6.4.2.4 *Chelates with oximes*

Amongst dioximes studied so far, only α-benzil dioxime has a sufficiently high distribution constant [log K_D(HA) $= 2.0$ for $CHCl_3$; p$K_a = 10.1$] to be used in EC. A 0.2% solution of α-benzoin oxime in chloroform has been used to separate zirconium and niobium (elution with 1 M HCl containing oxalic acid) from molybdenum(VI) (elution with conc. HNO_3)[56].

For the separation of copper from molybdenum(VI), 0.64 M 5,8-diethyl-7-hydroxydodecan-6-one oxime in toluene or 0.54 M 10-hydroxyeicosan-9-one oxime in toluene sorbed on Amberlyst XAD-2 has been recommended[57]. Copper was eluted with 0.5 M H_2SO_4 and 0.005 M H_2SO_4, and molybdenum with Na_2CO_3 (pH ≈ 11) and NH_3 (pH ≈ 12), respectively.

Iron(III) has been separated from copper on a cellulose powder column containing 22% Lix-64 (a mixture of 2-hydroxy-5-dodecylbenzophenone oxime and 5,8-diethyl-7-hydroxy-6-dodecanone oxime) in cyclohexane[58]. Iron was eluted with a mixture of 0.5 M $(NH_4)_2SO_4$ and 0.002 M NH_4HF_2, and copper with 1.5 M H_2SO_4.

6.4.2.5 *β-Diketonates*

The use of acetylacetone in EC is limited because of its solubility in an aqueous phase [log K_D(HA) $= 1.4$ and 0.7 for $CHCl_3$ and C_6H_6, respectively; p$K_a = 8.8$]; the kinetics of extraction with benzoylacetone [log K_D(HA) $=$ 3.15 for C_6H_6; p$K_a = 8.7$] or dibenzoylmethane [log K_D(HA) $= 5.3$ for C_6H_6; p$K_a = 9.3$] are rather slow. In this respect, dipropionylmethane [log K_D(HA) $= 2.28$ for $CHCl_3$; p$K_a = 10.0$] and isobutyrylacetone [log K_D(HA) $= 2.63$ for $CHCl_3$; p$K_a = 9.2$] seem to be the most promising agents of this group.

The selective separation of neptunium(IV) from most fission products was accomplished on a Porapak-Q column containing a 4 M solution of 2-thenoyltrifluoroacetone in *o*-xylene[59] [log K_D(HA) $= 1.5$; p$K_a = 6.2$]. Fission products were eluted from the column with 1 M HCl, and neptunium with 10 M HNO_3. The behaviour of various metals on paper impregnated with 2-thenoyltrifluoroacetone has been described by De and Bhattacharya[60].

Table 6.3 Separation of metals using *N*-benzoyl-*N*-phenylhydroxylamine (HBPHA) and neotridecanohydroxamic acid (HNDHA)

Metals separated	*Support*	*Stationary phase*	*Elution*	*Ref.*
Hf–Ta	Teflon	0.05 M HBPHA in $CHCl_3$	Hf: 2 M HCl + 0.1 M HF Ta: 2 M HF	53
Pa–Ta	,,	,,	Pa: 2 M HCl + 0.1 M HF Ta: 2 M HF	
Pa–Nb	,,	,,	Pa: 2 M HCl + 0.1 M HF Nb: 2 M HF	
Pa–Nb–Ta	,,	0.01 M HBPHA in $CHCl_3$	Pa: 6 M HCl + 0.05 M HF Nb: 2 M HCl + 0.4 M HF Ta: 2 M HF	
Re^{VII}–Tc^{VII}	Chromosorb W	0.1 M HBPHA in $CHCl_3$	Re: 5 M $HClO_4$ Tc: 0.05 M $HClO_4$	54
Th–Pa	Celite 545	0.05 M HBPHA in $CHCl_3$	Th: 4 M HCl Pa: 1 M $H_2C_2O_4$	20
U^{VI}–Th	,,	,,	U: 0.025 M HCl Th: 0.5 M HCl	
Co–Fe	,,	,,	Co: 0.01 M HCl + 3% H_2O_2 Fe: 0.1 M $H_2C_2O_4$	
Co–Cu–Fe	,,	,,	Co: 0.05 M HCl + 3% H_2O_2 Cu: 0.2 M HCl Fe: 0.1 M $H_2C_2O_4$	
Zn–Cu	,,	,,	Zn: 0.1 M $HClO_4$ + $NaClO_4$ (pH = 2.5) Cu: 0.1 M $HClO_4$	
Am^{III}–U^{VI}–Th	Microthene 710	0.3 M HNDHA in cyclohexane	Am: 1 M NH_4NO_3 (pH = 2.9) U: 0.03 M HNO_3 Th: 2 M HNO_3	55
Pu^{III}–Np^{IV}–Fe^{III}*	Microthene 710	0.3 M HNDHA in cyclohexane	Pu: 2 M HNO_3 Np: 0.1 M $H_2C_2O_4$ Fe: 6 M HCl	55
Np^{V}–Pu^{III}–Fe^{III}†	,,	,,	Np: 2 M HNO_3 Pu: 2 M HNO_3 + 0.15% hydroquinone Fe: 6 M HCl	

* In the presence of iron(II) sulphamate

† In the presence of iron(II) sulphamate and sodium nitrite as a redox system

6.4.2.6 *Dialkyl phosphates*

Dialkyl phosphoric acids dissolved in organic solvents are dimeric in form. If present in excess, they form with metals extractable chelates of the general formula $M(HA_2)_n$; at higher concentrations of mineral acids the formation of extractable mixed complexes may occur and for this reason the curve of dependence of the distribution ratio on the concentration of acid passes through a minimum.

The most important reagent of this group is di-(2-ethylhexyl) phosphoric acid (HDEHP) [$\log K_D(HA) = 3.42$ for n-octane; $pK_a = 1.4$]. HDEHP has found extensive application for the separation of tervalent lanthanides and actinides[6-17]. From the extraction constants of the lanthanum chelate ($\log K_{ex} = -2.6$) and lutecium chelate ($\log K_{ex} = 2.9$) with toluene as the organic solvent, it follows that the separation factor $\alpha(Lu/La)$ is higher than 3×10^5. The mean separation factor for two adjacent rare earths or transplutonium elements is *ca.* 2.5, which is higher than that obtained using extractants of other types.

The distribution ratio of 26 elements (Zn, Ge, $As^{III,V}$, Se^{IV}, Zr, Nb, Mo^{VI}, Tc^{IV}, $Ru^{III,IV}$, Rh^{III}, Ag, Cd, In, Sn^{IV}, $Sb^{III,V}$, Te^{IV}, I, Cs, Ce, Tb, Tm, Ac, Th, Pa^{V}, U^{VI} and $Np^{IV,VI}$) between HDEHP on Hostaflon-2 and hydrochloric acid (0–12 M) has been determined by batch experiments[29]. From the results obtained, procedures for the separation of fission products from irradiated uranium fuel have been worked out.

A survey of EC separations of various elements is given in Table 6.4[61-81]. EC Columns have also been used for gram-scale purification of ^{244}Cm[82], for the final purification of radioruthenium[83] and for the determination of ^{90}Sr in urine[84]. R_F data for various elements on thin layers of silica gel impregnated with 0.2 M HDEHP in benzene have been published by Muchová and Jokl[85]. The hydrolysis of americium(III) and curium(III) has been investigated on a paper supporting HDEHP[86]. Another reagent of this group, dibutyl phosphoric acid, forms with lanthanides chelates of high extraction constant ($\log K_{ex} = 1.3$ and 6.8 for lanthanum and lutecium, respectively; dibutyl ether as the organic solvent)[47] and, therefore, the elution of individual elements from the column is more difficult than in the case of HDEHP.

Batch experiments have showed recently that 1-methylheptyl phenylphosphonic acid is more effective for the separation of lanthanides than HDEHP[87].

6.4.2.7 *Chelates with other agents*

There are many chelating agents, still unexplored, which can probably be used as suitable stationary phases in EC, e.g. β-isopropyltropolone [$pK_a = 7.04$; $\log K_D(HA) = 3.37$ for chloroform], 4-benzoyl-3-methyl-1-phenylpyrazol-5-one [$pK_a = 4.1$; $\log K_D(HA) = 3.6$ for chloroform], 8-hydroxyquinoline [$pK_a = 9.7$; $pK_H = 5.0$; $\log K_D(HA) = 2.6$ for chloroform] and dibutyl dithiophosphoric acid [$pK_a = 0.22$; $\log K_D(HA) = 2.5$ for carbon tetrachloride][47].

Table 6.4 Separation of metals using di-(2-ethylhexyl) phosphoric acid (HDEHP)

Metals separated	*Support*	*Stationary phase*	*Elution*	*Ref.*
Ra–Th	Chromosorb W	HDEHP	Ra: 0.01–0.1 M HCl	61
			Th: remains in the column	
Sc–Ti	Celite	,,	Ti: 1.5 M H_2SO_4	62
			Sc: 12 M HNO_3 + 1 M HCl	
Y–Ho	Plaston CTFE	1 M HDEHP in Amsco	1.03 M HCl ($a = 2.02$)	63
			1.0 M HNO_3 ($a = 2.29$)	
La–Gd–Tb	Polytetrafluoroethylene	HDEHP	La, Gd: 0.8 M HCl	64
			Tb: 1 M HCl	
La–Ce^{IV}	Fluoroplast	,,	La: 6 M HNO_3	65
			Ce: 10% H_2O_2 in 4 M HCl	
Ce^{IV}–trivalent lanthanides	Chromosorb W	,,	Lanthanides: 5 M HNO_3 + 0.5 M $NaBrO_3$	66, 67
			Ce: 5 M HNO_3 + ascorbic acid	
Eu^{II}–trivalent lanthanides	Polyfluorocarbon PF-4	,,	Eu: 0.2 M HCl + 3 M NH_4Cl	68
			Lanthanides: HCl (conc.)	
Eu^{III}–Tb	Hostaflon C2	,,	Eu: 0.6 M HCl	69
			Tb: 4 M HCl	
Eu^{III}–Gd	Kel F	,,	1.35 M HCl ($a = 1.7$)	70
Eu^{III}–Tb	,,	,,	1.2 M HCl ($a = 2.75$)	
Ho–Er	,,	,,	2.42 M HCl ($a = 3.1$)	
Tm–Lu	,,	,,	5.5 M HCl ($a = 3.25$)	
Yb–Lu	,,	,,	5.75 M HCl	
Gd–Tb	Silica gel	,,	1.7 M HCl (60 °C)	71
Nd–Sm–Gd		,,	0.3 M H_2SO_4	72
Lanthanides–Ru	Kieselguhr	HDEHP	Ru: 0.1 M HNO_3	
			Ce, Nd: 0.3 M HNO_3	
			Pm, Sm, Eu: 0.4 M HNO_3	
U^{VI}–U^{IV}	Voltalef	5% HDEHP in xylene	4 M $HClO_4$ ($a = 4$)	74
	?	HDEHP	U^{VI}: HCl (conc.)	75
			U^{IV}: 6 M H_3PO_4	
U^{VI}–Th	?	,,	U^{VI}: HCl (conc.)	
			Th: 6 M H_3PO_4	

Table 6.4 *continued*

Metals separated	*Support*	*Stationary phase*	*Elution*	*Ref.*
Np^{V}–Np^{IV}–Np^{VI}	Fluoroplast 4	HDEHP	Np^{V}: 0.5–3 M HCl Np^{IV}: 0.3 M $H_2C_2O_4$ Np^{VI}: 7 M HCl	76
Am, Cm–U^{VI}	Kel F	,,	Am, Cm: 1.33 M HCl U: 12 M HCl	70
Cm–Bk, Es, Cf	Celite	,,	1 M 2-methyl-lactic acid (pH = 1)	77
Bk^{IV}–trivalent actinides	Flux-calcinated diatomaceous earth	,,	Actinides: 8 M HNO_3 + 0.2 M $NaBrO_3$ Bk: H_2O_2 + 4 M HNO_3	78
Tc–Mo	Hostaflon C2	,,	Tc: 1 M HCl (60°C) Mo: 12 M HCl (30°C)	79
Ga–In	Fluoroplast 4	,,	2 M HBr	80
Te–Sb	Hostaflon C2	,,	Te: 9 M HCl + $HClO_3$ (45°C) Sb: 0.4 M HCl (70°C)	79
Sb–Nb	,,	,,	Sb: 0.1 M HCl (50°C) Nb: 3 M HCl + 0.4 M HF (50°C)	81

6.4.3 Ion-association compounds

A quantitative treatment of the liquid–liquid distribution of ion-association compounds is extremely difficult and, therefore, optimum conditions for the EC separation can be found only from experimentally determined *D* values. The most important reagents of this group are long-chain amines and neutral organophosphorus reagents, which are liquids at room temperature and thus can be used as a stationary phase without organic diluents.

6.4.3.1 Associates with amines

The extraction by high molecular weight amines (M.W. 250–600) is considered to take place via the formation of ammonium salts (e.g. $R_3NH^+ Cl^-$) which are sparingly soluble in aqueous solutions, but freely soluble in organic solvents. The amine salt may undergo, under proper experimental conditions, an exchange reaction with anionic metal complexes (e.g. MCl_{n+1}^-). For EC purposes, primary [e.g. a mixture of primary amines with 18–22 carbon atoms with a mean M.W. of *ca.* 300 (Primene JM-T)], secondary [e.g. *N*-docenyl-(trialkylmethyl)amine with a mean M.W. of *ca.* 372 (Amberlite LA-1) or *N*-lauryl(trialkylmethyl)amine with a mean M.W. of *ca.* 374 (Amberlite LA-2)], tertiary [e.g. tri-n-octylamine (TOA), tri-n-laurylamine (TLA) or mixtures of tertiary amines with a mean M.W. of *ca.* 392 (Alamine 336)] or quarternary amines [e.g. tricaprylmethylammonium chloride with a M.W. of 475 (Aliquat 336)] can be used. The efficiency of the extraction usually increases in the order primary $<$ secondary $<$ tertiary $\sim$ quarternary.

The distribution ratios for 25 elements between tri-n-octylamine (TOA) supported on Hostaflon 2 and hydrochloric acid have been determined by Denig and co-workers[29]. *D* Values for most elements studied increased with increase of acid concentration (Se^{IV}, Te^{IV} and Ag are exceptions); however, in strongly acidic solutions some decreases of *D* values for several metals have been observed.

From liquid–liquid distribution studies, it follows that the separation factor depends more on the composition of the aqueous phase (the formation of anionic metal complex) than on the type of amine used. Conditions for the separation of various elements on EC columns are summarised in Table 6.5[35, 88-97].

High molecular weight amines have found extensive application in paper and thin-layer EC (for earlier applications see Refs. 12 and 17). Thus, the chromatographic behaviour of 28 metal ions has been investigated on thin layers of cellulose supporting Primene JM-T, Amberlite LA-2, Alamine 336 or Aliquat 336. It was found[98] that R_F values increased in most cases with an increase of concentration of hydrofluoric acid (0–20 M). R_F Spectra show many possibilities for the separation of individual metals (e.g. Mo, W, Re).

Hydrochloric acid has been applied for the separation of various elements on silica gel impregnated with 0.1 M TOA or Amberlite LA-1 in benzene[99], or on cellulose impregnated with Amberlite LA-1[42] or its mixture with tributylamine[100].

Table 6.5 **Separation of metals using *N*-lauryl(trialkylmethyl)amine (Amberlite LA-2), tri-n-octylamine (TOA), mixtures of C_8–C_{10} straight-chain tertiary amines (Alamine 336) and tricaprylmethylammonium chloride (Aliquat 336)**

Metals separated	*Support*	*Stationary phase*	*Elution*	*Ref.*
La–Th–Zr–U^{VI}	Silica gel	Amberlite LA-2	La, Y: 0.025 M H_2SO_4 + 1 M $(NH_4)_2SO_4$ Th: 0.1 M H_2SO_4 Zr: 4 M HCl U: 1 M $HClO_4$	88
Cu–Pb–Zn–Cd	Cellulose powder	,,	Cu, Pb: 2 M HCl Zn: 6 M HCl Cd: 10 M HCl	89
In–Cd	Hyflo Supercel	,,	In: 0.1 M HCl Cd: 10 M HCl	90
In–Sn^{IV}–Te^{VI}	,,	,,	In, Sn: 10 M HCl Te: 0.1 M HCl	
Sb^{V}–Sn^{IV}	,,	,,	Sb: 3 M HCl Sn: 10 M HCl	
W–Zr–Nb	Fluoroplast 4	0.5 M TOA in C_6H_6	W: HCl (conc.) Zr: 7 M HCl Nb: 2 M NH_4Cl + 0.2 M NH_4F	91
Mo–Zr–Nb	,,	,,	Mo: HCl (conc.) Zr: 7 M HCl Nb: 2 M NH_4Cl + 0.2 M NH_4F	
Nb–Re^{VII}	,,	,,	Nb: 2 M HCl Re: 11 M HCl	92
Nb–Fe^{III}	,,	,,	Nb: 2 M HCl Fe: 0 1 M HCl	
Nb–Sn^{IV}	,,	,,	Nb: 2 M HCl Sn: H_2O	
U–other elements	Styrene–divinylbenzene	TOA	Other elements: 6.6 M HCl or 6.3 M HNO_3 U: H_2O	93

Ni–Mn–Co–Cu–Fe–Zn–Cd	Silica gel	TOA in C_6H_6(1:1)	Ni: 7 M HCl Mn: 3.5 M HCl Co: 2 M HCl Cu: 0.6 M HCl Fe: 0.01 M HCl Zn: 0.3 M HCl + 0.5 M HNO_3 Cd: 0.5 M HNO_3	94
Ni–Co	Polyurethane foam	TOA	Ni: 8 M HCl Co: 1 M HCl	35
Al–Ga–Fe	Teflon	,,	Al: 2 M NH_4CNS (pH = 2.5) Ga: 0.1 M HCl Fe: 1.5 M HI + 2 M HCl	95
Cm–Cf	Celite 300	Alamine 336	11 M LiCl + 0.02 M HCl	96
V^{V}–Mo^{VI}–W^{VI}	Amberlist XAD-2	0.2 M Aliquat 336 in toluene	V: 0.15 M H_2SO_4 + 0.5% H_2O_2 W: 0.30 M H_2SO_4 + 0.5% H_2O_2 Mo: 4 M H_2SO_4 + 0.5% H_2O_2	97
Sn^{IV}–Mo^{VI}	,,	,,	Sn: 1 M H_2SO_4 + 0.5% H_2O_2 Mo: 4 M H_2SO_4 + 0.5% H_2O_2	

Forty three ions have been chromatographed on thin-layers of silica gel impregnated with Primene JM-T, Amberlite LA-1 and Alamine 336 using hydrobromic acid and acidified lithium bromide as eluents[101]. Hydroiodic acid generally gives similar results to hydrochloric or hydrobromic acids, whereas hydrofluoric acid systems differ considerably[42].

The use of 0.2–15 M nitric acid as eluent for *ca.* 40 ions on thin layers of silica gel impregnated with Alamine 336 S and Alamine 336 S oxide has also been reported[102].

The R_F spectra of 25 ions on silica gel supporting various amines using 0.01–5 M sulphuric acid or sulphate salts have been published by Brinkman and de Vries[103]. 0.5–1.0 M Sulphuric acid has been recommended[104] for the separation of rare earths, thorium, zirconium and uranium on silica gel loaded with Amberlite LA-2. The separation of adjacent rare earths can best be accomplished by a mixture of 0.025 M H_2SO_4/0.1–0.5 M $(NH_4)_2SO_4$ [104].

The thin-layer EC behaviour of 68 metal ions on cellulose impregnated with Amberlite LA-2-CNS^- has been systematically studied by Graham and Carr[105, 106] using ammonium thiocyanate (0.1–7 M) as the mobile phase. R_F Values generally increased with an increase in thiocyanate concentration (the formation of highly charged anionic complexes which are less extractable than singly charged anions). In the case of Pb, Tl^I and Mn^{II}, the R_F values passed through a minimum.

Recently, organic acids (lactic, tartaric, citric, nitrilotriacetic and ethylenediaminetetra-acetic acids) have been applied as eluting agents in thin-layer EC[85].

6.4.3.2 Associates with neutral organophosphorus reagents

Neutral organophosphorus reagents 'S' generally form extractable ion associates of the type MCl_nS_s or $M(NO_3)_nS_s$. The most important reagent of this group is tri-n-butyl phosphate (TBP). TBP is an oily liquid completely miscible with many organic solvents, but it has a low solubility in water (1.5×10^{-3} M at 25 °C) which is further reduced when TBP is diluted or when the aqueous solution contains salts[107].

From distribution studies it may be deduced[29] that the extraction increases with hydrochloric acid concentration up to the highest concentration (Ge, Zr, Nb, Sb^V, Pa, Np^{IV}) or passes through a maximum (Zn, Mo, Tc^{IV}, Ru^{IV}, Cd, In, Sn^{IV}, Sb^{III}, Te^{IV}, I, U^{IV} and Np^{VI}), whereas the D values of Se^{IV} and Ag decrease with an increase of acid concentration; $As^{III,V}$, Ru^{III}, Rh^{III}, Cs, lanthanides, Ac and Th are extracted only partially or not at all.

The optimal conditions for the separation of various metals are summarised in Table 6.6[33, 34, 36, 79, 81, 108-123]. EC columns have also been utilised for an investigation of the influence of the diluent for TBP[124] and for the group concentration of microimpurities[125].

Tri-n-octylphosphine oxide (TOPO) dissolved in cyclohexane has been used for EC separations (see Table 6.6)[44, 45, 126]. 0.8 M TOPO in butyl alcohol has been recommended for the separation of the lanthanides by thin-layer EC (elution with 1–7 M HNO_3)[127].

Tetrabutyl hypophosphoric acid has been used for the separation of the

Table 6.6 **Separation of metals using tri-n-butyl phosphate (TBP) and tri-n-octylphosphine oxide (TOPO)**

Metals separated	*Support*	*Stationary phase*	*Elution*	*Ref.*
Be–Zn–Mn	Fluoroplast 4	TBP	Mn: 3 M NH_4CNS Be: 0.75 M NH_4CNS Zn: H_2O	108
Ti^{IV}–Nb^{V}–Mo^{VI}	,,	,,	Ti: 9 M HCl Nb: 5 M HCl Mo: 1 M HCl or 15% tartaric acid	109
Zr–Nb	Hostaflon 2	,,	Zr: 6 M HCl Nb: 2.5 M HCl	79
Zr, Nb–Np	,,	,,	Zr, Nb: 8 M HNO_3 + $NaClO_3$ (50 °C) Np: 8 M HNO_3 + SO_2 (50 °C)	81
Th–Pa	,,	,,	Th: 9 M HCl Pa: 3 M HCl	79
Ta–various metals	Porous fluoroplast	,,	Various metals: 1 M HF + 1 M HNO_3 Ta: TBP	110
Ni–Fe	Polyurethane foam	,,	Ni: 4 M HCl Fe: 0.01 M HCl	36
Co, Cu–Fe	,,	,,	Co, Cu: 4 M HCl Fe: 0.01 M HCl	
Ni–Bi–Pd	,,	,,	Ni: 0.1 M $HClO_4$ + 3% thio-urea + 1% $NaClO_4$ Bi: 0.5 M $HClO_4$ Pd: H_2O	33, 34
Pd–other metals	Voltalef	,,	Pd: sorbed from 0.1 M $HClO_4$ + 0.1–5% thiourea + 1% $NaClO_4$; eluted with H_2O	111
Pd, Pt–Au	Daiflon	,,	Pd, Pt: 3.5 M HCl Fe: 1 M HCl Au: HNO_3 (conc.)	112
Rh–Pd, Pt–Ir–Au	Porasil	,,	Rh: 4 M H_2SO_4 Pd, Pt: 5 M HCl Ir: 0.1 M HCl Au: 14 M HNO_3	113

Table 6.6 *continued*

Metals separated	*Support*	*Stationary phase*	*Elution*	*Ref.*
Zn–Cd–In	Fluoroplast 4	TBP	Zn: 1 M LiBr–0.5 M HBr Cd: 1 M LiBr In: 0.05 M HBr	114
Cd–In	,,	,,	Cd: 0.8 M HCl (60 °C) In: H_2O	115
Zn–In	,,	,,	Zn: 0.8 M HBr In: H_2O satd. with TBP	116
In–Sb^{V}	,,	TBP–isoamyl alcohol (1:1)	In: 4 M HCl Sb: 10% thiourea–4 M HCl	117
Mn, Ni, Co, Cu, Zn–Sb	Silica gel	TBP	Mn, Ni, Co, Cu, Zn: 9 M HCl Sb: 2.5% tartaric acid + 1% KI	118
Bi–Po	Teflon	,,	Bi: 6 M HCl Po: 2 M HF	119
Bi–Pb	Voltalef	,,	Pb: 0.5 M $HClO_4$ + 0.5% thiourea Bi: 0.5 M $HClO_4$	120
Te–Se	Fluoroplast 4	,,	Se: 5 M HCl + Cl_2 Te: 1 M GCl	121
Pu^{III}–U^{IV}–U^{VI}	Voltalef	,,	4.8 M HNO_3 + 0.1 M hydrazine	122, 123
Ti, Sc–Th–Zr, Hf	Amberlist XAD-2	0.5 M TOPO in cyclohexane	Ti, Sc: 5 M HNO_3 Th: 12 M HCl Zr, Hf: 1 M HCl	45
Th–U^{VI}	Microthene 710	,,	Th: 5 M H_2SO_4 U: 1 M HF	44
Np–other elements	,,	0.1 M TOPO in cyclohexane	Np: 2 M HCl–chlorine water (1:1)	125

actinides in various oxidation states (e.g. Pu^{III}, Pu^{IV} and Pu^{VI}) or for their mutual separation (e.g. U, Np, Pu, Am, Cm and Cl) by paper or thin-layer EC (eluents: nitric or perchloric acids)[128-130].

6.4.3.3 *Associates with sulphoxides*

High molecular weight sulphoxides form extractable ion associates of the same type as phosphine oxides (e.g. $UO_2(ClO_4)_2{\cdot}S_4$ or $Pd(ClO_4)_2{\cdot}S_4$). A systematic study of the extraction properties of di-n-octyl sulphoxide (DOSO), bis(n-octylsulphinyl)methane (BOSM), bis(n-octylsulphinyl)ethane (BOSE) and *p*-tolyl sulphoxide (PTSO) has been carried out by Kennedy and Fritz[131]. Using paper EC, the interactions of these sulphoxides with some 50 metal ions were investigated in hydrochloric and nitric acids. DOSO and BOSM were also studied in perchloric acid and ammonium thiocyanate–perchloric acid mixtures. It was found that uranium(VI) (along with Pd, Au^{III} and probably also Hg^{II}) is rather selectively retained, on an Amberlist XAD-2 column containing 0.5 M DOSO in 1,2-dichloroethane, from a mixture of 1.0 M $LiClO_4$ and 0.025 M $HClO_4$, whereas other ions (including Zr, Th and lanthanides) pass through the column[132].

6.4.3.4 *Associates with other reagents*

A 1 M solution of 4-t-butyl-2-(α-methylbenzyl)phenol in carbon tetrachloride sorbed on Fluoroplast 40 has been recommended for the separation of caesium and rubidium ($\alpha = 12$ in 0.5 M NaOH)[133].

Amberlist XAD-2 impregnated with di-isopropyl ether (DIPE) has been used for the separation of Ga (5 M HBr), In (1 M HBr) and Tl^{III} (3 M HNO_3 followed by DIPE)[134].

The elution of In (10 M HCl), Fe (5 M HCl), Ga (2 M HCl) and Au^{III} (0.3 M NaOH) from a Fluoroplast column loaded with n-decyl alcohol has been utilised for the radiochemical determination of microimpurities of the above metals in arsenic[46].

Methyl isobutyl ketone sorbed on Amberlist XAD-2 has been recommended for the separation of Mo^{VI} (1 M HCl + 3 M H_2SO_4) from Sn^{IV} (0.1 M HCl in methanol)[45]. The same solvent was also used for the separation of Zr, Nb, Cr and Co (1 M HF + 2 M HCl) from Ta (30% H_2O_2)[135].

The R_F spectra of about 30 metal ions on thin layers of silica gel impregnated with tri-n-octylarsine oxide, tetramethylpyrazole and tetraphenylpyrazole have been published by Brinkman and co-workers[136, 137]. Hydrochloric acid and acidified lithium chloride were used as the mobile phase.

6.4.4 Mixed metal complexes

A synergic enhancement of the distribution of metals using a mixture of hydroxy group-containing chelating reagents and a neutral organophosphorus reagent (the formation of an adduct MA_nS_s) can be successfully used to perform EC separations and concentration in more acidic media. Although the optimum conditions for the separation can be calculated from known overall extraction constants[26], until now only a few papers have been published utilising this effect. Thus a saturated solution of hydroxamic acid in TBP sorbed on Fluoroplast 4 has been applied to the separation of V^{IV} from V^{V} [138]. The elution was carried out with 5 M and conc. phosphoric acid, respectively. A mixture of HTTA (1.3 g) and *N,N*-dibutyldiethylcarbamyl phosphate (1.3 ml) loaded on 10 g of Celite has been recommended for the separation of Eu–Cm (elution with acetate buffer, pH $\approx$ 3)[139] and of Cm–Cf ($a = 3.8$ in diluted nitric acid)[140]. A mixture of acetylacetone (60%) and TBP (40%) sorbed on Fluoroplast 4 has been used for the separation of indium (4 M HCl) from antimony(V) (10% thiourea in 4 M HCl)[117].

The behaviour of rare earths has also been investigated on paper impregnated with a mixture of HTTA and TOPO in carbon tetrachloride; hydrochloric acid was used as the eluent[141].

6.5 CONCLUSION

EC is a very efficient method for the separation of chemically similar elements. The optimum conditions for the separation can be calculated from known values of extraction constants or distribution ratios determined in batch liquid–liquid distribution studies.

Another possible use of EC, still insufficiently explored, lies in the preconcentration of elements in trace amounts as well as in macro amounts. In contrast to ion-exchange chromatography, this method allows concentration even from solutions of high ionic strength such as sea water, industrial waste waters, etc.

References

1. Carrit, D. E. (1953). *Anal. Chem.*, **25,** 1927
2. Siekierski, S. and Kotlinska, B. (1959). *Atomnaya Energ.*, **7,** 160
3. Siekierski, S. and Fidelis, I. (1960). *J. Chromatog.*, **4,** 60
4. Fidelis, I. and Siekierski, S. (1961). *J. Chromatog.*, **5,** 161
5. Green, H. (1964). *Talanta*, **11,** 1561
6. Katykhin, G. S. (1965). *Zh. Analit. Khim.*, **20,** 615
7. Kiba, T. (1966). *Bunseki Kagaku*, **15,** 991
8. Stronski, I. (1967). *Öster. Chem. Ztg.*, **68,** 5
9. Spěváčková, V. (1968). *Chem. Listy*, **62,** 1194
10. Cerrai, E. and Ghersini, G. (1970). *Advan. Chromatog.*, **9,** 3
11. Pointrenaud, C. (1970). *Chim. Anal. (Paris)*, **52,** 39
12. Brinkman, U. A. Th. (1971). *Progress in Separation and Purification*, Vol. 4, 241 (E. C. Perry and C. J. Van Oss, editors) (New York: Wiley-Interscience)
13. Katykhin, G. S. (1972). *Zh. Anal. Khim.*, **27,** 849
14. Stronski, I. (1973). *Isotopenpraxis*, **9,** 273
15. Braun, T. and Ghersini, G. (1975). *Extraction Chromatography* (Budapest: Akademiai Kiadó)
16. Eschrich, H. and Drent, W. (1971). *Eurochem. Technical Report ETR-271* (Mol, Belgium)
17. Brinkman, U. A. Th., de Vries, G. and Kuroda, R. (1973). *J. Chromatog.*, **85,** 187
18. Šebesta, F. (1970). *J. Radioanal. Chem.*, **6,** 41

19. Šebesta, F. (1971). *J. Radioanal. Chem.*, **7,** 41
20. Šebesta, F. and Lázníčková, A. (1972). *J. Radioanal. Chem.*, **11,** 221
21. Yakovlev, Yu, V. and Stepanets, O. V. (1970). *Zh. Anal. Khim.*, **25,** 578
22. Stepanets, O. V., Yakovlev, Yu, V. and Alimarin, I. P. (1970). *Zh. Anal. Khim.*, **25,** 1906
23. Alimarin, I. P., Yakovlev, Yu. V. and Stepanets, O. V. (1972). *J. Radioanal. Chem.*, **11,** 209
24. Yakovlev, Yu. V., Stepanets, O. V. and Sukhanovskaya, A. I. (1973). *Zh. Anal. Khim.*, **28,** 173
25. Starý, J. and Burcl, R. (1971). *Radiochem. Radioanal. Lett.*, **7,** 235
26. Starý, J. and Freiser, H. (1976). *Equilibrium Constants of Liquid–Liquid Distribution Reactions, Part IV* (*Oxford: Pergamon Press*)
27. Sillén, L. G. and Martell, A. E. (1964). *Chem. Soc. Special Publ.*, No. 17 (London: The Chemical Society)
28. Glueckauf, E. (1955). *Trans. Faraday Soc.*, **51,** 34
29. Denig, R., Trautmann, N. and Herrmann, G. (1970). *J. Radioanal. Chem.*, **5,** 223
30. Horwitz, E. P. and Bloomquist, C. A. A. (1972). *J. Inorg. Nucl. Chem.*, **34,** 3851
31. Pszonicka, M. and Siekierski, S. (1972). *Chim. Anal.* (*Warsaw*), **17,** 1321
32. Becker, R. and Hecht, F. (1973). *Microchim. Acta*, 624
33. Braun, T. and Farag, A. B. (1972). *Talanta*, **19,** 828
34. Braun, T. and Farag, A. B. (1972). *Anal. Chim. Acta*, **61,** 265
35. Braun, T., Huszár, E. and Bakos, L. (1973). *Anal. Chim. Acta*, **64,** 77
36. Braun, T., Bakos, L. and Szabó, Z. (1973). *Anal. Chim. Acta*, **66,** 57
37. Braun, T. and Farag, A. B. (1973). *Anal. Chim. Acta*, **65,** 115
38. Braun, T. and Farag, A. B. (1973). *Anal. Chim. Acta*, **66,** 419
39. Braun, T. and Farag, A. B. (1972). *Anal. Chim. Acta*, **62,** 476
40. Moskvin, L. N. and Tsaritsyna, L. G. (1970). *Radiokhimiya*, **12,** 730
41. Moskvin, L. N. and Tsaritsyna, L. G. (1970). *Radiokhimiya*, **12,** 737
42. Brinkman, U. A. Th. and de Vries, G. (1972). *J. Chem. Educ.*, **49,** 244
43. de Vries, G. and Brinkman, U. A. Th. (1972). *J. Chromatog.*, **64,** 374
44. Testa, C. (1970). *Anal. Chim. Acta*, **50,** 447
45. Fritz, J. S. and Latwesen, G. L. (1970). *Talanta*, **17,** 81
46. Kuzmin, N. M., Vlasov, V. S. and Bokova, T. A. (1972). *Zh. Anal. Khim.*, **27,** 1807
47. Starý, J. (1964). *The Solvent Extraction of Metal Chelates* (Pergamon Press: Oxford; (1966). Mir: Moscow)
48. Spěváčková, V. and Křivánek, M. (1970). *Radiochem. Radioanal. Lett.*, **3,** 63
49. Spěváčková, V. (1972). *Thesis*, Nuclear Research Institute, Rež, Czechoslovakia
50. Spěváčková, V. and Křivánek, M. (1974). *J. Radioanal. Chem.*, **21,** 485
51. Ueno, K., Yano, T. and Kojima, T. (1972). *Anal. Lett.*, **5,** 439
52. Woidich, H. and Pfannhauser, W. (1972). *Z. Anal. Chem.*, **261,** 31
53. Šebesta, F. and Pošta, S. (1973). *Radiochem. Radioanal. Lett.*, **14,** 183
54. Šebesta, F., Pošta, S. and Randa, Z. (1972). *Radiochem. Radioanal. Lett.*, **11,** 359
55. Delle Site, A. (1973). *J. Radioanal. Chem.*, **14,** 45
56. Yoshida, H., Yonezawa, C. and Gunji, K. (1970). *Bunseki Kagaku*, **19,** 818
57. Fritz, J. S. and Beuerman, D. R. (1972). *Anal. Chem.*, **44,** 692
58. Cerrai, E. and Ghersini, G. (1969). *Analyst*, **94,** 599
59. Wehner, H., Al-Murab, S. and Stoeppler, M. (1973). *Radiochem. Radioanal. Lett.*, **13,** 1
60. De, A. K. and Bhattacharyya, C. R. (1971). *Separ. Sci.*, **6,** 621
61. Šebesta, F. and Starý, J. (1974). *J. Radioanal. Chem.*, **21,** 151
62. Aly, H. F. and El-Haggan, M. A. (1971). *Mikrochim. Acta*, 4
63. Michelsen, O. B. and Smutz, M. (1971). *J. Inorg. Nucl. Chem.*, **33,** 265
64. Fahland, J., Lange, G. and Herrmann, G. (1970). *J. Inorg. Nucl. Chem.*, **32,** 3149
65. Levin, V. I., Tronova, I. N., Shlyagin, K. N. and Savel'ev, G. I. (1971). *Radiokhimiya*, **13,** 618
66. Krtil, J. and Bezděk, M. (1970). *Jaderná Energie*, **16,** 181
67. Krtil, J., Bulovič, V. and Maximovič, Z. (1971). *Jaderná Energie*, **17,** 10
68. Alimarin, I. P., Miklishanskii, A. Z. and Yakovlev, Yu, V. (1970). *J. Radioanal. Chem.*, **4,** 75
69. Riccato, M. T. and Herrmann, G. (1970). *Radiochim. Acta*, **14,** 107
70. Stronski, I. (1970). *Radiochim. Acta*, **13,** 25

71. Raichev, H. and Gesheva, M. (1973). *7th Radiochemical Conference, Marcánské Lázně*
72. Winget, J. O. and Lindstrom, R. E. (1971). *Separ. Sci.*, **6**, 573
73. Joon, K., denBoef, R. and deWit, R. (1971). *J. Radioanal. Chem.*, **8**, 101
74. Schmid, E. R. (1972). *Mikrochim. Acta*, 544
75. Luginin, V. A. and Tserkovnitskaya, I. A. (1970). *Radiokhimiya*, **12**, 898
76. Ushatskii, V. N., Preobrazhenskaya, L. D. and Omelchenko, M. G. (1972). *Radiokhimiya*, **14**, 892
77. Aly, H. F. and Latimer, R. M. (1970). *Radiochim. Acta*, **14**, 27
78. Overman, R. F. (1971). *Anal. Chem.*, **43**, 600
79. Denig, R., Trautmann, N. and Herrmann, G. (1970). *J. Radioanal. Chem.*, **6**, 57
80. Alimarin, I. P., Bolshova, T. A., Dolzhenko, T. G. and Kandybina, G. M. (1971). *Zh. Anal. Khim.*, **26**, 741
81. Denig, R., Trautmann, N. and Herrmann, G. (1970). *J. Radioanal. Chem.*, **6**, 331
82. Buijs, K., Maino, F., Müller, W., Reul, J. and Toussaint, J. Cl. (1971). *Angew. Chem.*, **83**, 766
83. Riccato, M. T. (1971). *Radiochim. Acta*, **15**, 3
84. Testa, C. (1970). *Energia Nucleare* (*Milan*), **17**, 320
85. Muchová, A. and Jokl, V. (1972). *Chem. Zvesti*, **26**, 303
86. Korotkin, Yu, S. (1973). *Radiokhimiya*, **15**, 676
87. Weaver, B. and Shoun, R. R. (1971). *J. Inorg. Nucl. Chem.*, **33**, 1909
88. Shimizu, T. and Ikeda, K. (1973). *J. Chromatog.*, **85**, 123
89. Leszko, M. and Karcz, W. (1970). *Zesz. Nauk. Uniw. Jagiellon,. Pr. Chem.*, **15**, 183
90. Stronski, I. (1970). *Radiochem. Radioanal. Lett.*, **5**, 113
91. Gibalo, I. M. and Mymrik, N. A. (1970). *Zh. Anal. Khim.*, **25**, 1744
92. Gibalo, I. M. and Mymrik, N. A. (1971). *Zh. Anal. Khim.*, **26**, 918
93. Beranová, H. and Tejnecký, M. (1972). *Coll. Czech. Chem. Commun.*, **37**, 3579
94. Neef, B. and Grosse-Ruyken, H. (1973). *J. Chromatog.*, **79**, 275
95. Shmanenkova, G. I., Mikheeva, L. M. and Elfimova, G. I. (1972). *Zh. Anal. Khim.*, **27**, 2161
96. Aly, H. F. and Abdel-Rassoul, A. A. (1972). *Z. Anorg. Allg. Chem.*, **387**, 252
97. Fritz, J. S. and Topping, J. J. (1971). *Talanta*, **18**, 865
98. Brinkman, U. A. Th., Sterrenburg, P. J. J. and de Vries, G. (1971). *J. Chromatog.*, **54**, 449
99. Muchová, A. and Jokl, V. (1971). *Chem. Zvesti*, **25**, 280
100. Cornett, F. W. and Gilbert, T. W. (1971). *Anal. Lett.*, **4**, 69
101. Brinkman, U. A. Th., de Vries, G. and Leene, H. R. (1972). *J. Chromatog.*, **69**, 181
102. Leene, H. R., de Vries, G. and Brinkman, U. A. Th. (1973). *J. Chromatog.*, **80**, 221
103. Brinkman, U. A. Th. and de Vries, G. (1971). *J. Chromatog.*, **56**, 103
104. Shimizu, T. and Ishikura, R. (1971). *J. Chromatogr.*, **56**, 95
105. Graham, R. J. T. and Carr, A. (1970). *J. Chromatog.*, **46**, 293
106. Graham, R. J. T. and Carr, A. (1970). *J. Chromatog.*, **46**, 301
107. Marcus, Y. and Kertes, A. S. (1969). *Ion Exchange and Solvent Extraction of Metal Complexes* (New York: J. Wiley)
108. Nga, N. K., Reshetnikova, L. P., Novoselova, A. V. and Bol'shova, T. A. (1972). *Vestnik Moskov. Univ., Ser. Khim.*, **27**, 342
109. Alimarin, I. P., Gibalo, I. M. and Lapenko, L. A. (1972). *Vestnik Moskov, Univ., Ser. Khim.*, **27**, 435
110. Goryanskaya, G. P., Kaplan, B. Ya., Kovalik, I. V., Merisov, Yu. I. and Nazarova, M. G. (1972). *Zh. Anal. Khim.*, **27**, 1498
111. Yusaf, M., Šulcek, Z. and Doležal, J. (1971). *Anal. Lett.*, **4**, 119
112. Akaza, I., Kiba, T. and Kiba, T. (1970). *Bull. Chem. Soc. Jap.*, **43**, 2063
113. Pohland, Ch. and Steele, T. W. (1972). *Talanta*, **19**, 839
114. Bolshova, T. A., Dolzhenko, T. G., Grishina, K. V. and Alimarin, I. P. (1973). *Zh. Anal. Khim.*, **28**, 259
115. Dolzhenko, T. G., Bolshova, T. A. and Alimarin, I. P. (1973). *Zh. Anal. Khim.*, **28**, 263
116. Alimarin, I. P., Bolshova, T. A., Dolzhenko, T. G. and Fernandes-Gomes, M. M. (1971). *Zh. Anal. Khim.*, **26**, 1213
117. Alimarin, I. P., Bolshova, T. A. and Bakhareva, G. A. (1973). *Zh. Anal. Khim.*, **28**, 1300

118. Shulepnikov, M. N., Shmanenkova, G. I., Shchelkova, V. P. and Yakovlev, Yu. V. (1973). *Zh. Anal. Khim.*, **28,** 608
119. Raichev, Kh, and Khalkin, V. (1970). *Radiokhimiya*, **12,** 778
120. Šulcek, Z. and Sixta, V. (1971). *Coll. Czech. Chem., Commun.*, **36,** 1561
121. Moskvin, L. N. and Tsaritsyna, L. G. (1970). *Radiokhimiya*, **12,** 187
122. Schmid, E. R. (1970). *Microchim. Acta*, 301
123. Schmid, E. R. (1970). *Monatsh. Chem.*, **101,** 1856
124. Woidich, H. and Pfannhauser, W. (1973). *Mikrochim. Acta*, 279
125. Ershova, N. I., Bolshova, T. A. and Alimarin, I. P. (1971). *Zh. Anal. Khim.*, **26,** 243
126. Santori, G. and Testa, C. (1973). *J. Radioanal. Chem.*, **14,** 37
127. Higashi, K. and Miyake, Y. (1971). *Osaka Kogyo Gijutsu Shikensho Kiho*, **22,** 282
128. Korotkin, Yu. S. (1971). *Radiokhimiya*, **13,** 137
129. Korotkin, Yu. S. (1973). *Radiokhimiya*, **15,** 625
130. Korotkin, Yu. S. (1973). *Radiokhimiya*, **15,** 671
131. Kennedy, D. C. and Fritz, J. S. (1970). *Talanta*, **17,** 823
132. Kennedy, D. S. and Fritz, J. S. (1970). *Talanta*, **17,** 837
133. Rais, J., Krtil, J. and Chotívka, V. (1971). *Talanta*, **18,** 213
134. Fritz, J. S., Frazee, R. T. and Latwesen, G. T. (1970). *Talanta*, **17,** 857
135. Yonezawa, Ch., Tamura, N. and Onishi, H. (1973). *Bunseki Kagaku*, **22,** 437
136. Leene, H. R., de Vries, G. and Brinkman, U. A. Th. (1971). *J. Chromatog.*, **57,** 173
137. Brinkman, U. A. Th., Taphoorn, J. H. and de Vries, G. (1973). *J. Chromatog.*, **84,** 407
138. Tserkovnitskaya, I. A. and Luginin, V. A. (1969). *Vestnik. Leningr. Univ., Ser. Fiz. Khim.*, **24,** 162
139. Aly, H. F. and El-Haggan, M. A. (1970). *Radiochem. Radioanal. Lett.*, **3,** 249
140. Aly, H. F. and Raieh, M. (1971). *Anal. Chim. Acta*, **54,** 171
141. Cvjetičanin, N. (1972). *J. Chromatog.*, **74,** 99

7
Particle Size Analysis

P. J. LLOYD and C. R. G. TREASURE
Loughborough University of Technology

7.1 INTRODUCTION

One of the major problems of reviewing a subject such as particle size analysis results from the wide diversity of techniques which have been developed. The first task has been therefore to classify the methods by the general principles involved, with sub-sections devoted to particular techniques. Some overlap is inevitable. In such a classification no place can be found for those papers which compare different methods or which review some of the techniques. Such papers will be considered in a general section following this introduction. Two further sections encompass many of the principles of the earlier classification but are areas of great interest in their own right and are considered separately. These are the important fields of aerosol sizing and on-line particle size analysis. The equipment described in these sections has been developed for these particular purposes, but the reader should refer to previous sections for other useful methods; for example, microscopy and light-scattering photometry are important techniques in aerosol size analysis. Similarly, for papers on particle shape the section on automatic microscopy should be read.

7.2 GENERAL REVIEW

An extensive review of most of the important methods of size analysis has been given by Alex[1-4]. A preliminary account described the mathematics of particle size distributions, various useful statistical and equivalent diameters, sampling and sample size for a given precision. Sample collection and preparation for microscopy is followed by accounts of optical and electron microscopes, and a number of scanning television microscopes. The problem of coincidence in stream counting is discussed and followed by discussions of the Coulter Counter, the acoustic counter and light-scattering counters for airborne and liquid-borne particles. Selected references (294) mainly cover the period 1960–70. Papers on particle size analysis published in the period 1970–71 have been reviewed by Alex, Koglin and Leschonski[5] (113 references). Two further papers providing introductions to particle size analysis, with surveys of the principles and measurements involved and accounts of the main types of sizing method, have been written by Leschonski[6,7]. Other review papers are those by Lankard and Niesz[8] on powder characterisation with reference to ceramic manufacture, and by Swift and co-workers[9] on methods and problems associated with the size analysis of fine-grained suspended sediments. Lauer[10] reviews sieving, sedimentation and counting methods while Scullion and Harris[11], in a paper on ancillary techniques, draw attention to the importance of preliminary examination of the particulate material, sampling, preparation and time effects.

Comparative methods for the size analysis of PVC latexes in the range 0.22–1.3 μm have been examined by Davidson and Collins[12]. The electron microscope gives results similar to those of other methods provided beam exposure and beam current are minimised. The authors consider that light scattering, sedimentation and fractional creaming provide more rapid and reliable methods.

7.3 PARTICLE STATISTICS

7.3.1 Particle shape

Shape is an important property of a particle, influencing not only the behaviour of a single particle, but also that of particulate masses. The problem has been to measure and define shape in as simple and unambiguous a manner as possible in order to establish correlations and to be able to predict the behaviour of non-spherical particles. The single figure sphericity factor of Wadell has been much used; Heywood's shape factors, linked to the basic geometrical form of the particle, can be used to derive correlations between the various 'sizes' of an irregular particle. Beyond this the concept of shape begins to lose meaning as recourse is made to such techniques as diffraction or the specification of a particle outline by a Fourier series.

Pahl, Schadel and Rumpf[13-15] in a three-part review of particle shape give four independent characteristics required to describe the geometry of particles: size, shape, roundness and roughness. After enumerating a number of the well known definitions of particle 'size', the authors collate methods of describing the shape and roundness of irregular particles. The review is extensive and is informative on the shape factors which have been defined and examined for their reproducibility as well as the types of problem for which they have been used (120 refs.). In a short paper, Deloye[16] briefly discusses particle size and shape, the use of shape coefficients according to the method of analysis, surface area and particle size distribution.

A statistical method for determining the sphericity of an assemblage of particles is described by Perez-Rosales[17]. Randomly packed particles are cemented by coloured plastic. The resulting solid is sliced into sections and analysed by means of a square grid. It can be shown that the sphericity of uniformly sized particles is a simple function of grain volume, number of grid-line/grain-perimeter intersections, total length of grid lines, number of mesh points lying within grain perimeters, total number of grid mesh points, and the photomicrograph magnification. Tests with known geometric shapes confirm the applicability of the method, but also the large number of intersections which have to be counted to obtain the asymptotic result. The method seems ideally suited to evaluation by automatic image analysis techniques.

For routine assessment of particle shape, direct examination of a representative sample is not suitable owing to the time required. Indirect methods, in which a bulk property of the material which is strongly influenced by particle shape is measured, have been developed. King and Dickin[18] have shown how a quick standard flow rate test can be used to evaluate the shapes of natural silica sands. The shape characteristic determined is the sphericity introduced by Riley and defined as

$$\left[\frac{\text{Diameter of circle inscribed in projected particle outline}}{\text{Diameter of circle circumscribing projected particle outline}}\right]^{\frac{1}{2}}$$

Because of the inadequacy of this function for describing angular shapes, the expressions developed are applicable only to natural silica sands with subrounded or sub-angular particles. Quartz sand was also the subject of measurement by Zolotar and Leonidov[19] investigating the effect of particle shape on pump wear due to conveyed suspensions. Apart from size, the authors measured the radii of curvature, the angular orientation and the number of the asperities. It was found that the number of asperities and the radii of curvature were normally distributed for each size, and that the orientation correlated with the radii of curvature. No explanation was put forward apart from the suggestion of random effects.

The gas atomisation of liquid metals was found by Rao[20] to result in the production of two main types of particle: spheres and ligaments. Shape was characterised by the mean length-to-diameter ratio of a number of particles in a given size range. These measurements were in general consistent with the conclusions from apparent density and flow-rate tests.

Airborne particles are sometimes classified according to their equivalent volume diameter (d_v) using the Coulter Counter, although their aerodynamic diameter (or Stokes diameter, d_s) is the more appropriate size. The relationship between these two size parameters was determined by Tomb and Corn[21] for some irregular particles by separating the particles according to their aerodynamic size in a Timbrell aerosol spectrometer (horizontal elutriator) and sizing the material collected at certain locations with a Coulter Counter. It was found that for coal and limestone d_v/d_s was approximately unity, while for an extreme shape, such as mica, d_v/d_s was approximately 1.3.

Electron micrographs illustrating the sizes and shapes of various raw materials for ceramics are given by Farges[22] with a short qualitative description in each case. The possibility of determining the shape of an object from a single view in scanning electron microscopy has been discussed by Olivei[23], who showed that it is feasible by exploiting the electron-photometry of the surface.

The use of an image analysis system (QTM-720) for the direct morphological characterisation of carbon black in the electron microscope has been described by Hess and co-workers[24]. The system provides quantitative size and non-dimensional shape information, and classification into spheroidal, ellipsoidal or fibrous shape types (subdivided into nine categories in all), using pattern recognition methods. By measuring projected lengths, Feret diameters at 45° spacing, projected areas, perimeters and calculating appropriate ratios, measures of anisometry, irregularity and orientation are obtained.

Instead of using a scanning system, as described above, for the detection and location of particles of particular size and shape on a slide, Naylor and Kaye[25] suggest the use of an optical spatial filtering system. The diffraction pattern from a transparency of a particle is subjected to holographic filtering and, as a result, a bright spot appears in an appropriate part of the output image plane, depending on the particle shape. Determination of the frequency and identification of particular particles (e.g. asbestos fibres) may thus be quickly carried out. These authors[26] have also investigated the use of optical character recognition techniques, in which the diffraction pattern formed by an object is interrogated by a rotating disc from which a sector has been removed. Fourier analysis of the resulting waveforms gives a spectrum which is characteristic of the shape.

Another method relying on optical diffraction has been developed by Ross[27] for monitoring low concentrations of particulate materials in suspension and determining separately the amounts of fibrous and non-fibrous material present, a capability which may be of great use in pollution monitoring. The technique involves analysing the diffraction patterns produced by a laser beam traversing the suspension with a rotating radial grating. Fibrous particles produce 'linear' diffraction patterns and more circular particles produce more circularly symmetric patterns. The rotating grating modulates the light flux which is detected and analysed into its a.c. and d.c. components, corresponding to fibrous and non-fibrous particles. Two limitations are that there should normally not be more than one particle in the measurement volume at one time, and that the size of the irregularly shaped particles is not determined.

In recent years, particle shape characterisation by a Fourier analysis of the particle projected area outline has been developed, the harmonic components of the Fourier series being identified with various geometrical shapes. Tilmann[28] has examined the effect on the analysis of orienting dispersed sand particles, so that the plane of maximum projected area (stable rest position) was at 45° to the usual horizontal plane and could be tilted in three different directions. The harmonic frequency distributions were tested statistically against the normal projection frequency distribution and no significant differences were found in 28 out of 30 comparisons. It was concluded that the shape information carried by the sand grains did not differ significantly between the four orientations so that no special care in ensuring the stable rest position on a slide would be necessary. Plainly this is a conclusion not to be applied indiscriminately to particles of any shape. In the application of the Fourier analysis method, it is necessary to locate the centroid of the particle. Naylor[29], in a note, cites earlier references to this method, stating that the results are not sensitive to small errors in the location of the centroid.

7.3.2 Particle size distribution

The common forms of particle size distribution functions, power law (Gates–Gaudin–Schuhmann), special exponential (Rosin–Rammler–Sperling–Bennett), normal (Gaussian) and logarithmic-normal are discussed by Rammler and Bahr[30, 31]. The second part of the article is concerned with the special graph papers which have been developed for plotting and evaluation of the distribution parameters.

In the derivation of moment measures of the size distribution of sediments, Hails and co-workers[32] have compared the mean size, standard deviation, skewness and kurtosis, calculated for various fractions of phi. They conclude, as theoretically expected, that the smaller the phi interval, the better. In practice the measures should be calculated at more than one phi interval for certainty. Svarovsky[33] shows how additional scales on logarithmic-probability graph paper enable the mode and the geometric standard deviation of a log-normal size distribution to be determined quickly and the size distribution function easily obtained. Using computed curves for the passage of aerosols having log-normal size distributions through channels, it is shown by Lekhtmakher[34] how more accurate values of the geometric mean size and the geometric standard deviation, together with an estimate of the error in these quantities, can be obtained. A mathematical investigation of the size distribution of ground mixtures when the components have log-normal or Rosin–Rammler size distributions is given by Reményi[35]. In a paper by Preining[36] on the application of information theory to particle size analysis, it is shown that the informational entropy associated with an experimentally determined size distribution is not only dependent on the size distribution itself, but also on the instrument used, especially its resolution and accuracy. The information contained in the size distribution is known only to an accuracy inherent in the method and described by a conditional probability function. Calculations of this information are used to evaluate three methods of aerosol sizing: optical counters, centrifuges and cascade impactors.

7.3.3 Sampling

The term sampling is taken here to include not only consideration of the statistics of sampling—to calculate, for instance, the size of sample required for a given precision of result—but also the design and realisation of the act of sampling itself, whether from a bulk powder or from a suspension.

7.3.3.1 *Size of sample*

Barbery[37] derives a simpler version of a formula due to Gy for sampling from a bulk for size analysis

$$\theta^2(\hat{P}_0) = \frac{f_0 \rho_0 d_{\frac{1}{2}}^3}{p_0} \times \frac{1}{m}\left[1 - \frac{m}{M}\right]$$

where $\theta^2(\hat{P}_0)$ = coefficient of variation of the estimated mass fraction in the size range 0, f_0 = an average shape factor for the size range, ρ_0 = an average particle density for the size range, $d_{\frac{1}{2}}$ = an average diameter for the size range, p_0 = mass fraction of the sample in the size range, m = mass of sample and M = mass of bulk. It is assumed that the sampling has been equiprobable. Another assumption is that the probability of a particle being sampled corresponds to the ratio (volume of particle: volume of the bulk material). This sampling model has been used by Pethö[38] to calculate the mass of sample required for the determination of density and particle size distribution.

For data which have a log-normal distribution, Hale[39] gives a formula to determine the sample size necessary in order that the geometric mean of the distribution can be calculated with a given precision and with a given probability. Hunt[40] also considers the same problem, again in connection with air pollution measurements. The formula derived is essentially the same as that of Hale, differing only in the factor which expresses, in terms of the logarithm of the geometric standard deviation, the confidence limits within which the geometric mean of the distribution is expected to lie with a given probability. Hale takes this factor to be the normal deviate corresponding to a specified level of confidence; Hunt correctly uses the 't' statistic associated with $(n - 1)$ degrees of freedom at the specified confidence level, where n is the number of data in the sample.

7.3.3.2 *Sampling operation*

In order to obtain representative samples of particulate suspensions for size analysis, a device has been developed by Burt *et al.*[41] which enables 100 ml of suspension to be divided into 10 ml samples. Tests indicated that the sampling reproducibility was significantly better than when samples were drawn from a stirred suspension by a syringe.

McBride and Sherman[42] describe a sampling technique used to determine the size distribution of zinc particles formed by condensation when zinc vapour was expanded supersonically with helium carrier gas through a nozzle. A portion of the nozzle flow was decelerated by mixing with quiescent air at matched pressure and the particles collected on a rotating cylinder which was shuttered in such a way that several discrete deposits could be obtained in a very short time. Electron microscopy indicated a median size of 140 Å and a narrow size distribution.

An effective long-term sampler for atmospheric aerosols has been developed by Heard and Whiffen[43]. A continuous series of 3 h samples, representing a filtered volume of 45 l, is formed on a strip of membrane filter which collects efficiently over the range 0.01–5 μm diameter. Aqueous droplets are removed by evaporation and the collected samples maintained at low relative humidity. The instrument can operate for periods up to 8 weeks.

The preparation of half-ton standard reference samples of five types of asbestos is described by Timbrell and Rendall[44]. Details of the blending and milling operations to produce respirable fibrous dusts with negligible contamination are given.

7.4 FIELD SCANNING

The observation and subsequent measurement of a large number of particles arranged in some field is an extremely important method of particle size analysis. This classification encompasses a wide range of techniques from simple optical microscopy to the use of laser holography. For simplification, the classification will be further broken down into the following sections: manual optical microscopy; automatic microscopy (equipment and setreology); electron microscopy; averaging methods; recording particles in motion.

7.4.1 Manual optical microscopy

The manually operated optical microscope has been used for the sizing of particles since the invention of the microscope. The method has been considered by some workers to be the only absolute method of particle size analysis, since each particle is measured by the operator. However, many different 'particle sizes' can be measured and comparison of the ease of measurement of each size by skilled and unskilled operators has been the subject of many studies. One of the latest has been by Nathan *et al.*[45], who have compared results obtained by the image shearing technique, the British Standard method using the globe and circle graticule, and by using a linear graticule. They concluded that unskilled operators always tended to oversize whilst skilled operators obtained greater consistency with the image shear method. However, their most important conclusion was that the errors in particle size distributions measured were larger than those currently allowed for. The accuracy of a determination has been further discussed by Hausen[46], who was interested in measuring accurately a few metal particles of interest in a matrix of unwanted material. The method consists of initially finding the average number of particles per field by examining in detail six fields. Total counts are then obtained by multiplying the average number of particles by the total number of fields examined over the whole mount. The accuracy of this method depends wholly on the six fields being statistically representative of the whole distribution. When this is so, a large number of particles are effectively counted, leading to accurate proportioning of components.

A novel technique for measuring the size distribution of aerosol sprays has been described by Fainerman and Sapiro[47]. The drops are allowed to fall on to filter paper treated with an indicator solution such that on contact with the liquid a colour change occurs. The edges of the spots are clearly defined and sizing is relatively easy in the range 90–1740 μm. A relation between the size of the stain and the liquid drop is given.

The use of a polaroid camera to take photomicrographs of particles for subsequent analysis has been described by Vialard-Goudou *et al.*[48]. The analysis of photomicrographs has also been considered by Ross[49], who gives methods for calculating a number of standard parameters from measurements made on photomicrographs of titanium dioxide particles by a Zeiss–Endter particle size analyser, a device for matching the diameter of a variable iris to the particle image. The conclusions made by this investigator are a little suspect as he examines only 465 particles in each analysis and does not mention the number of fields of view he has examined. His calculations of the confidence limits are certainly incorrect. White *et al.*[50] describe a simple modification to the Zeiss–Endter machine to make it suitable for data-logging.

7.4.2 Automatic microscopy

7.4.2.1 Equipment

Since the early 1950s, there have been many attempts to develop automated systems of microscopy in order to relieve the tedium of manual analysis, and to increase and maintain the precision of results over a larger period than is possible with an operator. The most successful instruments have consisted of a television camera which replaces the eye in viewing down the microscope stem and a small on-line computer to decipher the picture and produce the required information. Pillar[51] discusses the features required by a universal system for measuring and processing the characteristic geometrical and optical magnitudes of microscopic objects. Blifford and Gillette[52] also describe a general analysis system which is applied to both the electron and the scanning electron microscope, as well as optical processing equipment. The signals are recorded on magnetic tape and are processed off-line by a digital computer. Their analysis system will also classify particles according to shape by extracting features from the digitised picture. Gahm[53] continues the discussion with the Zeiss range of instruments used as examples. This discussion illustrates in particular the limitations of the use of television type scanners for high resolution requirements. The Quantimet 720 image analysis system has been described by Berry and Kelly[54]. Riediger[55] evaluated the Quantimet 720 for the routine analysis of thermal precipitator and conimeter samples. He found that the method gave satisfactory results if well defined optical and electronic conditions were observed, when the measurements were reproducible and without subjective errors. A bibliography of articles on automatic quantitative computers was published in 1973[56].

Bontemps[57] gives a general description of another instrument, the ΠMC, and its application to the characterisation of clays for the ceramic industry. Sobolev[58] describes a Russian instrument for the classifying of microparticles according to their dimensions. A television camera is used and the 'time equivalents' of chords of the particles above a predetermined limit are automatically counted. The distribution of the chords is obtained by resetting the limit. Gartner and Üelhoff[59] also describe a television scan system which not only sizes but also records local changes in luminance and therefore the temperature of radiant objects on the surface of copper melts. Vogelsgang[60] has compared the results obtained using the four image analysers and has found no deviations specific to a particular instrument, but difficulties due to depth of focus were common to all.

The principal disadvantage of television scan systems is that the resolution of particle surfaces and of phase variations within the particle profile is not very high. When such resolution is required, the television camera is replaced by a single photomultiplier and the optical system is stopped down until only a small area of the field of view is observed. The stage is now moved digitally and a resolution of 0.5 μm, the theoretical limit of the optical system, has been claimed[53,61]. Reiche[62] has also described a laser scan system in which the size of the laser beam is kept to a minimum and the amount of light is recorded with a photomultiplier. The system was found to be greatly superior to television scan systems.

7.4.2.2 Stereology

The development of image analysing computers has led many investigators to consider whether new descriptors of particle systems can be defined. An excellent paper by Coleman[63] outlines some of the difficulties. He considers assemblies of cubes and prisms and defines five different concepts of randomness. He considers also the distribution of intercepts through an assembly and shows how different moments of distributions can be calculated. This paper contains useful references and is a good introduction to the subject. More experimentally, Lewis *et al.*[64] determine the size distribution of spheres by a modification to the Saltykov method, which uses the distribution of areas obtained when a random plane cuts an isotropic bed of particles. The paper repeats early work published by Wicksell[65]. In a similar paper, Huska[66] relates the size distribution of sections to the size distribution by weight of fine sand. He does, however, discuss the effects of shape and density.

Dullien and Mehta[67] use a Quantimet instrument to obtain a chord distribution of sodium chloride particles mounted in oil, and the chord distributions of the void space in compacts by examining cut and polished sections after saturating the voids with Wood's metal. The authors back-calculated the chord distribution to give the original size distribution of the particles both by Shiel's method and by that due to Cahn and Fullman. In a later paper, Dullien and Dhawan[68], also using the Quantimet instrument, predict the size distribution of non-spherical objects from the chord distribution, but they presuppose knowledge of the shape function.

7.4.3 Electron microscopy

Particle size analysis with an electron microscope is beset with many problems, of which the most important is the preparation of a well-dispersed sample on the grid so that each particle can be seen separately. A particularly interesting paper by Porstendörfer and Heyder[69] details a study on latex particles. They prepared their grids either (i) by diluting with distilled water and permitting a drop to dry on a formvar membrane supported by a copper grid; or (ii) by spraying the diluted latex with compressed air or ultrasonics and permitting the aerosol formed to sediment on to the grids; or (iii) by allowing the airborne particles to dry and the latex particles to be collected by a thermal precipitator. For their material no difference in the size distribution due to the method of preparation was found. However, the authors did find that the mean size of the latex particles differed from the nominal size given by manufacturers by as much as 21% and that for latex particles of mean size greater than 0.5 μm the standard deviation of the distribution was also greater. The authors' investigations also showed that the most reliable results were obtained when the particles were not shadowed, when a minimum electron intensity was used, and when the microscope was used with cooled surroundings.

Kruglitskii *et al.*[70], who examined metal oxide powders with an electron microscope, favour both carbon replication and platinum shadowing. They calculated the specific surfaces of the oxides from their results and found that the values agreed favourably with those measured by nitrogen adsorption, using the BET method. Abdel-Azim and Vlajic[71] describe a novel method for obtaining good dispersed grids from particles in a solution. After ultrasonic dispersion, air is forced through a porous sintered filter until a separate foam layer is formed above the liquid. Prepared grids are placed in contact with the foam and a good dispersion obtained.

The use of liquids in an electron microscope is usually impossible because the high vacuum needed will evaporate most liquids. Larson *et al.*[72] describe the use of a wet cell for use in an electron microscope used in their studies of the formation of the spherical chromium hydrous oxide sol of particle size less than 0.5 μm. Their technique consists of retaining the liquid between two 400 mesh electrolytic copper screens and covering these with three films. They used formvar, nitrocellulose and silicon monoxide films and found that the cell formed would withstand the high vacuum at modest electron beam intensities. Using this technique, they studied the formation of the sol.

7.4.4 Averaging methods

There are many methods which do not yield a detailed particle size distribution but measure a parameter which is some function of the whole distribution. The simplest method is to measure the attenuation of a light beam which is a function of the projected area of the particles in the beam provided the concentration is not so high as to cause multiple scattering. Such an instrument is called a turbidity meter when the particles are suspended in a liquid and a nephelometer when an aerosol is considered. Alternatively, the diffraction pattern of the field of particles on a microscope slide can be examined to give another average of the particle size distribution. This technique has been found very useful in the routine analysis of microscope slides on which particles have been dispersed by one of the aerosol samplers such as the thermal precipitator.

7.4.4.1 Turbidimetry

Witmer and Gollan[73] describe a turbidity meter which has been developed for measuring the concentration of crude oil in the ballast water in oil tankers. They obtained a calibration curve which was independent of the type of crude oil and also compared results with the particle size distribution obtained by photographing the oil droplets as they passed through a specially developed flow cell. Pierce and Cowan[74] describe the use of a spectrophotometer as a turbidity meter for the measurement of an acrylic epoxy ester and modified alkyd resins in the particle size range 60–890 Å. Their paper also considers the factors which influence the size of the anionic electrodeposition micelles.

7.4.4.2 Nephelometry

A nephelometer is an instrument which by measuring the extinction coefficient of an aerosol at a number of different wavelengths, usually four, can yield a number of parameters of great importance for monitoring the atmosphere. These parameters include visibility, spectral scattering coefficients, total aerosol mass and a value for the exponent of the assumed power law function for the aerosol size distribution. Heintzenberg and Quenzel[75] present a general theory for calculation of the scattering coefficient, which is independent of a specific measuring system. They also considered the problem of the calibration of nephelometers by scattering media with differently shaped scattering functions, and the error in the determination of the scattering coefficient when the shape of the scattering functions of the media are ignored. Their theory is applied to two designs of nephelometer, that of Ruppersberg[76] and that of Charlson[77].

The latter instrument has been used extensively in an investigation into the Los Angeles survey. Charlson[77] gives the background to the study and the general conclusions. Ensor *et al.*[78] and Thielke *et al.*[79] give greater detail. These workers, using the Minnesota Aerosol Analysing System, developed by Whitby and co-workers, were able to record at the same time the size distribution of the aerosol and the measurement of the extinction coefficient at four distinct wavelengths: 365 ± 15, 436 ± 5, 546 ± 5 and 576 ± 15 nm, in a Charlson integrating nephelometer. Thus a unique series of 212 measured size distributions were obtained for each of which Mie light scattering computations were made for direct comparison of the calculated extinction coefficient with the empirical value. The measured results were all slightly higher than those calculated but the differences were well within the accuracy of the nephelometer method. In the paper by Thielke *et al.*[79] the comparison was continued. The authors found that broad wavelength band scattering ($\Delta\lambda \approx 100$ nm at 500 nm) correlated well with monochromatic scattering, and that if the aerosol was assumed to follow a power law size distribution, the exponent correlated well with that of an assumed power law of wavelength dependence of scattering. Good agreement was obtained between the volume concentration and the measured light scattering coefficients.

Two papers have been published concerning the absolute calibration of nephelometers. Karabegov[80] proposes the use of suspensions of pyrex glass particles which he had found to remain stable for two or more years. Horvath[81] proposes the use of opal glass as the calibration standard, using colour and neutral density filters to simulate intensity and wavelength dependence. Heintzenberg and Quenzel[82] have considered the effect on the determination of the scattering coefficients due to the loss of large particles by deposition in the sampling tube. They develop quantitative relationships between the largest diameter of particle present and the scattering law coefficient for a given power law size distribution and conclude that any loss of the larger particles will cause little error in the calculated scattering coefficients. Cohen[83] criticises the mathematics of an earlier paper by Grasse[84] and shows that following certain assumptions the particle size distribution in seven intervals can be calculated directly and not by the method proposed.

7.4.4.3 Other methods

Another method of measuring an average parameter of a particle array is the Talbot Diffraction Size Analyser (DISA). Essentially the diffraction pattern formed by an array of particles on a microscope slide is observed and the average projected area deduced. Du Toit and Murray[85] have described comparative tests on three such instruments used for the automatic assessment of samples of different minerals obtained with a thermal precipitator. Results show that great care must be taken in the initial calibration, since, although consistent results were obtained with each instrument, a one-to-one relationship between any pair of instruments was not obtained.

7.4.5 Methods for particles in motion

The ideal method of measuring the particle size distribution of particles suspended in a gas or moving with high speed is to record instantaneously the particles or droplets by photography or holography. By this method uncertainties due to sampling can be avoided and volatile drops can be measured. Clark and Dombrowski[86] have made a critical assessment of the commonly used weighting procedure used in the analysis of photographic records of sprays. This method uses a double flash technique to determine drop velocities and then weights the count for each size by a factor equal to the normalised velocity of the size. The authors claim that this could be in error because the periodic nature of drop formation can lead to unsteady state conditions, especially near the nozzle, and thus unpredictable errors. McCreath *et al.*[87] describe a technique for measuring the velocities and size of particles in flames. The paper pays close attention to the selection of film and filters. A thorough review of the methods available for measuring the particle size of droplets has been presented by Bürkholz[88]. He reviews the use of sampling, of measuring the light scattered from individual particles and of droplet holography.

Gross and Peter[89] have described a complete holographic system for measurement of particle size distribution and density of aerosol particles in a model of a human respiratory system. The principal advantages of the holographic system are emphasised. It is possible to obtain as a hologram a record of particles dispersed throughout a volume and then by using a laser beam to re-create the original three-dimensional image and so make measurements of particle size at leisure. Bexon[90] describes a novel method for determining the magnification generated in the in-line method of holography, thus ensuring that a correct assessment of particle size is achieved.

The determination of the particle size and distribution of very small particles is a continuing problem facing the analyst who would prefer an alternative method to the electron microscope. Hinds and Reist[91,92] have described a method using lasers in which the Doppler shift in frequency of the scattered light due to monodisperse particles is used to yield a velocity distribution and the particle size for concentrations in the range 10^3–10^6 particles per cm^3. For polydisperse suspensions it is necessary to assume a two-parameter analytical function describing the size distribution and to derive the parameters from the observations. This method yields a powerful tool for *in situ* measurements of very fine aerosols in the size range 0.03–3 μm diameter. Gollub *et al.*[93] have also used Doppler shift to record the growth of droplets in a cloud chamber as a function of time.

Reagan and Herman[94] suggest three methods in which laser radar (lidar) might be used to obtain particle size distribution data remotely over a large volume of the atmosphere. The three methods suggested are: to use the solar radiometer, monostatic lidar or bistatic lidar techniques.

7.5 SEDIMENTATION METHODS

7.5.1 General

Methods of particle size analysis which depend on the differing settling velocities of particles according to their sizes rely on two principal assumptions: that the particles are discrete, and that their velocities, and hence positions after some settling period, can be determined with the aid of a mathematical expression, such as Stokes' law. There is no easy solution to the problem of finding suitable suspending liquids and dispersing agents for previously untested materials apart from experience with similar samples or the use of compilations of published data. Most organisations active in particle size analysis no doubt have made their own lists. Some years ago Welwyn Hall Research Association circulated privately a list of some 100 materials and appropriate dispersing agents, but one double the size has now been published by the Akademie der Wissenschaften of East Germany[95] and should be most useful.

Moore and Orr[96] have considered the influence of Brownian motion on sedimentation analysis by comparing actual analyses with theoretical distributions derived from microscopy and various diffusion equations. They suggest that the effect of Brownian motion can be neglected down to sizes of 0.2 μm, provided that the mean value of the ratio of Brownian to settling motion is less than *ca.* 0.20. An extensive series of graphs presenting the results of size analyses with the Cahn sedimentation balance and the Martin

Sweets centrifugal photosedimentometer over the range 0.1–150 μm has been given by Klingler[97]. The materials investigated ranged from baby powder to zinc and included Arizona road dust. The instruments are discussed in Sections 7.5.2.2 and 7.5.3.

7.5.2 Gravitational methods

7.5.2.1 Incremental methods

Leroux[98] discusses the pipette method of Andreasen and the conditions which must be fulfilled for the application of Stokes' law. The preparation of the samples is described, with particular reference to soil samples and the importance of dispersion. The systematic errors of the method are not considered. With the necessary precautions taken it is proposed that the method can be used to size to 0.1 μm with satisfactory accuracy (this assumes no convection during 8 days settling with 12 sample extractions). A similar extension of the Andreasen method to sub-micron sizes is reported by Gate[99] for the size analysis of kaolins. In the event, a homogeneous suspension in a shallow cylinder is sampled by a small pipette fitted with a perspex tip having four radial sampling intakes. By sampling at 2 mm below the free surface, an analysis to 0.5 μm requires a little over two hours sedimentation. The results compare favourably with centrifugal sedimentation data.

As an alternative to the hydrometer method, much used for soil size analysis, but which requires 11 h to obtain data at the 2 μm size, Aitein and Poulin[100] have devised a captive diver method for determining the density of a suspension at any particular depth and time. This is found from the apparent loss in weight of a small diver suspended by a fine wire from a balance. In spite of the disturbances which must be caused by the introduction and removal of the diver at each reading, the results agree well with the standard hydrometer method and the method requires only 2 h for an analysis from 80 to 2 μm.

Pezzetta[101], in a report on the size analysis of silt-clay sediments to 1 μm, reports on the falling drop method for determination of the concentration of samples extracted from a settling suspension. The method consists of timing the fall of a standard drop of each sample through an immiscible organic liquid. Calibration with drops of sodium chloride solutions of known densities enables the density of the sample drop to be estimated and the solids concentration calculated. Pezzetta reports no loss of accuracy in the size data, compared with the standard method of drying and weighing. The report also contains a large number of data tables, including settling times for quartz particles, calculated in $\frac{1}{4}$ phi size class intervals, in the temperature range 20(0.2)–30 °C.

The determination of the solid concentration at a specific depth in the settling suspension without the disturbing presence of a foreign body, or the removal of a sample of the suspension, is plainly desirable and opens the way to continuous recording. The extinction of a beam of light traversing the sedimentation cell and the application of the Lambert–Beer law has often been used for size analysis, usually with assumptions regarding the extinction coefficient of the particles and the relation of Stokes' diameter to the projected area diameter. Polyakov and co-workers[102] describe a photosedimentometer used for the analysis of tungsten powders. The paper is short and not very informative. The output of the photocell is apparently differentiated and a logarithmic recording of optical density plotted against time. The determination and calculation of the size distribution of a powder is said to take 5–10 min, for a size range of 0.5–75 μm. The details of how this is achieved are not revealed.

The variability of the extinction coefficient can be reduced by using white light and collecting both transmitted and diffracted light; the time of analysis can be reduced by using a scanning technique so that the effective measurement depth is continuously reduced during sedimentation of the suspension. Both these modifications have been used by Allen[103], as well as assessment of the concentration of the solids by the absorption of an x-ray beam, which is not particle size dependent. Brief descriptions and comparative results for these instruments, developed in 1969–70 and previously reported, and for a centrifugal version of the x-ray sedimentometer, are given in the paper. A patent[104] describing Allen's gravitational and centrifugal equipment using x-ray absorption has been published. An apparatus similar to that of Allen is the Miromeritics 5000 Sedigraph, which has been available for some time. Kunst[105] reports on the machine from the user's viewpoint. His main problem concerns the elimination of tiny air bubbles in the sedimentation cell, the effect of which is very noticeable when trying to size below 0.5 μm. Base-line drift, observed after 1 or 2 days, was attributed to minute distortion of the cell window. Comparative tests showed good agreement with the normal Andreasen method, and reproducibility tests gave consistent results. The analysis of a clay sample from 50 to 0.5 μm requires 40 min; a built-in computer enables the size distribution to be plotted directly by a recorder.

7.5.2.2 Cumulative methods

The most common cumulative sedimentation method in use is that known as the sedimentation balance. Over the years the systematic errors associated with the method have been well documented and a number of designs and techniques have been developed to reduce these errors. Keienburg[106] describes the use of an automatic recording sedimentation balance for the size analysis of cement and discusses the determination of the gradient of the sedimented weight *versus* time curve by approximating the curve with a linear or quadratic expression. The choice of expression is not so important as the ratio of the times (t_{i-1}, t_i, t_{i+1}) at which the sedimented weights are found for calculation of the slope of the curve at time t_i. The mean and coefficient of variation obtained from the weight frequency distribution curve are to be preferred for characterising the distribution, rather than endeavouring to fit standard size distribution functions.

One of the largest sources of error in the method arises from the loss of sediment which, because of concentration gradients or the 'pumping' action of the balance pan, is not collected, so that the ultimate settled weight is not that expected. Bürkholz[107] shows both theoretically and experimentally that the loss for a pan suspended in a vessel of greater diameter depends on the gap width and on the particle size. He describes an improved sedimentation vessel containing an insert which almost touches the pan and is almost of the same diameter. He shows that the losses in this type of vessel are small, so that the weight of fraction still suspended at any particular time may be determined from the initial suspended weight, and the analysis terminated at a convenient time. Scott and Mumford[108] consider the same problem and show that instead of complicated sedimentation vessels a simple mathematical technique is sufficient to yield the 'correct' end-weight. It is assumed that the final approach to the unknown asymptote, represented by complete sedimentation, is a first-order exponential function:

$$w_t = w_\infty[1 - \exp(-at)]$$

where w_t is the weight deposited on the pan at time t, w_∞ the final settled weight, a a constant. A plot of values of $w_{(t+\Delta t)}$ against w_t, with constant Δt, will give a straight line which intersects the line of unit slope, $w_t = w_{(t+\Delta t)}$, at the point (w_∞, w_∞). Values of the end-weight found by this method, and used to calculate the particle size distributions of various silica powders, have given results in excellent agreement with those found by using the Leschonski modification of the Sartorius balance sedimentation vessel. Comparable results were obtained with calcite powders[109].

Shelnutt[110] in a detailed report describes a centrifugal method and the use of the Cahn sedimentation balance. The sedimentation accessory was modified by a pan support lever which held the pan against the lower end of the sedimentation column and prevented loss of material during mixing. It was found that practically all the material was collected on the pan eventually. A Cahn electrobalance was also used by Gibbs[111] in an investigation of the accuracy of settling tubes used for the analysis of geological sediments. Samples of glass spheres (2 μm to 1 mm) were introduced at the tops of the sedimentation tubes by a layering technique. It was concluded that for the best compromise of accuracy, resolution and time taken, tubes with settling lengths of 140 cm and diameters not less than 13 cm should be used. Similar work by Anan[112] on the size analysis (2 to 0.0625 mm) of coastal sand samples has led to the development of a settling tube 20 cm in diameter and with a 200 cm fall distance. In effect, the two-layer technique is used by admitting the 20 g sample to the tube by means of a 'Venetian blind' gate. The balance pan is suspended from a force transducer so that the weight of sediment can be continuously recorded.

The size distribution of the droplets of one liquid dispersed in another is determined by Grossman[113] from observations of the volume of the dispersed phase collecting at the bottom of a sedimentation column as a function of time. In this case the volume of the sediment is not neglected as in the normal sedimentation balance method. The expression used for calculating the size distribution is

$$\text{Fractional mass (volume) less than size } R = \frac{La - l + ut}{L - l + ut}$$

where L = total length of settling column, a = total volume fraction of dispersed phase, l = height of settled column of dispersed phase at time t, $u = dl/dt$, and R = size of droplet which has settled the distance $(L - l)$ in time t.

A somewhat similar exercise has been carried through by Lebedev *et al.*[114] for analysis of an emulsion in which the disperse phase is lighter than the continuous phase. A small quantity of the emulsion is injected into the bottom of a column filled with the continuous phase liquid. The droplets rise from their line start position and are collected under a hood. The increased buoyancy causes the hood to rise and an attached vane obscures a beam of light. The apparatus is adjusted for linearity between hood displacement and photoresistor output. The recorded curve is thus a cumulative volume or weight distribution plotted to an inverse size squared scale.

7.5.3 Centrifugal methods

The development of centrifugal methods followed naturally from the desire to analyse finer materials without the delay arising from the use of gravitational methods and errors arising from the effects of Brownian motion. In recent years the disc centrifuge has been much used and Shelnutt[110] describes the use of the Martin Sweets centrifugal photosedimentometer (a variant of the original Kaye disc centrifuge), using the two-layer technique. A set of linear equations approximating to Rose's generalised extinction curve was used in the computations to provide a 'correction' for the size-dependent light scattering. Although frequency distributions with and without the correction showed differences, the cumulative distribution curves were very similar. The injected suspension concentration is given in the text as 0.03% by volume maximum, but the examples quoted used concentrations of 0.075% and 0.15% without comment. The presence of streaming was not specifically observed.

Some doubts about size distributions determined with the two-layer technique and a beaker type centrifuge have been expressed by Zwicker[115]. Although the MSA Analyser was found to have a high reproducibility, it did not always reveal changes in size distribution found by other methods. In tests with standard glass beads and ground alumina, samples drawn from the sediment, the bulk suspension and the feed layer did not show the expected size separation on which the theory of the method is based. From ancillary observations it is suggested that density streaming is the major cause of the poor separation, although no adequate explanation has been found of why reasonable agreement with other methods occurs when the analysed material has a log-normal (or nearly so) size distribution.

One way of avoiding the possibility of density streaming with the two-layer technique is to use an initially homogeneous suspension. Owing to radial dilution, the resulting equations are, however, analytically intractable and recourse is made usually to approximate solutions. Bagness and co-workers[116] have been using this homogeneous suspension technique with pesticides and assessing suspension concentrations at the sampling radius of a disc centrifuge by light absorption. No attempt, however, has been made to analyse the resulting plot, but the ratio (E_t/E_s) is taken as the cumulative weight distribution, where E_t and E_s are the optical densities at time t and time zero, respectively. This is plainly incorrect and, if anything, would bear more resemblance to a cumulative surface distribution rather than one by weight. Nevertheless, the pseudo-distributions produced are claimed to be in reasonable agreement with electron micrograph results for narrowly sized silica spheres. The restricted size range may well have been responsible for the apparent agreement. The speed and good reproducibility of this method suggest uses in quality control situations.

Instead of extracting samples for concentration determination or using inferential methods, the sedimentation balance technique has been applied directly to beaker-type centrifuges. An apparatus is described in detail by Bürkholz[117]. With a rotational radius of 30 cm and a fall distance of 1.5 cm the particles in the initially homogeneous suspension settle, in an almost constant force field with almost constant velocity. The mass of the particles deposited on a collecting plate is determined from the movement of the plate, which is measured by an inductive transducer and recorded. The recorded graph, on which displacement is directly equivalent to sediment weight, is analysed and a cumulative distribution obtained in the same way as for a gravity sedimentation balance. The extreme particle sizes which can be determined are *ca.* 0.02 and 9.6 μm. The sample weight is small (10–50 mg), but the method appears very reproducible.

A centrifugal sedimentation balance is also described by Tarasov[118]. The float supporting the collecting plate is located in a separate chamber and its position, depending on sediment weight, determined by a light beam. The particle size range covered is from 0.05 to 20 μm. From the drawing of the sedimentation cell it would appear that there may be some loss of material for the same reasons as in the earlier gravitational sedimentation balances. This problem would not arise with the instrument described by Bürkholz since in that case the collecting pan covers the total cross-section of the vessel.

Scholtan and Lange[119] describe the use of an ultracentrifuge to determine the size distribution of polymer latices, by measuring the optical density at a given radius as a function of time. Three corrections to the apparent size distribution, calculated as for gravitational sedimentation, and with a uniform extinction coefficient, are discussed and applied: the influence of concentration on particle size calculated by Stokes' law, the effect of dilution due to the radial motion of the particles, and the effect of particle size on the scattering of light. The corrected concentration ratios are computed from a series of terms involving the observed optical density, calculated specific turbidities, and the times for various particle sizes to clear the light beam. The method depends on calculation of the initial concentration of the smallest size in order to find the concentration of the next larger size, and so on. This represents a solution of the problem of the centrifugal photosedimentometer using a homogeneous suspension. In the case when the concentration can be directly assessed, e.g. by x-ray absorption, the basic integral equation relating the concentration and size distribution function has, until now, been solved by an approximation method.

Svarovsky and Friedova[120] have now described a method in which the unknown size distribution function is assumed to follow the three-parameter equation described by Harris. The parameters are found by means of a direct search curve-fitting technique applied to the measured concentrations. Kamack[121] has now found a useful exact solution of the concentration–size distribution relationship, for centrifuges using homogeneous suspensions, in the form

$$F(D_m) = \exp a\left\{c(D_m) - \int_0^{\infty} h(x)c\left\{D_m \exp(-x)\right\} \mathrm{d}x\right\}$$

where $F(D_m)$ = weight fraction less than size D_m, $a = 2 \ln (r/r_0)$, with r the radius at the sampling point and r_0 the radius at the free surface of suspension, D_m = largest size particle to settle from r_0 to r in time t, $c(D_m)$ = normalised concentration at r at time t, and $h(x)$ = function depending on apparatus constant 'a'. Graphs of $h(x)$ for various values of 'a' are given in the paper.

Unlike the earlier approximate solution, the integration interval Δx is independent of the particle sizes at which $F(D_m)$ is to be evaluated and may be taken as small as is required for accuracy.

7.6 STREAM SCANNING METHODS

Into this classification are placed all those instruments in which a stream of particles passes through a sensing volume. A signal is produced which may be used for registering the passage of a particle and of which the magnitude is some function of the size of the particle. The Coulter Counter is one such instrument. The particles are immersed in an electrolyte and are drawn through a small orifice, thus momentarily changing the resistance of the orifice. This change is monitored by the voltage change arising from the constant current maintained between two electrodes. The light-scattering photometers in which a particle passes through a light beam and the scattered light is monitored are also stream scanning instruments. Inseparable from the practical operation of this type of equipment is the problem of coincidence, that is the possibility that more than one particle may be in the sensing volume at the same time.

7.6.1 Coulter Counter

As one of the most significant modern advances in particle size analysis, the Coulter Counter has continued to be the subject of much development. Apart from the many patents taken out by Coulter Electronics, there have been significant papers on the development of the principle of operation. Schultze and Thom[122] in a paper aimed at the counting of platelets in whole blood samples describe the application of 'hydrodynamic focusing' in which a stream of particles suspended in electrolyte are directed through the centre of the orifice surrounded by an annulus of clean electrolyte. One of the two electrodes is situated in the clean electrolyte. The authors claim that this technique cuts down the 'noise' to a level expected by modern electronics and thus extends downwards the smallest size measurable by a single orifice.

They also describe improved electronics which analyse the electronic pulse produced into 256 channels. Interesting plots are presented which show the sensitivity by giving distributions obtained from latices of 2.49 μm particle size, which demonstrate the presence of singlets, doublets and triplets. Bebyakov[123] discusses the magnitude of errors caused by changes in the resistance of the electrolyte and of concentration on the calibration in the Russian equivalent of the Coulter Counter. Three techniques are proposed to overcome the difficulties.

The problem of coincidence has been discussed further by Bader *et al.*[124]. Coincidence occurs when more than two particles arrive in the orifice at the same time and is of great importance, especially when high concentrations are used. In the use of the instrument, it is therefore of some importance to know at what concentration the coincidence effect becomes important. Bader *et al.* have discussed the problem with reference to the number of particles found in small volumes, a model used by other workers. However, since the Coulter Counter is a stream scanning instrument, to subdivide the stream is somewhat arbitrary and the model falls short of the true coincidence phenomenon.

As with many methods of particle size analysis, the techniques used are a very important factor in obtaining good results. Especially welcome are the substandard latices described by Horfield[125]. Use of these will enable many operators to avoid the pitfalls arising from the use of pollens. Horfield's paper also includes a discussion of calibration techniques. Lines[126] emphasises that a result is only as good as the sample on which the analysis is made. He presents a paper describing the difficulties and precautions found necessary when comparing two models of the Coulter Counter. He obtained a reproducibility of better than ± 1.0 wt.%, which compared favourably with the instrumental variation of less than ± 0.5 wt.%. It is particularly important to ensure that the particles are kept uniformly dispersed throughout the suspension. Nienow and Ang[127] discuss the use of different types of stirrers and their application. They finally suggest a system with an angled bottom to the 'beaker' which prevents particles gathering in a 'dead spot'. This is especially important for particles greater than *ca.* 50 μm in size. Maguire *et al.*[128] present a note on calculating the settings on a Coulter Counter, Model A, in order to obtain minimum 'noise' at the lowest setting. Barclay[129] discusses the operational techniques of the Coulter Counter Model T and its limits of interpretation.

Another measure of the success of the Coulter Counter is its application in many fields. McCave[130] describes techniques he has developed for the measurement of fine to coarse sands up to 1000 μm using 10:90 saline:glycerol mixture as an electrolyte. Since this mixture has a viscosity of 178 cP at 20 °C, the difficulty of keeping particles in suspension is reduced. Marshall and Meakin[131] discuss the application of a Model T Coulter Counter to dissolution rate determinations of insulin crystals, with particular reference to the possibility of using the technique without non-aqueous electrolytes or the saturated solution methods for analysing water soluble materials. Baudet[132] has compared the distributions of kaolin of particle size less than 20 μm obtained by Coulter Counter and by other methods. After correction for particle shape, good agreement was obtained between analyses. Baudet found that the Coulter Counter had many advantages for the analysis of clays since it measured a property of the particle which was independent of density and shape of the constituent materials.

An ingenious system has been described by Steinkampf *et al.*[133] which characterises individual particles, as they pass through in a stream, by volume using the Coulter principle, by light scattering and by fluorescent techniques. This system can also be used to separate individual particles. Another development of the Coulter principle has been proposed by Fedrick, Bell and Hanson[134]. They cause a gas at reduced pressure to become conducting by passing a current through it, and pass the glow discharge through a small orifice. Particles to be characterised are also passed through the orifice and cause a change in the discharge current which has been shown to be proportional to particle size. The paper is very preliminary and the effects of change in current with particle size need to be further investigated, but the technique appears very promising.

7.6.2 Single particle light-scattering photometers

The counting and sizing of individual particles in gas streams by measurement of individual light scattering has long been an important method of particle characterisation. Nikiforova and Sedunov[135] have published an important paper which investigates four possible sources of error in this type of aerosol counter. These sources are (1) non-uniformity of illumination of working volume; (2) coincidence and the effects of two or more particles in working volume; (3) photodetector noise and errors in calibration; and (4) errors in quantisation, or in how the size distribution is divided.

In similar vein, Oeseburg[136] examines the influence of the aperture of the optical system on the response curve of an aerosol counter. He also concludes that the optimum collection angle is 20–120° inclusive.

Schehl *et al.*[137] use the cavity of a He–Ne laser to obtain uniform illumination over a very small sensitive volume so that measurements of high concentrations are possible, in excess of 10^6 particles l^{-1}. They collect the scattered light over an angular range of 7.5–172.5° and claim good resolution in the range 0.2–1.0 μm. Wims and Myers[138] have also introduced a laser light scattering photometer which measures the intensity of light scattered from particles suspended in a liquid as a function of angle in the range 0.1–170°. The measurement of the molecular weight of a polystyrene standard in butan-2-one demonstrates the versatility and high sensitivity of the photometer. Cluck *et al.*[139] describe a portable battery-operated airborne particle counter and analyser utilising forward scattering at 135°. No resolution data are given in the paper. Another novel counter[140] has been introduced in which a particle is suspended in an electric field while a complete scattering diagram is plotted. Whilst the instrument is cumbersome for particle size distribution determination, the technique may have other applications.

Finally, a technique which almost defies classification but which may have uses in other fields has been described by Holland and Sawers[141]. The technique was developed to determine the particle size of silver compounds used in photographic emulsions. The particles are dispersed on a filter paper and covered with an electrolyte in which a silver electrode is immersed. The current to a second electrode, which is a 10 μm platinum wire, is monitored. When this probe touches a particle it is reduced and the current to the probe is changed. This current change is proportional to the volume of the particle. The filter paper is rotated and new particles contacted. The authors claim that the method works down to a particle volume of 10^{-2} μm^3.

7.7 SIZE SEPARATION METHODS

A sample may be classified into fractions of limited size ranges mechanically by sieving or by the interaction of a fluid and the particles.

7.7.1 Sieving

Sieving is perhaps the Cinderella of all methods commonly used for particle size analysis. It remains the most important method for particles above 200 mesh (56 μm), but is applicable down to 5 μm when electroformed micromesh sieves are employed. The importance of the method is seen in the continued publication of testing standards. During the period of this review, two have been published: the first[142] from South Africa is concerned with specification of test sieves, whilst the second[143] from Poland details rules for obtaining a good analysis. It is a great pity that much of the early work on sieving has been disregarded and needs to be rediscovered. A short paper by Kiff[144] emphasises the precautions needed to obtain reliable results. He particularly mentions the need for calibration, the need not to exceed the critical load and the need for controlled sieving time. These points are considered by Janke[145], who presents empirical formulae for calculating the load on the sieve as a function of $\Delta\phi$ (the phi interval between sieves), the mesh size and the acceptable limit on a sieve of mesh equivalent to 1.00 mm. An equation for sample size is presented in like manner. Such an approach has little to recommend it outside the limited range of marine sediment analysis in which the author is working. A better approach is that by Rose and English[146], who propose that the quantity of material in the near-mesh range (1.1–1.0 of the mesh size) is responsible for variations in the rate of sieving. If much of this material is present, then the residue on a sieve, even after a very long time of sieving, may not be the true result. Their explanation of this phenomenon is plausible: that these 'near-mesh particles' blind the mesh and thus do not permit further particles to pass through. If the load on the sieve is kept below a critical value, which will depend on the relative numbers of the near-mesh particles and which can be calculated using a semi-empirical formula, then the rate of sieving is rapid and the residue approaches the correct proportion. Although the concept of critical loading has been discussed before, its functional dependence on the size distribution of the material will be of importance in operation and design.

West and Dumbleton[147] consider that wet sieving of soils is an improvement over the British Standard (BS 1371) method of wet sieving first on a fine sieve and then drying and completing the analysis by dry sieving, since the drying process can lead to agglomeration. Comparison of analyses obtained by the wet and dry methods are given. Carter[148] presents a method of wet sieving with precision microsieves which he claims is both inexpensive and capable of high reproducibility. A stack of micromesh sieves can be used. Lanes[149] has compared micromesh sieves with square and round openings and finds the distribution obtained depends on the shape of the openings. This result is not new, having been used by the manufacturers of road stone as a method of differentiation by shape for many years. He was also able to use micromesh sieves of 10–15 μm aperture on an Alpine air jet sieve provided the dusts did not have a tendency to agglomerate.

Niedek[150-152] has proposed to shorten the total time for wet sieving by estimating the residues quantitatively from the attenuation of a traversing light beam. He claims very rapid determination which could be used for on-line control. Bauer *et al.*[153] describe a new method for the continuous classification of material by sieving. Reasonably high efficiencies were obtained at particle sizes above 75 μm, but below that size efficiency was very poor.

7.7.2 Classification methods

Classification, or elutriation, methods are widely used in industry for the production of various grades of powders. Small-scale equipment has been developed and applied for the purpose of particle size analysis. An advantage of the method is that usually a significant quantity of the material is processed so that sampling errors are reduced and the fractions are available for further investigations. A disadvantage is that ideal separation of the various particle sizes is never achieved: there is always some misclassified material which is seldom divided equally between the two product fractions. The fractional weight of material in, say, the finer fraction will, therefore, differ from the fractional weight of feed material which is less than the separating size. Developments in classifiers in recent years have been aimed at improving the operating conditions and consequently the sharpness of separation so that improved correlation with other methods of size analysis may be obtained. Leschonski[154] discusses the problems of separation in fluid flows, particularly in air classifiers, with reference to the counter-flow gravity separator and the cross-flow separator, previously developed for laboratory size analysis. Schalter and Lapple[155] briefly describe centrifugal classifiers and the measurement of their performance with reference to the Donaldson classifier and its behaviour in the fractionation of plastic powders.

The use of a fairly standard arrangement of a three-column elutriator with column diameters in a ratio of 1:2:4, and a constant liquid flow through the columns in succession, is described by Follmer and Beavers[156]. The apparatus was used to produce silt fractions and size analysis data. There was no evidence on how well the separation was carried out, but there seems nothing in the design which would lead to sharper separations than are obtained in other similar equipment.

Laurenty[157] has patented a device for sorting crystals according to size.

In its simplest form the equipment consists of a tank with two inlets and two outlets. The suspension of crystals enters by the upper inlet and fresh liquor by the lower inlet. Inside, elutriation occurs as the crystals settle against a current of the fresh liquor rising through a set of angled vanes towards the larger upper outlet. Alternative forms in which centrifugal force fields are developed are also described. No operational data are given.

In another patent, Brociner and Owen[158] describe a centrifugal elutriator which consists essentially of a horizontally rotating disc centrifuge. The distance apart of the walls (h) and the distance from the axis of rotation (r) are related by the formula

$$hr^2 = \text{constant}$$

If elutriating liquid is introduced at the periphery at a constant flow rate, the inward elutriating velocity u at any radius counter-balances the settling velocity of particles of size x, given by $u = krx^2$, where k is a constant depending on operating conditions. Particles smaller than x are therefore carried with the overflow over a weir at the central port in the lower face of centrifuge. Since an infinitely long time is required theoretically to elutriate all particles of size x and smaller, and the elutriation time must be limited, a correction procedure has been devised. If Δx, say 0.1 μm, is the difference between size x and the size which is completely elutriated in the experimental time t, and this is assumed to be divided into size fractions of $0.1\Delta x$, each with a uniform size distribution, an expression can be found for the fraction of particles between x and (x − 0.1 μm) elutriated in time t. A similar expression for another elutriation time, say $2t$, allows determination of the fraction of particles in the range (x − 0.1 μm) and correction of the experimental data. The equipment has been used for elutriation in the range 0.1–5 μm.

The use of an electrostatic field for size analysis of an aerosol by effectively measuring the mobilities of the charged particles was reported by Whitby *et al.* some time ago. A paper by Turetskii *et al.*[159] reports on the size separation of an iron powder, ranging up to 0.42 mm in size, in a corona-type chamber separator. Three stages were necessary to achieve an acceptable separation. This could have been due to the experimental conditions, since the process was being developed industrially. However, it suggests that electrostatic methods of size analysis are not likely to be so effective as fluid classification methods.

7.8 SURFACE AREA DETERMINATION

The surface areas of lamp phosphors and associated materials have been determined by King and O'Hare[160] using four different methods—Knudsen flow permeametry, nitrogen adsorption, air permeametry and calculation from Coulter Counter data. Agreement was obtained between the gas adsorption and air permeametry methods when the median particle size exceeded *ca.* 11 μm. For materials with smaller median sizes the Knudsen technique provided a better measure of external surface. Estimates of internal surface and of sphericity (by comparison with surface areas computed from

Coulter data) were also made. For speed and quality control, Harris and Scullion[161] investigated the correlation of Coulter counts at one size-level to the specific surface of a wet-milled product determined by air permeability. It was concluded that, for the closely controlled milling process, the simple counting technique was acceptable as a particle size control test.

In two papers, Rammler and Bahr[162, 163] review graphical methods which have been devised for deriving specific surface values from particle size distribution data and give a method for the determination of specific surface when the size analysis follows the Gaudin–Schumann, Rosin–Rammler, normal or log-normal distributions. It is shown that in each case the product of the idealised specific surface (for spherical particles of unit density) and the particle size parameter of the distribution is a function only of the distribution parameter (exponent or standard deviation). Graphs of these functions enable the idealised specific surface to be obtained.

Relying on the photo-extinction method of assessing surface area, English[164] showed that with care, and by examining at least 10 samples, small differences in the apparent specific surface of two similar powders could be detected.

The Micromeritics MIC-2200 surface area analyser essentially measures the volume of nitrogen gas adsorbed by the sample at liquid nitrogen temperature, at the single relative pressure point of 0.3. The error introduced by assuming a zero intercept of the BET plot is partly corrected by a volume adjustment. To assess the errors arising from this procedure, calibration was carried out by Harvard and Wilson[165] with the SCI/IUPAC/NPL standard powders and other materials. The errors were generally within 10%, but were greater for the highly microporous materials.

7.9 AEROSOL SIZE ANALYSIS

7.9.1 General

With the developing interest in environmental pollution, methods of sampling and size analysis of aerosols are of increasing importance and form the main part of this section. An editorial article[166] discusses present concern over particles less than 2 μm in size and reviews briefly methods already existing and being developed for sampling and sizing these particles, in readiness for expected legislation. Because of the suspect efficiency of the dry filters used with the normal stack sampling equipment, alternative methods of determining the dust mass depend on measuring the change in resonant frequency of a quartz crystal with mass of deposited dust, or the attenuation of a beam of β-particles. Inertial impaction devices are used for sizing, but small cyclones and a classifier with helical channels are under development. Light scattering counters are discussed and the use of the thermal precipitator for collection and sizing of particles of 0.1 μm and smaller. A system under study for this purpose is the diffusion battery with a condensation nuclei counter monitoring the numbers of inlet and outlet particles.

Whitby *et al.*[167] describe the Minnesota Aerosol Analysing System (MAAS) used in the Los Angeles smog project. The instrument system consisted of three aerosol sensors operated in parallel, and which measured the aerosol size distribution from 0.0035 to 6.8 μm diameter automatically. One main sampling system was used to supply the MAAS and other instruments being used to obtain comparative data; it was designed to minimise diffusion losses in the sampled aerosol. The Whitby Aerosol Analyser is fully described; the electric mobility classification was improved by increasing the dilution of the aerosol. Improved calibration was also obtained by the production of monodisperse aerosols below 0.02 μm size, and accuracy was increased by determination of the fraction of particles charged and the average charge as a function of particle size.

The optical particle counter (Royco) was modified by using a filtered air sheath which reduced the concentration of particles in the sensing zone, enabled more uniform light-intensity to fall on the narrow stream of particles, and thus produced an improvement in resolution. The apparatus was calibrated with latex aerosols. For the very smallest particles, number counts were obtained with a condensation nucleus counter (General Electric). Subsequent experiments suggest that the number counts obtained were *ca.* 1.75 times too large.

The use of pore filters ('nuclepore' filters and gold membrane filters) for the analysis of aerosols in the lower stratosphere is advocated by Spurney and Lodge[168]: classical diffusion batteries, because of their weight, would be unsuitable.

Measurements of indoor aerosols were made with a light scattering counter[169] in a controlled environment. The results demonstrated the effect of differing activities and the effect of the filtration systems and recirculation on the aerosol size spectra.

The particulate matter emitted from cars is a complex mixture of inorganic salts, soot and carbonaceous material. Elaborate sampling systems are required for measurement and characterisation of such particles. A review of apparatus and results has been given by Habibi[170]. The apparatus involved comprises the electrostatic precipitator, total exhaust filter and impactor units, which have all been used for the monitoring of lead emissions. An integrating nephlometer was used to measure the light scattering due to air suspended exhaust particles and a correlation found between the scattering and the lead and carbon concentrations.

For the calibration and standardisation of aerosol sizing equipment, known experimental aerosols are required and 'Dow' polystyrene spheres are widely used to provide closely monodisperse aerosols of spherical particles. Electrical mobility measurements by Megan and Wells[171] of spheres in the range 0.05–1.0 μm radius have shown that the measured mobility of particles less than 0.05 μm radius is less than the theoretical value. It is concluded that a layer of stabiliser is present which substantially increases the radius of the 0.05 μm spheres to an effective airborne radius of 0.058 μm. An investigation of radioactively labelled fused clay aerosols by Kotrappa and Wilkinson[172] has shown that the radioactivity is proportional to the cube of the particle size, i.e. that the specific activity is constant. Only for one of the aerosols tested was the specific activity found to be size dependent, owing to the effect of heat treatment on the ruthenium labelled clay.

7.9.2 Inertial methods

7.9.2.1 Impaction

Since May's first development of the cascade impactor, many variations of this device have been produced. Arising from the awareness of the effects of air pollution on health, high volume cascade impactors have been made in order to provide sufficient size-fractionated aerosol samples for subsequent examination. Two papers describe such equipment: Burton *et al.*[173] report on the field testing of an impactor sampling at 20 ft^3 min^{-1} (540 l min^{-1}). This uses jet plates 12 in (30 cm) in diameter, each with 300 jets. Including the final filter, an aerosol can be fractionated into five size ranges from 0.01 to 7 μm or larger. Comparative sampling tests with other impactors gave satisfactory results. It was found that the glass fibre collection medium with a pH of 11 adsorbed atmospheric acid gases and led to erroneous results, probably due to the formation of sulphate particulates. Glass fibres with an adjusted pH of 6.5 eliminated most of this effect. The high volume impactor of Gussman *et al.*[174] samples at 30 ft^3 min^{-1} (810 l min^{-1}). In this design each of the four separation stages consists only of a single slit and impact plate. This results in better collection efficiencies, less variability in position of particles in the jet than in a comparable round jet, and a more compact design. Calibration was carried out with laboratory-produced aerosols; no comment on the use of the binderless fibre glass collection medium was made.

The 20 stage impactor of Berner[175] comprises four 5-stage cascade impactors (each *ca.* 8 cm in diameter) which are coupled together and are operated in parallel, the throughput of each being in the range 10–28 l min^{-1}. The jets are formed by a ring of holes drilled in each jet-plate and impinge on the annular impact plates. By arranging that the intervals between the separation characteristics of each stage are logarithmically equal, the measurement errors remain the same in all the ranges of the particle size spectrum. For adhesion of the particle layer to the impact plate, and prevention of bounce and re-suspension, coating the surface of the baffle with vaseline was found effective. A six-stage impactor with baffles modified to allow weighing and direct determination of the impacted material gave more precise results. Comparative tests indicated that the impactors gave reliable dust concentration values and comparable frequency distributions.

For the sampling of town air and to obtain thin uniform deposits suitable for optical assessment, Biles *et al.*[176] devised an instrument using the third and fourth stages of a standard Casella cascade impactor and oscillating slides. The deposit covers an area of *ca.* 15 × 14 mm.

A method of determining the cumulative size distribution of aerosols from multi-stage cascade impactor data has been described by Picknett[177]. From the collection efficiency–particle size characteristics of the impactor stages, the material density, and the relative masses deposited on each stage and in the final filter, it is possible to synthesise a cumulative size distribution. This is equivalent to the $(N + 1)$ monodisperse aerosols which would theoretically give the observed results with the N-stage impactor. The calculation is best routinely carried out by computer. Tests with a variety of model distributions show that with accurate data the median diameter and interquartile diameters are determined to within 5 and 15%, respectively.

The stage constants of cascade impactors are usually expressed in terms of spheres of unit density. Kotrappa[178] shows how calibration curves made with irregular particles can be reconciled by the use of experimentally established shape factor relationships.

The labour of measuring aerosol particles deposited on a slide is reduced if the particles are ordered with respect to size so that only counting is required. Zebel[179] has described an impactor which deposits particles with aerodynamic diameters between 0.5 and 5 μm according to their size on a single stage. Concentric flows of clean air and aerosol are accelerated and aerodynamically focused in a shaped round nozzle. The calibration derived from a theoretical evaluation agrees qualitatively with the empirical calibration.

The behaviour of large particles in impactor devices has been considered in two papers. Yuu and co-workers[180] have discussed the effect of gravitation on the stage separation efficiency of a cascade impactor and propose a method of size analysis taking the gravitational effect into account. Zebel[181] has calculated the trajectories of the coarser particles which have bounced and shows that they could reach the outer boundaries of the plate before coming to rest. In contrast, the finer particles are separated almost uniformly over the plate surface.

7.9.2.2 Classification

The deposition of aerosol particles according to size and the direct application of centrifugal force derived from rotating systems are the distinguishing features of conifuges and later developments generally referred to as aerosol spectrometers. Poynting[182] describes a miniature conifuge which is battery-operated and completely portable. It will sample unit density particles between 1 and 8 μm diameter at a sampling rate of 25 cm^3 min^{-1}. The size range can be altered by the use of interchangeable orifices which control the internal air recirculation.

Owing to the radial component of the airflow in the annular conical ducts of conifuges, the Coriolis forces of the rotating airflow have a tangential component which accelerates the airflow to an angular velocity relative to the cone. This limits the range of total airflow rates suitable for the proper operation of the instrument. The influence of the Coriolis forces has been recalculated by Stöber and Mercer[183, 184] and a value of 9000 suggested as the maximum allowable Reynolds number of the tangential airflow in conifuges for stable flow.

The use of the conifuge for the determination of the size distribution of therapeutic aerosols has been described by Hauck[185] and an estimate made of the resolving power of the instrument.

Following Gerber's[186] extensive assessment of the Goetz aerosol spectrometer in 1971, in which he refuted some of the doubts which had been cast on it, Horvath[187] briefly reports on the similar evaluation of a slightly different spiral aerosol spectrometer. He points out that high accuracy with respect to particle size obtained during calibration with monodispersed particles does not automatically give high accuracy with polydispersed aerosols. The use of a microscope for counting may lead to a false maximum in the size distribution because of decreasing visibility of the sub-micron particles. Also the differential of the empirical concentration function is required which, for reasonable accuracy, would require *ca.* 10^6 particles to be counted at each location, rather than the usual procedure of counting 400–1000 particles and drawing by hand a smooth curve through the data points.

Stöber and Boose[188] have theoretically investigated the effect of developing flow profiles on the deposition patterns of monodisperse aerosols in apparatus such as the horizontal elutriator and the Goetz centrifuge. They only found small effects due to the developing flow, but in the case of the Goetz centrifuge differences between experiment and theory are due to a slow double vortex flow resulting from external forces, particularly Coriolis forces. In spite of Gerber's paper, they suggest that inhomogeneous deposits of monodisperse aerosols are to be expected with the Goetz centrifuge, which may impair the evaluation procedure.

A simplified version of the Stöber and Flachsbart spiral particle separator is described by Kotrappa and Light[189]. An aerosol sample is drawn into an expanding spiral duct through an inlet at the axis of the rotor and is entrained in a laminar stream of clean air. The particles are deposited according to their aerodynamic sizes along a 46 cm collection foil lining the outer wall of the duct. The smallest particles are retained on a filter at the end of the duct. Large samples can be collected without significant change in calibration or resolution characteristics. Normally, particles of 0.5 μm and larger are collected on the foil; separation of particles to a lower limit of 0.2 μm is said to be possible by appropriate operation. Stöber[190] comments on this machine, in particular doubting the usefulness of the radially expanding spiral duct, and the possibility of successful operation at 10 000 r.p.m. and of successful separation of 0.2 μm particles. Kotrappa[191] in his reply attributes reduced rotor weight and the exit filter to the expanding duct and reiterates that the machine is relatively simple and can be used routinely.

A modification of the Stöber aerosol spectrometer was constructed by Moss and co-workers[192], the main changes being a deeper channel and operation without the coolant housing at speeds less than 6000 r.p.m. in order to obtain distinct deposition patterns when sampling monodisperse aerosols. The apparatus has been used not only for size distribution determinations but also for density determinations.

An alternative method to the microscopic examination of the particles deposited on the foil of a spiral centrifuge is described by Porstendörfer[193]. He labelled aerosol particles with radioactive products and calculated the size distribution from the distribution of activity along the foil.

Although a few papers on aerosol beams have been published in the past ten years, no instrumental developments have yet been reported. Dahneke and Flachsbart[194] describe the generation of aerosol beams, aerosol beam focusing and aerosol beam spectrometry. A beam is generated when an aerosol is expanded from normal pressure through a nozzle or capillary into a vacuum chamber. The aerosol particles follow rectilinear trajectories at velocities of *ca.* 350 m s^{-1}. Focusing of the beam to minimise the solid angle describing it is done by injecting the aerosol into the centre of a laminar flow of clean air before traversing the nozzle. Unsymmetrical expansion of the beam due to jet design results in a varying deflection from the nozzle axis for the differing particle sizes. The deposits on a plate placed opposite the nozzle are, then, distributed according to particle size. Two possible applications of this technique are discussed: the isolation of small known size ranges by a slit, with subsequent counting by the scattered light technique, or passage of the isolated sample into an oven for vaporisation and analysis.

Iinoya and others[195] describe an instrument reminiscent of the cascade impactor for the measurement of the concentration and size distribution of a mist. Coaxial streams of the mist aerosol and clean gas impinge on each other: the smaller particles are transported radially, the larger particles continue on their course. The separation size lies in the range ten to several hundred microns.

7.9.3 Other methods

The free atmospheric aerosol is of great importance in meteorology and a number of condensation nuclei counters have been developed for the determination of the concentration of Aitken and condensation nuclei. A self-contained automatic and portable apparatus is described by Cabrol *et al.*[196]. To obtain the supersaturated water vapour environment necessary for droplets to form on the nuclei, the high pressure method is employed. After initial isothermal compression of the air sample, the adiabatic expansion has a volumetric ratio of 1.36, which allows a minimum detectable nucleus radius of 7×10^{-8} cm. Droplets are detected by the attenuation of a light beam, readings being obtained every minute. A theoretical calibration curve, taking account of thermal imbalance, gives an extinction *versus* number concentration plot in good agreement with an experimental calibration obtained with a 'General Electric' nucleus counter. The sensitivity of the apparatus is of the order of 100 particles per cm^3. Because of deviations from the classical shape of the size distribution in natural aerosols, it is important to know the smallest particle radius detectable in condensation nuclei counters. An equation relating this lower limit to direct particle counts and total droplet volume has been derived by Walter and Jaenicke[197].

An interesting device has been proposed by Matteson and Keng[198] for the size determination of an aerosol in the sub-micron range by the use of thermophoresis. The aerosol is to be passed upward through two separated porous metal plates. The upper plate is to be heated and the lower plate cooled so that a temperature gradient exists between them. Particles moving through the lower plate with the gas will come to an equilibrium position between the plates where the thermal and drag forces on the particles are equal in magnitude. The equilibrium stopping distance depends on the particle size, with the larger particles being the closer to the upper plate. When sufficient particles have been stratified in order of size, optical attenuation or radiation absorption techniques could be used to determine the size distribution. The range of application of the device is given approximately by $0.06 < \lambda/r < 2.5$, where λ is the mean free path of the gas molecules, and r is the particle radius.

A review of methods (101 references) for the determination of size and concentration of liquid aerosols below 10 µm in size has been made by

Williams and Hedley[199]. Excluding methods which involve separating the particles from the suspending gas, they discuss the photographic and holographic techniques and give an extensive discussion of methods dependent on the scattering of light by particles.

For the size analysis of spray droplets, mainly in the range 75–250 μm, Janerus and co-workers[200] have devised a sedimentation method. Droplets from all parts of the spray can be made to enter a settling chamber in which an evenly distributed transverse air stream deposits the drops according to their size. The floor of the chamber is divided into channels in each of which a small size range of drops is collected. Calibration is by using paraffin wax sprays and determining the sizes of the spheres formed from the droplets. The distribution of each droplet size over the floor of the chamber is also determined. The droplet size distribution of a spray is then calculated from an initial postulated distribution by an iterative process in which discrepancies between the calculated and experimental mass distribution over the floor of the settling chamber are used to provide corrections. The greatest systematic error would appear to lie in the calibration procedure in which hot wax droplets and solid spheres are assumed to behave in the same way. Indeed, it was observed that the distances travelled downstream by the cooling droplets were some 10% less than the calculated distances for the spheres. The reproducibility of the method appeared to be good.

The distribution of droplet sizes in a Venturi scrubber was determined by Akbrut and Kropp[201], both for the spray in still air, and for droplets moving in the gas flow, by equipment in which the drops could be collected on slides coated with a mixture of transformer oil and vaseline. The consistency of this mixture was such that no appreciable deformation of a droplet was said to occur when it landed on a slide. The finest drops were *ca.* 50 μm in size.

Two pieces of equipment designed to operate automatically for the determination of the mass concentration of an aerosol use essentially the same principles. In both, the particulate matter is isolated by drawing the aerosol through a moving filter tape. The mass of material on the tape is determined in one device from the attenuation of a transmitted beam of light[202]. Calibration is, of course, necessary for each type of aerosol for which it is used. In the other equipment, the mass of material is derived from the absorption of β radiation before and after particle deposition on the tape[203]. For the sampling of stack gases an oscillating probe is used through which a gas flow some 20% greater than the isokinetic value is maintained. This procedure is intended to avoid the problems of sampling at one fixed point, of having to determine the correct sampling velocity at each position of the probe, and of not collecting the more massive particles by sampling at less than the isokinetic velocity. The variability of results is such that the mean of ten readings is within 10% of conventional gravimetric methods.

7.10 ON-LINE PARTICLE SIZE ANALYSIS

The search for the more efficient use of particle production or handling systems has shown the need for reliable methods of on-line particle size analysis, either for continuous monitoring or for automatic control. The methods employed may encompass many of the principles outlined in the previous sections, but it is convenient to consider all of the on-line methods together. Most methods do not measure a detailed size distribution but some parameter which is dependent on the distribution. The Automatic Particle Size Monitor described by Hathaway[204] uses the attenuation of an ultrasonic beam to yield both a solids concentration and a particle size parameter. The sensor consists of two pairs of transmitting and receiving transducers operating at two different frequencies such that one pair measures the fraction of solids in the slurry, whilst the second pair measures a parameter proportional to the fraction of solids and the size distribution. No theory is given for the device but it seems to work well on mill pulps, which usually have the advantage of a size distribution function which does not change in shape with mill operating conditions. This means that a single parameter is sufficient tc determine the size distribution. The success of the instrument is further illustrated in papers by Diaz and Musgrove[205] and Webber and Diaz[206], who describe installations. A similar device, which uses the attenuation of x-rays at two different frequencies by the slurry, has been described by Carr-Brion and Jenkinson[207].

Another interesting device has been described by Cornillault[208] and Meric[209]. The suspension of particles is passed through a cell, the cross-section of which is slit-shaped. The cell is illuminated by laser light perpendicularly to the longest dimension and the resulting far-field diffraction pattern observed. If the suspension were monodisperse, then the diffraction pattern would be a set of light and dark rings. For wider size distributions the pattern is less distinct but the resulting radial intensity distribution is a function of the size distribution and particle shape. By measuring the intensities as a function of radius, using a disc with slots, the authors claim to be able to obtain a seven-point size distribution. Stark[210] was critical of the original design and has suggested some improvements. A similar device has been described by McSweeney and Rivers[211], who measure the intensities using annuli of glass fibres. The light to each annulus can then be measured by a single photocell, thus enabling the whole intensity distribution to be measured simultaneously.

Another device has been reported by Beck *et al.*[212] and Stanley-Wood *et al.*[213] which measures the electrical conductivity of a suspension of particles and relates it to a size distribution. When particles are present in a flowing turbulent fluid, they try to follow the turbulent eddies. Large particles, because of their higher inertia, tend to follow only the larger eddies, whilst small particles will follow all eddies. It is possible, therefore, using autocorrelation techniques, to relate the size of particles flowing in the fluid to the scale of turbulence. The system relies on measurements with particles of known size to determine calibration constants.

Acknowledgement

The authors would like to record their appreciation of the services of Mr. P. Foxcroft, Information Officer of the Particle Science and Technology Information Service (PSTIS), who provided the references and articles on which this review is based.

References

1. Alex, W. (1972). *Aufbereit.-Tech.*, **13,** 105
2. Alex, W. (1972). *Aufbereit.-Tech.*, **13,** 168
3. Alex, W. (1972). *Aufbereit.-Tech.*, **13,** 639
4. Alex, W. (1972). *Aufbereit.-Tech.*, **13,** 723
5. Alex, W., Koglin, B. and Leschonski, K. (1972). *Fortschritte Verfahrenstechnik*, Vol. 10, 189 (Dusseldorf: VDI-Verlag)
6. Leschonski, K. (1972). *Ullmanns Encyklopädie der Technischen Chemie*, Vol. II (Wienheim: Verlag Chemie)
7. Leschonski, K. (1973). *Chem.-Ing.-Tech.*, **45,** 8
8. Lankard, D. R. and Niesz, D. E. (1971). *Charact. Ceram.*, 349
9. Swift, D. J. P., Schubel, J. R. and Sheldon, R. W. (1972). *J. Sediment. Petrol.*, **42,** 122
10. Lauer, O. (1971). *Sprechsaal*, **104,** 780
11. Scullion, H. J. and Harris, J. E. C. (1973). *Proc. Soc. Anal. Chem., Anal. Div. Chem. Soc.*, **10,** 108
12. Davidson, J. A. and Collins, E. A. (1972). *J. Colloid Interface Sci.*, **40,** 437
13. Pahl, M. H., Schädel, G. and Rumpf, H. (1973). *Aufbereit.-Tech.*, **14,** 257
14. Pahl, M. H., Schädel, G. and Rumpf, H. (1973). *Aufbereit.-Tech.*, **14,** 672
15. Pahl, M. H., Schädel, G. and Rumpf, H. (1973). *Aufbereit.-Tech.*, **14,** 759
16. Deloye, F. X. (1972). *Bull. Soc. Fr. Céram.*, **95,** 3
17. Perez-Rosales, C. (1972). *J. Sediment. Petrol.*, **42,** 975
18. King, G. J. W. and Dickin, E. A. (1972). *Mater. Constr.*, **5,** 85
19. Zdotar, A. I. and Leonidov, L. D. (1973). *Zavod. Lab.*, **39,** 328
20. Rao, P. (1973). *Ph.D. Thesis* (Drexel University)
21. Tomb, T. F. and Corn, M. (1973). *Amer. Ind. Hyg. Ass. J.*, **34,** 13
22. Farges, P. (1972). *Bull. Soc. Fr. Céram.*, **95,** 19
23. Olivei, A. (1972). *Optik.*, **36,** 454
24. Hess, W. M., McDonald, G. C. and Urban, E. (1973). *Rubber Chem. Tech.*, **46,** 204
25. Naylor, A. G. and Kaye, B. H. (1972). *J. Colloid Interface Sci.*, **39,** 103
26. Kaye, B. H. and Naylor, A. G. (1972). *Pattern Recognition*, **4,** 195
27. Ross, A. R. L. (1973). *Opt. Laser Tech.*, **5,** 13
28. Tillmann, S. E. (1973). *J. Sediment. Petrol.*, **43,** 867
29. Naylor, A. G. (1972). *Nature Phys. Sci.*, **236,** 17
30. Rammler, E. and Bahr, A. (1972). *Chem. Tech.*, **24,** 345
31. Rammler, E. and Bahr, A. (1972). *Chem. Tech.*, **24,** 738
32. Hails, J. R., Seward-Thompson, B. and Cummings, L. (1973). *J. Sediment. Petrol.*, **43,** 889
33. Svarovsky, L. (1973). *Powder Tech.*, **7,** 351
34. Lekhtmakker, S. O. and Ruzer, L. S. (1972). *Kolloid. Zh.*, **34,** 805
35. Reményi, K. (1973). *Epitoanyag*, **25,** 326
36. Preining, O. (1972). *J. Aerosol Sci.*, **3,** 289
37. Barbery, G. (1972). *Inst. Mining Met. Trans. C*, **81,** 784
38. Pethõ, S. (1973). *Epitoanyag*, **25,** 313
39. Hale, W. E. (1972). *Atmos. Environ.*, **6,** 419
40. Hunt, W. F. (1972). *J. Air. Pollution Control Assoc.*, **22,** 687
41. Burt, M. W. G., Fewtrell, C. A. and Wharton, R. A. (1973). *Powder Tech.*, **7,** 327
42. McBride, D. D. and Sherman, P. M. (1972). *A.I.A.A.J.*, **10,** 1058
43. Heard, M. J. and Wiffen, R. D. (1972). *Atmos. Environ.*, **6,** 343
44. Timbrell, V. and Rendell, R. E. G. (1971/72). *Powder Tech.*, **5,** 279
45. Nathan, I. F., Barnett, M. I. and Turner, T. D. (1971/72). *Powder Tech.*, **5,** 105
46. Hausen, D. M. (1972). *Trans. Soc. Mining Eng., A.I.M.E.*, **252,** 328
47. Fainerman, V. B. and Sapiro, V. S. (1972). *Ind. Lab.*, **38,** 1230
48. Vialard-Goudou, A., Gimonet, M. and Fontaine-Bienfait, J. (1972). *Lab. Pharm. Probl. Tech.*, **20,** 34
49. Ross, W. D. (1973). *Filtr. Separ.*, **10,** 587
50. White, E. T., Batstone, R. J., Day, J. and Wright, P. G. (1971/72). *Powder Tech.*, **5,** 127
51. Piller, H. (1973). *Advan. Opt. Electron. Microsc.*, **5,** 95
52. Blifford, I. H., Jr. and Gillette, D. A. (1973). *Microscope*, **21,** 121
53. Gahm, J. (1973). *Advan. Opt. Electron. Microsc.*, **5,** 115

54. Berry, D. J. and Kelly, T. (1973). *Powder Tech. Publ. Ser. No. 2* (London: Powder Advisory Centre)
55. Reideger, G. (1972). *Staub.*, **32,** 11
56. Anon. (1973). *Bibliography* (Cambridge: Imanco Ltd.)
57. Bontemps, P. (1972). *Bull. Soc. Fr. Céram.*, **95,** 77
58. Sobolev, A. L. (1972). *Zavod. Lab.*, **38,** 970
59. Gärtner, K. J. and Üelhoff, W. (1973). *Optik.*, **37,** 15
60. Vogelsgesang, R. (1972). *Staub.*, **32,** 26
61. Anon. (1973). *Powder Met. Int.*, **3,** 152
62. Reiche, S. (1973). *Optic*, **37,** 50
63. Coleman, R. (1973). *Metallography*, **6,** 103
64. Lewis, H. D., Walkers, K. L. and Johnson, K. A. (1973). *Metallography*, **6,** 93
65. Wicksell, S. D. (1925). *Biometrika.*, **17,** 84
66. Huska, Von G. (1972). *Tonind-Z. Keram. Rund.*, **96,** 57
67. Dullien, F. A. L., and Mehta, P. N. (1971/72). *Powder Tech.*, **5,** 179
68. Dullien, F. A. L. and Dhwan, G. K. (1973). *Powder Tech.*, **7,** 305
69. Porstendörfer, J. and Heyder, J. (1972). *J. Aerosol. Sci.*, **38,** 141
70. Kruglitskii, N. N., Kupricnko, P. I., Simurov, V. V. and Vasilenko, V. P. (1972). *Sov. Powder Met. Metal Ceram.*, **11,** 741
71. Abdel-Azim, M. S. and Vlajic, M. (1972). *J. Amer. Ceram. Soc.*, **55,** 535
72. Larson, R. I., Fullam, E. F., Lindsay, A. D. and Matijevic, E. (1973). *A.I.Ch.E.J.*, **19,** 602
73. Witmer, F. E. and Gollan, A. (1973). *Environ. Sci. Tech.*, **7,** 945
74. Pierce, P. E. and Cowan, C. E. (1972). *J. Paint Tech.*, **44,** 61
75. Heintzenberg, J. and Quenzel, H. (1973). *Atmos. Environ.*, **7,** 509
76. Ruppersberg, G. H. (1967). *I.A.L.A.Bull.*, **31,** 11
77. Charlson, R. J. (1972). *J. Colloid Interface Sci.*, **39,** 240
78. Ensor, D. S., Charlson, R. J., Ahlquist, N. C., Whitby, K. T., Husar, R. B. and Liu, B. Y. H. (1972). *J. Colloid Interface Sci.*, **39,** 242
79. Thielke, J. F., Charlson, R. J., Winter, J. W., Ahlquist, N. C., Whitby, K. T., Husar, R. B. and Liu, B. Y. H. (1972). *J. Colloid Interface Sci.*, **39,** 252
80. Karabegov, M. A. (1972). *Measure. Tech.*, **15,** 291
81. Horvath, H. (1973). *Atmos. Environ.*, **7,** 521
82. Heintzenberg, J. and Quenzel, H. (1973). *Atmos. Environ.*, **7,** 503
83. Cohen, A. (1972). *Appl. Optics*, **11,** 1657
84. Grassl, H. (1971). *Appl. Optics*, **10,** 2534
85. du Toit, R. S. J. and Murray, N. M. (1972). *J. Mine Vent. Soc. S. Africa*, 213
86. Clark, C. J. and Dombrowski, N. (1973). *J. Aerosol. Sci.*, **4,** 27
87. McCreath, C. G., Roett, M. F. and Chigier, N. A. (1972). *J. Phys. E.*, **5,** 601
88. Bürkholz, A. (1973). *Chem.-Ing.-Tech.*, **45,** 1
89. Gross, J. and Peter, P. (1973). *Aerosol Report*, **12,** 64
90. Bexon, R. (1973). *J. Phys. E*, **6,** 245
91. Hinds, W. and Reist, P. C. (1972). *J. Aerosol Sci.*, **3,** 501
92. Hinds, W. and Reist, P. C. (1972). *J. Aerosol Sci.*, **3,** 515
93. Gollub, J. P., Chabay, I. and Flygare, W. H. (1973). *Appl. Optics*, **12,** 2838
94. Reagan, J. A. and Herman, B. M. (1972). *A.I.A.A.J.*, **10,** 1401
95. Bernhardt, C. (1973). *Flussigkeiten und Zusatzstoffe für die Sedimentationsanalyse* (Frieberg: Forsch. Aufbereit.)
96. Moore, D. W. and Orr, C. (1973). *Powder Tech.*, **8,** 13
97. Klingler, G. A. (1972). *Report T73-02532* (Ohio: Wright-Patterson Air Force Base)
98. Leroux, A. (1972). *Bull. Soc. Fr. Céram.*, **95,** 53
99. Gate, L. F. (1972). *Proc. Soc. Anal. Chem., Anal. Div. Chem. Soc.*, **9,** 225
100. Aïten, P. C. and Poulin, C. (1972). *J. Mats.*, **7,** 113
101. Pezzetta, J. M. (1972). *Tech. Report.*, **15,** WIS-SG-72-215 (University of Wisconsin)
102. Polyakov, B. I., Ralevskii, P. Kh. and Rumyantsev, V. K. (1973). *Sov. Powder Met. Metal. Ceram.*, **12,** 342
103. Allen, T. and Svarovsky, L. (1972). *Proc. Int. Conf. Ind. Measure Contr. Radiation Techn.* (*I.E.E. Conf. Pub.* **84**), 7
104. Allen, T. (1973). *Brit. Pat.* 1 327 044
105. Kienst, F. D. (1973). *Coulter Counter User's Conference, London*

106. Keienberg, R. (1973). *Zement-Kalk-Gips.*, **26,** 281
107. Bürkholz, A. (1972). *Staub.*, **32,** 15
108. Scott, K. J. and Mumford, D. (1971/72). *Powder Tech.*, **5,** 321
109. Scott, K. J. (1972). *South African Council for Scientific & Industrial Research, CEG Memo 72/15*
110. Shelnutt, J. W. (1971). *Report ARL 71-0134* (Aerospace Lab., U.S.A.F)
111. Gibbs, R. J. (1972). *J. Sediment. Petrol.*, **42,** 141
112. Anan, F. S. (1972). *Tech. Teport* **3**-*CRC* (Coastal Research Center, University of Massachusetts)
113. Grossman, G. (1972). *Ind. Eng. Chem. Proc. Des. Develop.*, **11,** 537
114. Lebedev, N. M., Zhuchkov, N., Debalk, V. P. and Pankova, E. D. *Ind. Lab.*, **38,** 791
115. Zwicker, J. D. (1972). *Powder Tech.*, **6,** 133
116. Bagness, J. E., Ottaway, A. V. and Young, B. W. (1972). *Proc. Soc. Anal. Chem., Anal. Div. Chem. Soc.*, **9,** 83
117. Bürkolz, A., Richter, W. and Muschelknautz, E. (1973). *Chem.-Ing.-Tech.*, **45,** 680
118. Tarasov, V. L. and Khodakov, G. S. (1972). *Kolloid. Zh.*, **34,** 957
119. Scholtan, W. and Lange, H. (1972). *Kolloid. Z. Z. Polym.*, **250,** 782
120. Svarovsky, L. and Friedova, J. (1971/72). *Powder Tech.*, **5,** 273
121. Kamack, J. H. (1972). *J. Phys. D*, **5,** 1962
122. Schultz, J. and Thom, R. (1973). *Med. Biolog. Eng.*, **11,** 447
123. Bebyakov, A. N. (1972). *Measure. Tech.*, **15,** 1218
124. Bader, H., Gordon, H. R. and Brown, O. B. (1972). *Rev. Sci. Instrum.*, **43,** 1407
125. Harfield, J. (1973). *Coulter Users Conference, London*
126. Lines, R. W. (1973). *Powder Tech.*, **7,** 129
127. Nienow, A. W. and Ang, H. M. (1972). *Lab. Pract.*, **21,** 495
128. Maguire, B. A., Seamay, R. J. and Halpin, R. K. (1972). *Powder Tech.*, **6,** 61
129. Barclay, L. M. (1973). *Coulter Users Conference, London*
130. McCave, I. N. (1973). *Coulter Users Conference, London*
131. Marshall, K. and Meakin, B. C. (1973). *Coulter Users Conference, London*
132. Baudet, G. (1972). *Bull. Soc. Fr. Céram.*, **95,** 31
133. Steinkampf, J. A., Fulwyler, M. J., Coulter, J. R., Hiebart, R. D., Herney, R. D. and Mullaney, P. F. (1973). *Rev. Sci. Instrum.*, **44,** 1301
134. Fedrich, N. A., Bell, A. T. and Hansen, D. N. (1972). *Environ. Sci. Tech.*, **6,** 1117
135. Nikiforova, N. K. and Sedunov, Y. S. (1972). *J. Aerosol. Sci.*, **3,** 441
136. Oeseburg, F. (1972). *J. Aerosol. Sci.*, **3,** 307
137. Schel, R., Ergun, S. and Headrick, A. (1973). *Rev. Sci. Instrum.*, **44,** 1193
138. Wims, A. M. and Myers, M. E. (1972). *J. Colloid Interface Sci.*, **39,** 447
139. Cluck, J. W., Adams, W. S. and Moroz, W. J. (1973). *ISA Trans.*, **12,** 208
140. Anon. (1973). Application Note API-3-101 (Santa Barbara, Calif.: Science Spectrum Inc.)
141. Holland, A. B. and Sawers, J. R. (1973). *Photo. Sci.*, **17,** 295
142. South African Bureau of Standards. (1971). *SABS*, 197
143. Polish Standards Committee. (1971). *PKN C-04501*
144. Kiff, P. R. (1973). *Proc. Soc. Anal. Chem., Anal. Div. Chem. Soc.*, **10,** 114
145. Janke, N. C. (1973). *J. Sediment. Petrol.*, **43,** 518
146. Rose, H. E. and English, J. E. (1973). *Trans. Inst. Chem. Eng.*, **51,** 14
147. West, G. and Dumbleton, M. J. (1972). *Transport Road Res. Lab. Rep.*, LR 437
148. Carter, J. M. (1972). *Trans. Inst. Mining Met. C*, **81,** C113
149. Lauer, O. (1972). *Aufbereit.-Tech.*, **13,** 422
150. Niedek, E. A. (1973). *Verfahrenstechnik*, **7,** 149
151. Niedek, E. A., Lipp, H. M. and Mieling, B. (1973). *Verfahrenstechnik*, **7,** 280
152. Niedek, E. A., Lipp, H. M. and Mieling, B. (1973). *Verfahrenstechnik*, **7,** 285
153. Bauer, H., Cordes, H. and Haas, H. (1972). *Aufbereit.-Tech.*, **13,** 428
154. Leschonski, K. (1972). *Aufbereit.-Tech.*, **12,** 751
155. Schaller, R. E. and Lapple, C. E. (1972). *J. Paint Tech.*, **44,** 85
156. Follmer, L. R. and Beavers, A. H. (1973). *J. Sediment. Petrol.*, **43,** 544
157. Laurenty, F. (1972). *U.S. Pat.* 3 688 901
158. Brociner, R. E. and Owen, D. (1973). *Brit. Pat.* 1 313 753
159. Turetskii, Ya. M., Gristan, E. L. and Konorov, M. I. (1972). *Res. Powder Met.*, **2,** 42
160. King, W. G. and O'Hare, B. P. (1973). *Coulter Users Conference, London*

161. Harris, J. E. C. and Scullion, H. J. (1973). *Coulter Users Conference, London*
162. Rammler, E. and Bahr, A. (1971). *Verfahrenstechnik*, **5,** 483
163. Rammler, E. and Bahr, A. (1972). *Verfahrenstechnik*, **6,** 312
164. English, J. E. (1972). *Brit. Chem. Eng. Process Tech.*, **17,** 715
165. Harvard, D. C. and Wilson, R., (1973). *Coulter Users Conference, London*
166. Iammartino, N. R. (1972). *Chem. Eng.*, **79,** 32
167. Whitby, K. T., Liu, B. Y. H., Husar, R. B. and Barsic, N. J. (1972). *J. Colloid Interface Sci.*, **39,** 136
168. Spurny, K. and Lodge, J. P. (1972). *Staub*, **32,** 18
169. Lum, R. M. and Graedel, T. E. (1973). *Atmos. Environ.*, **7,** 827
170. Habibi, K. (1973). *Environ. Sci. Tech.*, **7,** 223
171. Megaw, W. J. and Wells, A. C. (1972). *J. Aerosol. Sci.*, **3,** 1
172. Kotrappa, P. and Wilkinson, C. J. (1972). *J. Aerosol. Sci.*, **3,** 167
173. Burton, R. M., Howard, J. N., Penley, R. L., Ramsay, P. A. and Clark, T. A. (1973). *J. Air Pollution Control Assoc.*, **23,** 277
174. Gussman, R. A., Sacco, A. M. and McMahon, N. M. (1973). *J. Air Pollution Control Assoc.*, **23,** 778
175. Berner, A. (1972). *Staub.*, **32,** 1
176. Biles, B., Brown, C. and Ellison, J. McK. (1973). *Atmos. Environ.*, **7,** 101
177. Picknett, R. G. (1972). *J. Aerosol. Sci.*, **3,** 185
178. Kotrappa, P. (1973). *J. Aerosol. Sci.*, **4,** 47
179. Zebel, G. (1972). *Staub.*, **32,** 6
180. Yuu, S., Yukawa, T. and Iinoya, K. (1972). *J. Chem. Eng. Jap.*, **5,** 285
181. Zebel, G. (1973). *Staub*, **33,** 102
182. Poynting, R. (1972). *J. Phys. E*, **5,** 52
183. Stöber, W. and Mercer, T. T. (1972). *Environ. Sci.*, **6,** 1091
184. Stöber, W. and Mercer, T. T. (1969). *Report UR-3490-122*, Rochester University, New York
185. Hauck, H. (1972). *J. Aerosol Sci.*, **3,** 31
186. Gerber, H. E. (1971). *Atmos. Environ.*, **5,** 1009
187. Horvath, H. (1973). *Atmos. Environ.*, **7,** 1003
188. Stöber, W. and Brose, C. (1973). *Atmos. Environ.*, **7,** 119
189. Kotrappa, P. (1972). *Rev. Sci. Instrum.*, **43,** 1106
190. Stöber, W. (1973). *Rev. Sci. Instrum.*, **44,** 1149
191. Kotrappa, P. (1973). *Rev. Sci. Instrum.*, **44,** 1438
192. Moss, O. R., Ettinger, H. J. and Coulter, J. R. (1972). *Environ. Sci. Tech.*, **6,** 614
193. Porstendörfer, J. (1973). *J. Aerosol. Sci.*, **4,** 345
194. Dahneke, B. and Flachsbart, H. (1972). *J. Aerosol Sci.*, **3,** 345
195. Iinoya, K., Makino, K., Toyama, S. and Goto, A. (1973). *Kagaku Kogaku*, **37,** 858
196. Cabrol, C., Lopez, A., Chapius, A. and Fontan, J. (1972). *J. Aerosol. Sci.*, **3,** 281
197. Walter, H. and Jaenicke, R. (1973). *Atmos. Environ.*, **7,** 939
198. Matteson, M. J. and Keng, E. Y. H. (1972). *J. Aerosol Sci.*, **3,** 45
199. Williams, I. and Hedley, A. B. (1973). *Rep. Prog. Appl. Chem.*, **57,** 635
200. Janerus, I., Rosenblad, G. and Hansson, B. (1973). *J. Inst. Fuel*, **46,** 178
201. Akbrut, A. I. and Kropp, L. I. (1972). *Thermal Eng.*, **19,** 118
202. Polydorova, M. (1972). *Staub*, **32,** 8
203. Anon. (1973). *Chem. Eng.*, **80,** 44
204. Hatherway, R. E. (1972). *Coal, Gold, Base Min. S. Africa*, 49
205. Diaz, L. S. and Musgrove, P. M. (1973). *Trans. Soc. Mining Eng. A.I.M.E.*, **254,** 42
206. Webber, C. C. B. and Diaz, L. S. (1973). *Can. Mining. J.*, **94,** 36
207. Carr-Brion, K. G. and Jenkinson, D. A. (1972). *US Pat.* 3 666 943
208. Cornillault, J. (1972). *Appl. Optics*, **11,** 265
209. Meric, J. (1972). *Bull. Soc. Fr. Céram.*, **95,** 67
210. Stark, H. (1972). *Appl. Optics*, **11,** 1655
211. McSweeney, A. and Rivers, W. (1972). *Appl. Optics*, **11,** 2101
212. Beck, M. S., Lee, K. T. and Stanley-Wood, N. G. (1973). *Powder Tech.*, **8,** 85
213. Stanley-Wood, N. G., Lee, K. T. and Beck, M. S. (1973). *Proc. Soc. Anal. Chem., Anal. Div. Chem. Soc.*, **10,** 282

Index